Matthew Landrus

Leonardo da Vinci's Giant Crossbow

Matthew Landrus

Leonardo da Vinci's Giant Crossbow

With 75 Images

Author

Matthew Landrus
Wolfson College
Oxford University
Oxford OX2 6UD
UK
e-mail: matt@mail.wolf.ox.ac.uk

ISBN 978-3-662-50168-9   ISBN 978-3-540-68918-8 (eBook)

DOI 10.1007/978-3-540-68918-8

Production: Heather King
Typesetting and layout: Büro Stasch · Bayreuth (stasch@stasch.com)

Printed on acid-free paper

9 8 7 6 5 4 3 2 1

springer.com

For Mallica

Foreword

Leonardo was both repelled by violence and fascinated by war. One of his "prophecies" tells as that,

> Animals will be seen on earth who always fight amongst themselves with great destruction and by their fierce limbs a great proportion of trees in huge forests will be laid low throughout the universe... Nothing will remain on or under the earth or in the water that will not be persecuted, dislocated and despoiled... Why, earth, do you not open and precipitate them into the fissures of your great cataracts and caverns, and no longer exhibit to heaven such a cruel and despicable monster. (CA 1033va)

The "monster" wreaking such destruction is of course man himself.

Leonardo's "prophecies" belong to a literary genre, and were designed to entertain. However, reading this and other pieces in which he writes about man's wanton destruction of himself and all the creations of nature, we can sense his personal feelings. Why, then, did he devote so much time to the designing of weapons of massive destruction?

The obvious answer is that it was his job. He was employed at the Milanese court by Ludovico Sforza as one of his military engineers and was expected to deliver designs and expertise – more readily perhaps than he delivered paintings. In his capacity as an expert on cartography, weapons and fortifications, he subsequently joined the entourage of the notorious Cesare Borgia who was rampaging in central Italy on behalf of the pope. Leonardo also served in this capacity for the city of Florence and for his later French employers. However, his engagement with military engineering went beyond what was necessary just to do his job. He became deeply engaged in the way that weapons presented the possibility of harnessing force and deploying the "elements of machines" on a scale that was unthinkable in any other domain of his activity. War-like patrons were continuously attracted to the construction of decisive new weapons and were willing to invest sums of money far beyond any they would spend on the Arts.

Leonardo was much concerned with scale. He saw basic continuities in the design of everything in the universe, whether made by god or man and however large or small. The rules of design worked across all scales. This was not just a question of statics, but involved the behaviour of forces. As he knew from his engagement with mediaeval physics, theory tells us that if you propel something with twice the force the performance of the projectile ought to be doubled. A bigger canon with more gunpowder should accordingly send its missile proportionately further than a small one. Or it should send a proportionately bigger missile the same distance. Bigger bangs promised ever bigger blows. A crossbow 20 times the size of one carried by a soldier should in theory produce twenty times the power. But he came to realise that what book theory predicted and what actually happened were two different things. He was, as he repeatedly stressed, a man of "experience" and not of "book learning". He came to acknowledge that in the material world of real things, the simple law of the multiplication of cause and effect did not work quite as stated.

He did consider the neat rule that that the distance a missile from a crossbow would travel would be proportionate to the length of withdrawal of the bowstring, but saw that the bending of the arms of the bow threw this simple geometry out of line. He looked instead to the interior angle subtended by the string at the point of its withdrawal. There must, according to Leonardo's beliefs, be a mathematical law at work somewhere in the system. He was the first modern engineer to attempt to apply the geometrical mathematics of the laws of motion to the design of machines. His military designs were intended to operate, as he claimed of his flying machine, "in accordance with natural law".

He was also aware, as someone who knew about materials and worked with master fabricators who knew far more about the performance of wood than any modern engineer, that increasing scale brought with it new constructional problems. You could not simply make the arms of a crossbow thicker and thicker and longer and longer. A very thick arm of unitary construction would fracture in its outer layers. He thus looked to laminates and components that could slide over each other.

The main drawing at which Matt Landrus is looking with notable intensity tackles these and other of the problems that go with making a weapon of huge proportions in an entirely new way. He shows that the drawing is not some tidy version of a sketched fantasy that could be presented to a patron as a vapid piece of visual boasting. It was the culmination of incredibly concentrated design procedures for the whole and the parts. Only a fraction of the preparatory drawings survive. The process of drafting was meticulously controlled in scale, not least using a series of inscribed lines and compass points that are only visible in a close study of the paper's surface under raking light. Only by such means can we unravel the "archaeology" of the drawing and reconstruct Leonardo's design process.

The result of Landrus's researches is that we can now see Leonardo's finished demonstration of the giant crossbow as a new kind of drawing that embeds an enormous amount of precise, measured and measurable information within its graphic compass. Some of Leonardo's colleagues, like the Sienese engineer Francesco di Giorgio, one of whose manuscripts Leonardo owned, made wonderfully convincing drawings of machines in their full perspectival glory. But their results were very pictorial, designed to show the viewer (above all the patron) what the devices would look like, but they were not engineering drawings in the modern sense, produced to provide precise guidance to the fabricators. Leonardo in his finished drawing of the giant crossbow is working his way towards the kind of descriptive geometry devised by Gaspard Monge in the 18th century. The trick in such a drawing is to render the machine in its full spatial array but also to embody precise information about the dimensions and the construction of its parts. Matt Landrus, through a meticulous and systematic examination of the drawing in its context, shows for the first time the extraordinary degree to which Leonardo pulled off this trick.

Martin Kemp
Oxford, September 2009

Acknowledgements

As this book has taken shape alongside other projects related to Leonardo's proportion theories, I have had the fortune of advice and encouragement from a large number of friends and colleagues, some of whom I have thanked – although inadequately – elsewhere in the book. I am especially grateful to Martin Kemp, whose generous support of my work at Oxford University and thereafter made this book possible. He has offered essential recommendations on the original manuscript of this book, and has kindly written a preface. I would also like to thank Pietro C. Marani and James Bennett for their thoughtful advice on the original manuscript. At the outset of my work on Leonardo and ever since, I have had the exceptional, gracious support of Dario Covi. In recent years, J. V. Field, Claire Farago, and Francis Ames-Lewis have been especially supportive. I also wish to thank Thomas Ditzinger, my Editor at Springer, for his insights, guidance and patience. And for her assistance with the editorial process at Springer, I wish to thank Heather King. I am also very grateful to Armin Stasch for his perspicacious work on the index and other essential preparations of the book.

Initial research was possible with the help of a Mary Churchill Humphrey Centenary Memorial Scholarship from the University of Louisville. For the support of this scholarship and other resources, I am particularly indebted to John W. Shumaker and David Jones. In the UK, David and Pam Emms were most gracious and generous hosts as trustees of the scholarship. I am grateful to the University of Oxford's Hulm University Fund for awarding financial support for the reproductions.

Studies of the *Giant Crossbow* and associated drawings required an analysis of metal stylus marks on drawings at the various collections. At the Biblioteca Ambrosiana, I had the benefit of support from Vittorio Bergnach, Cesare Pasini, Pier Francesco Fumagalli, and Gianfranco Ravasi. I am grateful to Catherine Whistler for her help at the Ashmolean Museum. At the Windsor Royal Library, Martin Clayton and Jane Roberts kindly assisted with a large number of drawings. At the National Art Library of the Victoria and Albert Museum, I had the support of Andrew Russell, Rowan Watson, John Meriton, Jan van der Wateren. Michael Boggan kindly provided support at the British Library. At the Uffizi's Gabinetto Disegni e Stampe, Annamaria Petrioli Trofani, Lucia Monaci, and Massimo Pivetti were very helpful. At the Biblioteca Trivulziana, I had the generous support of Camilla Gavazzi, Ivanoe Riboli, and Giovanni Columbo. Christopher Baker was most helpful at the Christ Church Picture Gallery. At the Galleria dell'Accademia, Giovanna Nepi Sciré and Annalisa Perissa provided generous support. Bruce Barker-Benfield offered exceptional guidance at the Bodleian Library. Carmen Bambach and Mary Zuber kindly assisted with arrangements at the Metropolitan Museum in New York.

Library staff, some who are mentioned further above, assisted considerably with research. I am grateful to the staff at the University of Oxford Bodleian, Sackler, History of Art Department, and Modern History Libraries, and at other libraries in Oxford Colleges, such as Wolfson, All Souls, Christ Church, Magdalen, Merton and Trinity. My thanks also to the staff of the Biblioteca Leonardiana, Biblioteca Berenson, Biblioteca Medicea Laurenziana, Biblioteca Nacional in Madrid, Bibliothèque Nationale de France, Biblioteca Nazionale Centrale in Florence, and to staff of the art and rare books libraries at the British Museum, Brown University, the Centre d'Études Supérieures de la Renaissance, Harvard University, Institute de France, Kunsthistorisches Institut in Florence, Princeton University, University of Louisville, and especially the Warburg Institute. For access to rushes of the ITN Factual Television "Dream Machines" documentary I would like to thank Emily Roe.

For generous assistance at various stages of my research I am grateful to Susana Barreto, Mary Bergstein, Thomas Buser, Vanessa Daubney, Paola DeMatte, Gillian Eastmond, Frank Fehrenbach, Edward Ford, Paolo Galluzzi, Stella Gambling, Gail Gilbert, James Grubola, Jeffrey Johnson, Jay Kloner, Jean Liebel, Stephanie Maloney, Jorge Menezes Oliveira, Isabella Nardi, Johannes Nathan, Malcolm Oster, Carlo Pedretti, Donald Preziosi, Marina Prosperi, Michael Rocke, Gervaise Rosser, Vicky Sanderlin-McLoughlin, Albert van der Schoot, Patricia Trutty Coohill, Kim Veltman, Kim Williams, and Frank Zöllner. At the University of Oxford History of Art Department, I have had the support of Linda Whiteley, Marius Kwint, Vicky Brown, Nicola Henderson, Rachel Woodruff, Pamela Romano, Laura Iliffe, and Naomi Collyer. Within the Modern Histoy Faculty, Hubert Stadler and Louise Parkinson were most supportive. Individuals at Wolfson College (Oxford) that I would like to thank are too numerous to note here, though I should like particularly to thank Ben Simpson, Stephen Gower, Jan Scriven, Marjory Szurko, Martin Francis, and Sir Gareth Roberts. I would also like to thank my family for their generosity, including especially Kit Carter, Carol Irwin, Archie Irwin, Mary Weilage, Charles Weilage, Joan Landrus, Harvey Landrus, Peggy Landrus, Radhica Ganapathy, Lotica Ganapathy, and Gaurav Chawla. Most of all, I am grateful to my amazing wife, Mallica, who for the past thirteen years has inspired and encouraged me with her exceptional sense of humour, patience, and the occasional kick to remind me of a deadline. She has shared my interest in events circa 1500, as her research centers on that period in South Asian studies. Her support has been crucial to the completion of this book.

Matthew Landrus
Providence, March 2010

Contents

Introduction 1

Part I
The Idea: Locating the Giant Crossbow and Its Components 3

Chapter 1
Proposal to Ludovico Sforza, CA 1082r [391ra] 15
Locating Leonardo, c. 1480–1484 19
The Political Climate in Florence, c. 1480 25

Chapter 2
Dating the Crossbow 31
Manuscript B 32
War with Venice 33
A Programme for Siege and Defence Strategies 35

Chapter 3
A Real Project 43
Studies and Measurements 43
Leonardo's Sources 45

Part II
Making It Work: Functionality of the Giant Crossbow Design 59

Chapter 4
The Armature's Design and Proposed Construction 65
Bat Wing Form, Laminated Beams and Blocks 71
Materials: Willow, Iron, Hemp, and a Stone 72

Chapter 5
Proportional Design and Operation of the Armature 79
Proportional Form 82
The Motion of Missiles Cast ... Will Be in Proportion
to the Angle of the Propelling String 86

Chapter 6
The Lower Carriage, Side Poles, and Wheels 95
Lower Carriage 98
The Wheels 104

Chapter 7
The Upper Carriage 109
Proportional Consistency 109
Trigger Mechanism 113
The Crossbowman 117

Chapter 8
Mechanical Draughting Techniques 119
Page Format 125
The Sequence of Marks 128

Conclusion 139

Appendix A
Measurements 141
The Measurements and Proportions of Leonardo's Giant Crossbow 141
Lengths 144
Weights 144

Appendix B
Drawings That Were Closely Associated with the Giant Crossbow Project 145

Glossary
Crossbow-Related Terms Used by Leonardo 147

Bibliography 149
Manuscripts and Archival Sources 149
Printed Primary Sources 149
Printed Secondary Sources 150

Index 169

List of Figures

Fig. I.1. Leonardo da Vinci, Giant Crossbow,
Codex Atlanticus f. 149br [prev. 53vb], Biblioteca Ambrosiana, Milan 5

Fig. I.2. Leonardo da Vinci, A vertical bent-beam springald,
Codex Atlanticus f. 181r [64r], Biblioteca Ambrosiana, Milan 6

Fig. I.3. Leonardo da Vinci, Cannon factory,
Windsor Castle Royal Library, f. 12647, Windsor 6

Fig. I.4. Leonardo da Vinci, Studies of a springald, cannon, and crossbows,
Codex Atlanticus f. 149ar [53va], Biblioteca Ambrosiana, Milan 9

Fig. I.5. Illustration of the woodcut of a vertical bent-beam springald, the
original illustration attributed to Matteo Pasti, in: Roberto Valturio,
De re militari, 1472, Bodleian Library MS Douce 289, unfoliated, Oxford 9

Fig. I.6. Illustration of the woodcut of a cannon with pulleys and carriage, the
original illustration attributed to Matteo Pasti, in: Roberto Valturio,
De re militari, 1472, Bodleian Library MS Douce 289, unfoliated, Oxford 10

Fig. I.7. Illustration of the woodcut of a torsion spring springald, the original
illustration attributed to Matteo Pasti, in: Roberto Valturio,
De re militari, 1472, Bodleian Library MS Douce 289, unfoliated, Oxford 11

Fig. I.8. Leonardo da Vinci, Preparatory sketch for the Giant Crossbow,
Codex Atlanticus f. 147av [52va], Biblioteca Ambrosiana, Milan 13

Fig. 1.1. Leonardo da Vinci, A giant bent-frame springald,
Codex Atlanticus f. 145r [51vb], Biblioteca Ambrosiana, Milan 16

Fig. 1.2. Leonardo da Vinci, Preparatory sketches for the Giant Crossbow,
Codex Atlanticus f. 147bv [52vb], Biblioteca Ambrosiana, Milan 19

Fig. 1.3. Leonardo da Vinci, Illustration of a screw and ratchet winding mechanism,
Codex Atlanticus f. 148ar [53ra], Biblioteca Ambrosiana, Milan 21

Fig. 1.4. Leonardo da Vinci, Illustration of a large repeating sling,
Codex Atlanticus f. 159br [57br], Biblioteca Ambrosiana, Milan 23

Fig. 1.5. Leonardo da Vinci, Giant repeating crossbow,
Codex Atlanticus f. 182br [64vb], Biblioteca Ambrosiana, Milan 25

Fig. 1.6. Leonardo da Vinci, Trigger mechanisms,
Codex Atlanticus f. 1048br [376rb], Biblioteca Ambrosiana, Milan 27

Fig. 1.7. Leonardo da Vinci, Giant repeating crossbow,
Codex Atlanticus f. 1070r [387r], Biblioteca Ambrosiana, Milan 29

Fig. 2.1. Illustration of Leonardo's flying machine wing on Manuscript B,
MS B.N. It. 2037, f. 74r, Institut de France, Paris 31

Fig. 2.2. Illustration of Leonardo's drawing of the bottom side of a crossbow on
Codex Trivulzianus N 2162, p. 99 [f. 54r], Castello Sforzesco, Milan 32

Fig. 2.3. Leonardo da Vinci, Large sling engine,
Codex Atlanticus f. 182ar [64va], Biblioteca Ambrosiana, Milan 37

Fig. 2.4. Leonardo da Vinci, Preparatory sketch for the Giant Crossbow,
among other engineering studies, Codex Atlanticus f. 57v [17ra],
shown as originally attached to 69ar [22ra] on the right,
Biblioteca Ambrosiana, Milan 40

Fig. 2.5. Illustration of Leonardo's studies of the techniques used to join boards,
Codex Atlanticus f. 90r [33ra], Biblioteca Ambrosiana, Milan 42

Fig. 3.1. Illustration of a cheiroballista 49

Fig. 3.2. Illustration of an 11th century Byzantine diagram
of a Roman style cheiroballista, possibly copied from
Heron's Belopoiika; MS Codex Parisinus graec. 2442,
f. 76v, Bibliothéque Nationale, Paris 50

Fig. 3.3. Illustration of an anonymous painting of a ballista fulminalis,
De rebus bellicus, Bodleian Library, Oxford 51

Fig. 3.4. Illustration of a woodcut of the components of a springald style
battering ram in Conrad Kyeser's Bellifortis,
Universitätsbibliothek, MS 63, f. 80v, Universitätsbibliothek, Göttingen ... 54

Fig. 3.5a. Detail of the trigger mechanism on Codex Atlanticus f. 149br,
with superimposed arrows that show directions of the components 55

Fig. 3.5b. Illustration of a great crossbow and incendiary missile in the Löffelholz
manuscript, MS lat. 599 f. 34v, Bayerische Staatsbibliothek, Munich 55

Fig. 3.6. Illustration of Francesco di Giorgio Martini drawing of a dart smacking
device, MS Salluzziano 148, f. 61v (detail), Biblioteca Reale, Turin 56

Fig. II.1. Author's photograph of a reconstruction of Leonardo's Giant Crossbow
by ITN Factual Television, a view of the filming on Salisbury Plain
in September 2002 60

Fig. II.2. Author's photograph of a reconstruction of Leonardo's Giant Crossbow
by ITN Factual Television, now located at the Royal Armouries,
Fort Nelson, Hampshire, UK 60

Fig. II.3. Author's photograph of a reconstruction of Leonardo's Giant Crossbow
by ITN Factual Television, now located at the Royal Armouries,
Fort Nelson, Hampshire, UK 60

Fig. II.4. Author's photograph of a reconstruction of Leonardo's Giant Crossbow
by ITN Factual Television, now located at the Royal Armouries,
Fort Nelson, Hampshire, UK 62

Fig. II.5. Author's photograph of a reconstruction of Leonardo's Giant Crossbow
by ITN Factual Television, now located at the Royal Armouries,
Fort Nelson, Hampshire, UK 62

Fig. II.6. Author's photograph of a reconstruction of Leonardo's Giant Crossbow
by ITN Factual Television, now located at the Royal Armouries,
Fort Nelson, Hampshire, UK 62

Fig. 4.1. Illustration of Leonardo's drawing of a great crossbow,
Codex Atlanticus 142r [51rb], Biblioteca Ambrosiana, Milan 66

Fig. 4.2. Illustration the appearance of the Giant Crossbow's basic features
as they might look from a viewpoint directly above 67

Fig. 4.3. Detail of the central portion of the upper carriage,
near the 100 point stone ball, on Codex Atlanticus f. 149br 75

Fig. 4.4. Illustration of a typical stonebow (a stone throwing crossbow),
from R. Payne-Gallwey, The Book of the Crossbow, London:
Longmans, Green and Co., 1903, p. 157 76

Fig. 4.5. Illustration of Leonardo's drawing of an explosive dart
on Windsor folio 12651r 77

Fig. 4.6. Illustration of Leonardo's drawing of an explosive dart
on Manuscript B folio 50v 77

Fig. 5.1. Superimposition white lines over CA 149ar-br,
illustrating the crossed arc method Leonardo likely used
to determine the position of the armature 82

Fig. 5.2. Superimposition of black and white lines over CA 149br,
showing the proportions of the span with respect to a hexagon
inscribed in a circle 84

Fig. 5.3. Illustration of Leonardo's diagram of the proportional draw strengths of a crossbow on Codex Atlanticus f. 78r [27rb] 87

Fig. 5.4. Illustration of Leonardo's diagram of the proportional draw strengths of a crossbow on Codex Madrid I, f. 51r 87

Fig. 5.5. Superimposition of white lines over CA 149br, illustrating three possible spans of the armature with respect to the likely positions of the ropes at those positions 88

Fig. 5.6. Superimposition of black lines over CA 149br, showing the extent to which the armature would have extended forward before it had been bent, treated, and attached to ropes 90

Fig. 6.1. Detail of the front end of the crossbow on CA 149br 95

Fig. 6.2. Leonardo da Vinci, Cannon and carriages, Codex Atlanticus f. 154br [55vb], Biblioteca Ambrosiana, Milan 97

Fig. 6.3. Superimposition of black lines over CA 149br, illustrating the carriage proportions 98

Fig. 6.4. Superimposition of black lines and measurements over CA 149br illustrating its dimensions 101

Fig. 6.5. Diagram of the appearance of the upper carriage at the elevation of fourteen braccia, at an angle of approximately 16° 102

Fig. 6.6. Albrecht Dürer, Landscape with Cannon, Metropolitan Museum of Art, New York 103

Fig. 7.1. Superimposition of white lines over CA 149br, illustrating adjustments of the first phase of diagonal carriage metal stylus lines, with an outline of 30°/60°/90° triangle template 110

Fig. 7.2. Detail of the central portion of the upper carriage, near the crossbowman, on Codex Atlanticus f. 149br 112

Fig. 7.3. Superimposition of blue dots and red and green circles CA 149br, illustrating directions of a divider when plotting the armature curves 113

Fig. 7.4. Illustration of Leonardo's drawing of spanning bench, from Uffizi, no. 446r; Gabinetto dei Disegni e delle Stampe degli Uffizi, Florence 116

Fig. 8.1. Diagram of the crossed arc method 119

Fig. 8.2. Illustration of the Giant Crossbow, with notes about its scale and proposed dimensions 120

Fig. 8.3. Superimposition of white lines over CA 149br, illustrating the possible directions of divider when estimating the position of the armature 121

Fig. 8.4. Photograph of the sectors supposedly used by Brunelleschi to build the Florence Cathedral, Cathedral of Sta. Maria del Fiore, Florence 121

Fig. 8.5. Superimposition of white lines over CA 149br, illustrating the possible directions of divider when estimating the position of the armature 122

Fig. 8.6. Superimposition of white lines over CA 149br, illustrating the position of a metal stylus incision and possible directions of a divider when estimating the position of the armature ... 123

Fig. 8.7. Superimposition of white lines over CA 149br, illustrating the position of a metal stylus incision and possible directions of a divider when estimating the position of the armature ... 124

Fig. 8.8. Superimposition of white lines over CA 149br, illustrating the positions of a metal stylus incisions and possible directions of a divider when estimating the position of the armature ... 125

Fig. 8.9. Superimposition of white lines over CA 149br, illustrating the positions of a metal stylus incisions and possible direction of a divider 126

Fig. 8.10. Superimposition of white lines over CA 149br, illustrating the positions of metal stylus incisions 127

Fig. 8.11. Superimposition of white lines over CA 149br,
illustrating the positions of a metal stylus incisions and possible
directions of a divider when estimating the position of the armature ... 128

Fig. 8.12. Superimposition of white lines over CA 149br,
illustrating the positions of metal stylus incisions 129

Fig. 8.13. Superimposition of white lines over CA 149br,
illustrating the positions of metal stylus incisions 130

Fig. 8.14. Superimposition of white lines over CA 149br,
illustrating the positions of metal stylus incisions 131

Fig. 8.15. Superimposition of white lines over CA 149br,
illustrating the positions of metal stylus incisions 132

Fig. 8.16. Superimposition of white lines over CA 149br,
illustrating the positions of metal stylus incisions 133

Fig. 8.17. Superimposition of white lines over CA 149br,
illustrating the positions of metal stylus incisions 134

Fig. 8.18. Superimposition of white lines over CA 149br,
illustrating the positions of metal stylus incisions 135

Fig. 8.19. Superimposition of white lines over CA 149br,
illustrating the positions of metal stylus incisions 136

Fig. 8.20. Superimposition of white lines over CA 149br,
illustrating the positions of metal stylus incisions 137

Fig. 8.21. Illustration the positions of metal stylus incisions
that originally appeared on CA 149br 138

Abbreviations

Abbreviation	Leonardo's drawings and writings	Approximate dates
A	Manuscript A (2172 and 2185), Institut de France, Paris	1492
AR	Codex Arundel 263, British Library, London	1478–1518
ASH	Ashmolean Museum, Oxford	1478–1508
B	Manuscript B (2173 and 2184), Institut de France, Paris	1483–1490
C	Manuscript C (2174), Institut de France, Paris	1490–1491
CA	Codex Atlanticus and assorted sheets, Biblioteca Ambrosiana, Milan	1478–1518
CC	Christ Church College, Oxford	1489–1498
CF	Forster Codices I, II, and III (804.AA.0150–0152), Victoria and Albert Museum, London	1487–1494
CL	Codex Leicester, Collection of Bill and Melinda Gates, Medina, WA	1507–1510
CM	Madrid Codices I and II (8937 and 8936), Biblioteca Nacional, Madrid	1490–1505
CT	Codex Trivulzianus (N 2162), Castello Sforzesco, Milan	1485–1489
CU	Codice Urbinate Latinus 1270, Biblioteca Apostolica Vaticana, Vatican City	1490–1518
D	Manuscript D (2175), Institut de France, Paris	1508
E	Manuscript E (2176), Institut de France, Paris	1513–1514
F	Manuscript F (2177), Institut de France, Paris	1508
G	Manuscript G (2178), Institut de France, Paris	1510–1515
H	Manuscript H (2179), Institut de France, Paris	1493–1494
I	Manuscript I (2180), Institut de France, Paris	1497–1499
K	Manuscript K (2181), Institut de France, Paris	1503–1508
L	Manuscript L (2182), Institut de France, Paris	1497–1502
M	Manuscript M (2183), Institut de France, Paris	1497–1500
LOU	Musée du Louvre, Paris	1470–1500
T	Codex on the Flight of Birds (Cod. Varia 95) and assorted drawings, Biblioteca Reale, Turin	1485–1505
UF	Gabinetto Disegni e Stampe, Uffizi, Florence	1478–1480
V	Gallerie dell'Accademia, Venice	1487–1494
W	Windsor Castle Royal Library, Windsor	1487–1513
	Common references	
br	*braccio, braccia*	
Cod.	codex	
MS(S)	manuscript(s)	

Introduction

Although Leonardo's *Giant Crossbow* (Fig. I.1) is one of his most popular drawings, it has been one of the least understood. Most scholars have traditionally dismissed its purpose as a mere recreational drawing of a fanciful object, albeit a quintessential example of engineering draftsmanship. As for the design, there has been no adequate appraisal of Leonardo's approach, thanks in part to the generally considered implausibility of the drawing's practical function. *The following discussion* offers the first in-depth account of this drawing's likely purpose and its highly resolved design.

At issue will be the context of Leonardo's invention with an examination of the extensive documentary evidence, a short history of the great crossbow and ballista, the first accurate translation of the text and the technical specifications, and a detailed analysis of his design process for the crossbow, from start to finish. Dozens of preparatory drawings, along with the recent discovery of nearly invisible metal stylus preparatory incisions under the ink of the *Giant Crossbow* drawing, are evidence of Leonardo's intent to offer engineers and other viewers a thorough design of the massive machine. Possibly intended for a treatise on military engineering, the *Giant Crossbow* is one of the earliest dimetric illustrations, and happens to be the earliest surviving illustrated machine that is proposed to scale. To make sense of these new discoveries, the following study will also address a strategy that had been at the core of Leonardo's working philosophy: proportional method. An analysis of the *Giant Crossbow* project, will show how he used a consistent approach to ⅓ proportions throughout the design and drawing process. He employed this kind of proportional strategy at the start of almost every important project: for theories about physical laws (statics and dynamics), for civil and military engineering proposals and treatises, or for pictorial commissions such as the *Adoration of the Magi* and the *Last Supper*. In fact, recent discoveries about the *Giant Crossbow* reveal its importance to his approaches to both art and science (according to his, rather than our, interpretations of these terms). Proportion theory and proportional method assisted his technical drafting strategies in ways that will lead us to critique the traditional descriptions of these proto-scientific approaches as mere intuitive guesswork. For example, new evidence will emerge in the present study of his use of Archimedean and Euclidean geometry to determine the crossbow's primary dimensions, and therefore the appropriate form for its natural force and movement. This is a rare example of a direct influence of a specific Archimedean formula in Leonardo's work for the presentation drawing of a treatise. Thanks to this proof of his knowledge of geometry, evidence of his studies of impetus and force, and thanks to the highly polished and complex nature of the *Giant Crossbow* design, a new date for the drawing will be considered in the present book, associating the drawing with his activities and draughting capabilities between 1483 and 1493.

A question left unanswered in Leonardo scholarship has been: did *he* consider this machine worth building? Although some of the drawings by him and other Renaissance engineers are undoubtedly "paper" engineering, the number of preparatory studies for the Giant *Crossbow* presentation drawing and the functioning of its components indicate that he was developing something that would have resembled a practical machine on paper, as well as at full scale. No evidence exists with regard to the actual construction of this crossbow. Regardless of this, Leonardo's extensive crossbow and siege engine related drawings offer sophisticated approaches—by 15th century standards—to representing solutions for great crossbow construction. This is one reason for the 2002 reconstruction of the machine by ITN Factual for the documentary, *Leonardo's Dream Machines*. The present study will address the limitations of the ITN project as part of the survey of Leonardo's theoretical plans for a treatise illustration of a fully functional giant crossbow and the proportional methods used for its design.

Part I

The Idea: Locating the Giant Crossbow and Its Components

This section will offer an initial analysis of the Giant Crossbow drawing and a study of documents related to its historical context. The drawing of this machine (Fig. I.1) on Codex Atlanticus folio 149br [previously 53vb], and its various preparatory sketches and related studies, reveal Leonardo's impressively thorough approach to the drawing, as well as his proportional techniques for machine design and construction.

First, a description of the drawing: it has autograph illustrations and written statements alla mancina (from right to left), on paper folio 149br [53vb], located in the collection of 1 119 sheets of drawings known as the Codex Atlanticus (CA, see Fig. I.1), located at the Biblioteca Ambrosiana in Milan. Folio 149b is the lower half of folio 149, measuring 204.1 mm on the left side, 278.6 mm along the top side, 209 mm on the right side, and 282.1 mm along the bottom. There is no watermark on the leaf. The verso (back side) is blank, with the exception of raised diagonal and horizontal lines resulting from sharp metal stylus incisions pushed through from the other side. These are visible behind the crossbow carriage, trigger hammer handle and two straight lines between armature ends. Illustrations on the recto (front side) include a giant crossbow, two trigger mechanisms, and a worm screw and gear mechanism, all drawn with pen and dark brown ink with the addition of brown wash. A large number of metal stylus incisions are also visible on the drawing (Fig. 8.20), including four pinholes, a small number of freehand metal stylus marks near the crossbowman, and the majority of straight lines, drawn with the help of a straight edge. Compared to other leaves in the Codex Atlanticus, CA 149br is relatively unfaded. A comparison between the CA folios and a group to which they previously belonged, now located at the Windsor Royal Library, indicates that the former group has faded or degraded more than the latter group. Large portions of both sets were part of a mixed set before Pompeo Leoni edited them around 1582–1590.[1] The condition of the paper of CA 149br is relatively fresh and unmarked. The outer edges of CA 149br are glued to the inside edges of a rectangular aperture inside a large acid-free sheet of paper, the left side of which is stitched into the binding of the second leather volume of the Codex Atlanticus.[2]

---

[1] Pompeo Leoni acquired and edited numerous manuscripts and drawings from the son of Francesco Melzi around 1582–1590. D. Luigi Gramatica (1919) *Le memorie su Leonardo da Vinci di Don Ambrogio Mazenta*, Milan, pp 39–40; notes 27 and 31. I have studied both collections in person and have found the unbound sheets at Windsor generally in better condition than the Atlanticus sheets. For histories of the manuscripts and drawings see: Carlo Pedretti (ed) (1957) *Leonardo da Vinci, fragments at Windsor Castle from the Codex Atlanticus*, London: Phaidon Press, pp 11–18. Kate Steinitz (1948) *Manuscripts of Leonardo da Vinci, their history, with a description of the manuscript editions in facsimile*, Los Angeles. Kenneth Clark (1935/1969) *A catalogue of the drawings of Leonardo da Vinci in the collection of His Majesty the King at Windsor*, Phaidon, Cambridge, revised with C. Pedretti, p ix. A. H. Scott-Elliot (1956) *The Pompeo Leoni volume of Leonardo drawings at Windsor*, Burlington Magazine 98(634):11–17.

[2] It would be beneficial to unbind these sheets. The practise of opening the large leather volumes causes the pages to rub against one another. Around the edges of some sheets, especially those prepared with a bone ground, dust has settled, rubbed from the original drawings.

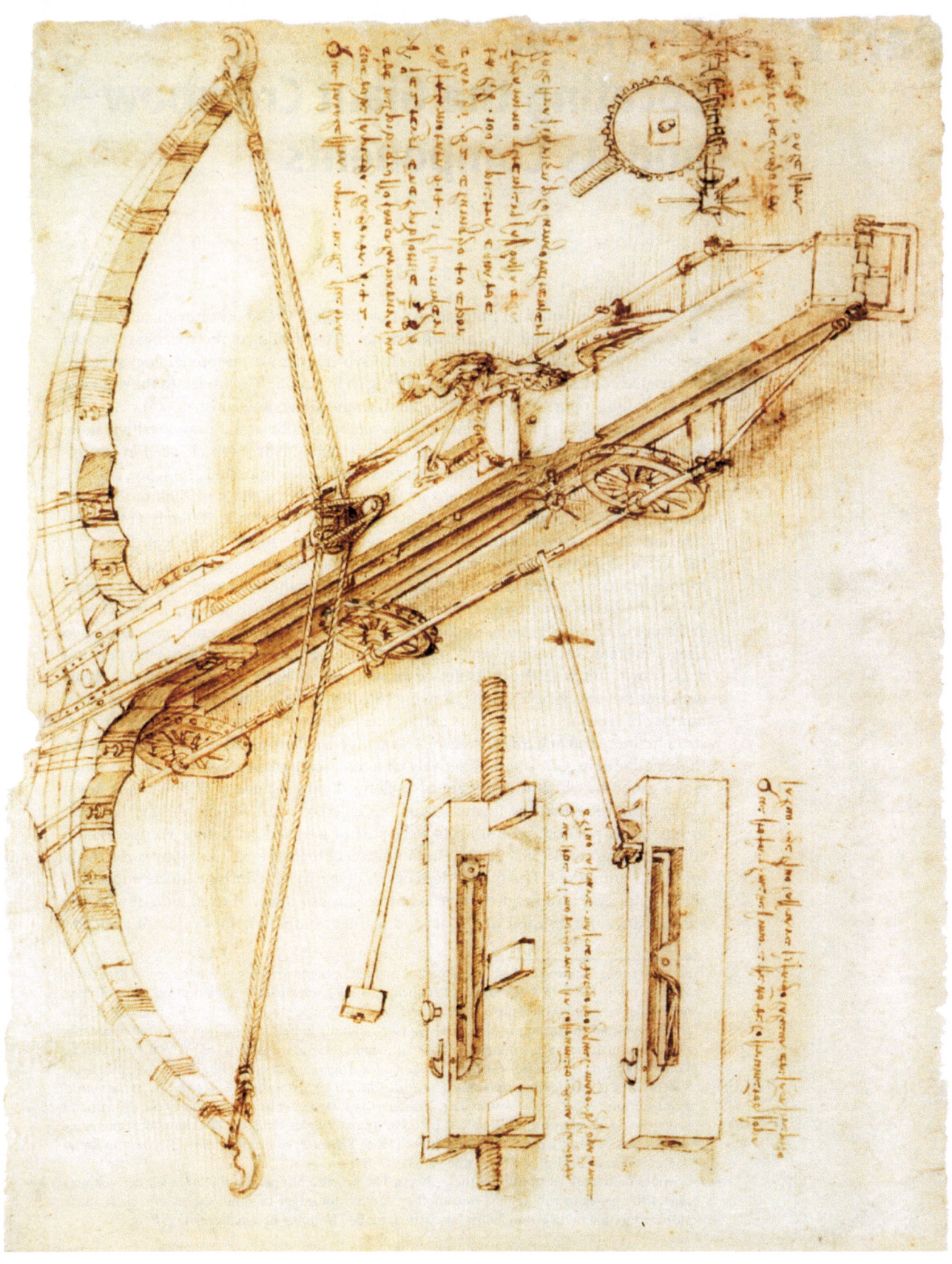

Leonardo produced the drawing sometime between 1483 and the early 1490s as evidence of his skills as an engineer, and for inclusion in a planned treatise on military engineering that would update Roberto Valturio's (1405–1475) famous treatise, De re militari (On the Military Arts, c. 1451–1460).[3] First printed in Verona in 1472, the Latin treatise of twelve books was immediately popular and widely distributed, following the popularity of manuscript copies in the 1450s and 1460s. It is the first book printed with illustrations of a technical nature. Leonardo possibly owned the Italian language version of the treatise, published in Verona by Boninus de Bonninis in 1483. Compare for example Leonardo's series of siege weapons—Figs. I.2, I.3 and I.4 (top centre)—with copies of the woodcut illustrations in *De re militari*—Figs. I.5, I.6 and I.7. For the drawing to be convincing as a treatise illustration and diagram, as in an engineer's diagram or schematic, Leonardo apparently strove to make it look like a serious military proposal, for reasons discussed below.

That the drawing might be a serious military proposal goes against some of the previous scholarship. For example, Jean Liebel—one of the most knowledgeable historians of the great crossbow—writes that,

> The sizes of some of these great crossbows, which a poet called *arbalestes grandesimes*, 'huge crossbows' (Gay 1928: 419), has undoubtedly been exaggerated by some authors who go so far as to attribute to them a span of 5 or 10 m! Leonardo da Vinci (1974a edn, fos 147v, 149r) also amused himself by drawing two [crossbows], of which he said that the bow of one was 14 *braccia*, that is 8.40 m, and the other 42 *braccia*, that is 25 m between the end of each bow (*nelle istremita di ciascuno braccia*)! More reasonably, it can be estimated that the span of the biggest crossbows was, at the most, between 1.6 and 2 m.[4]

It is clear in this statement, and in Liebel's examples that follow it, that there is no evidence of an actual great crossbow before the 16th century that exceeded an armature width of two metres. Whereas this is accurate, one can argue that Leonardo's intention for his design was not mere self-amusement, as per the evidence that will follow. The *Giant Crossbow* was a well-planned project, more thoroughly considered than some of the "paper engineering" examples in contemporary manuscripts. Proposed in this case was not only a dangerous machine—to build, operate, or for one to confront at the business end—but also an engine that the lord of Milan, Ludovico Sforza, would have been compelled to appreciate, as one of the most informed and respected patrons of military engineers in Europe.

Moreover, the second sentence of Liebel's statement mistranslates the note on CA 149br, "*ha di montata braccio* 14," as referring to the bow's width. This phrase actually specifies the bow's "elevation" or "draw"(*montata*), stating that: "it has an elevation [or draw] of 14 *braccia*." With regard to the bow width, the instruction is clear on that folio that the "crossbow opens at its arms… 42 braccia" (*apre nelle sue braccia… b* 42).[5] Without easy access to a reliable translation of the CA 149br text in 1998, Liebel had to rely on his experience as a military engineering historian to draw the natural conclusion

---

[3] I am grateful to Professor Bambach for her advice at the 2007 College Art Association meeting in New York on the likelihood of a later date for the drawing, perhaps as late as 1494, considering its likely associations (stylistic and otherwise) with Leonardo's early treatise programme. At the Metropolitan Museum I presented a paper on the metal stylus marks and other preparatory characteristics of the drawing.

[4] Jean Liebel (1998) *Springalds and great crossbows*, Juliet Vale (trans), Royal Armouries, Leeds, p 25. This is a reference to 147av [52va] and 149br [53vb]. He makes a similar comment about Leonardo's designs for "his own amusement" on CA 147v and 149br, p 31. The reference to "Gay 1928" is: V. Gay (1928) *Glossaire archéologique du Moyen Age et de la Renaissance*, vol. 2, B. Stein (ed), Picard, Paris, p 419. The reference to "1974a edn." is: Leonardo da Vinci (1974) *Il codice atlantico*, Giunti, Florence, folios 147av [52va] and 149br [53vb].

[5] In December of 2003, I wrote to Jean Liebel for his advice on the translation and he kindly replied with a confirmation that he had studied the original drawing and that he drew conclusions from the evidence available at the time. I am very grateful for his supportive letters.

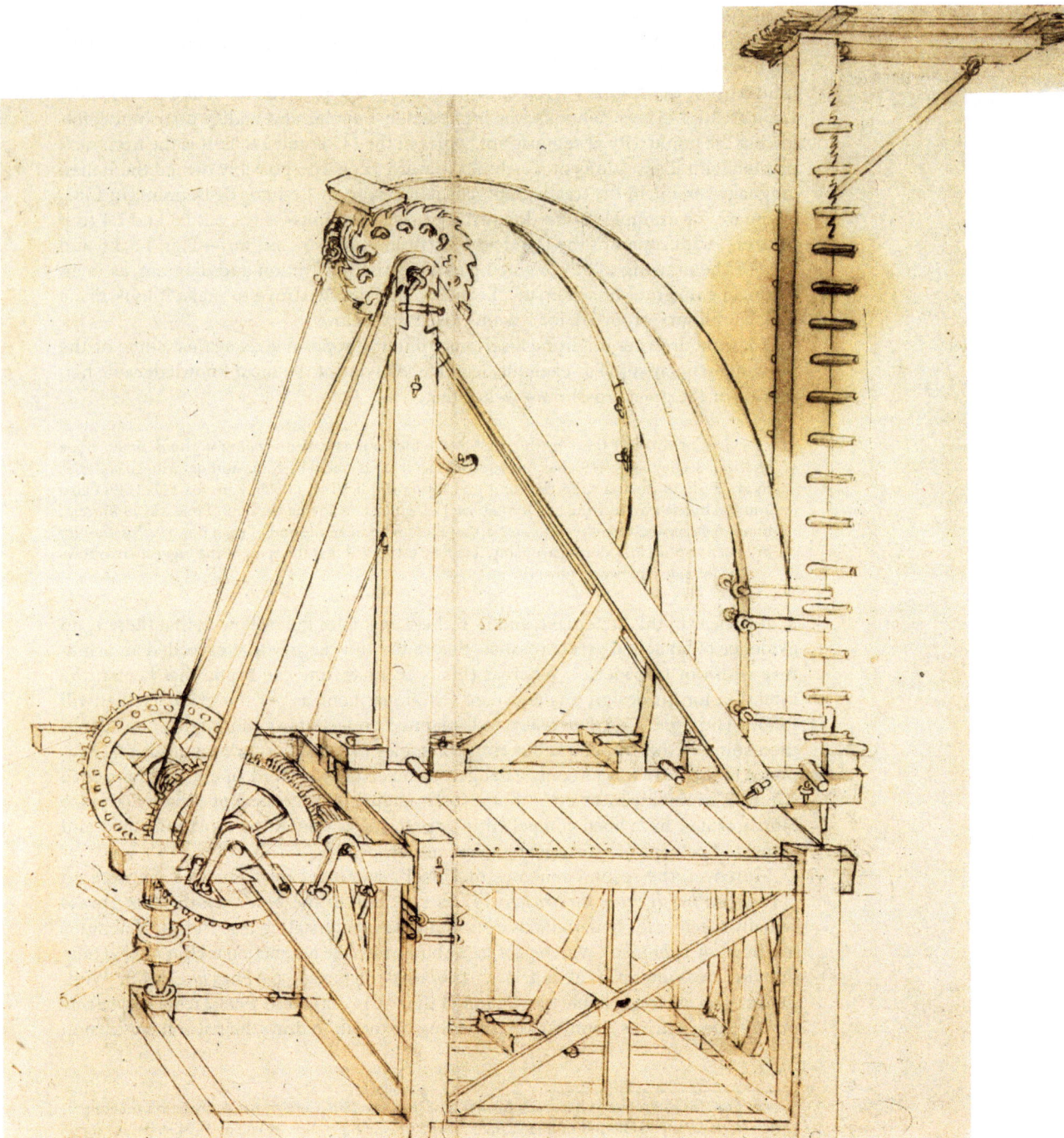

Fig. I.2. Leonardo da Vinci, A vertical bent-beam springald, Codex Atlanticus f. 181r [prev. 64r], Biblioteca Ambrosiana, Milan

that there appeared to be no prima facie evidence that the *Giant Crossbow* was anything more than a fun concept drawing. This had been a general academic consensus about the drawing until very recently. It will be shown in the following discussion, however, that Leonardo seriously considered how a 42 *braccia*-wide bow should be illustrated in accurate detail, and that his interest in CA 149br and 147av (Fig. I.8) was not for self-amusement.

Fig. I.3. ▶
Leonardo da Vinci, Cannon factory, Windsor Castle Royal Library, f. 12647r, Windsor

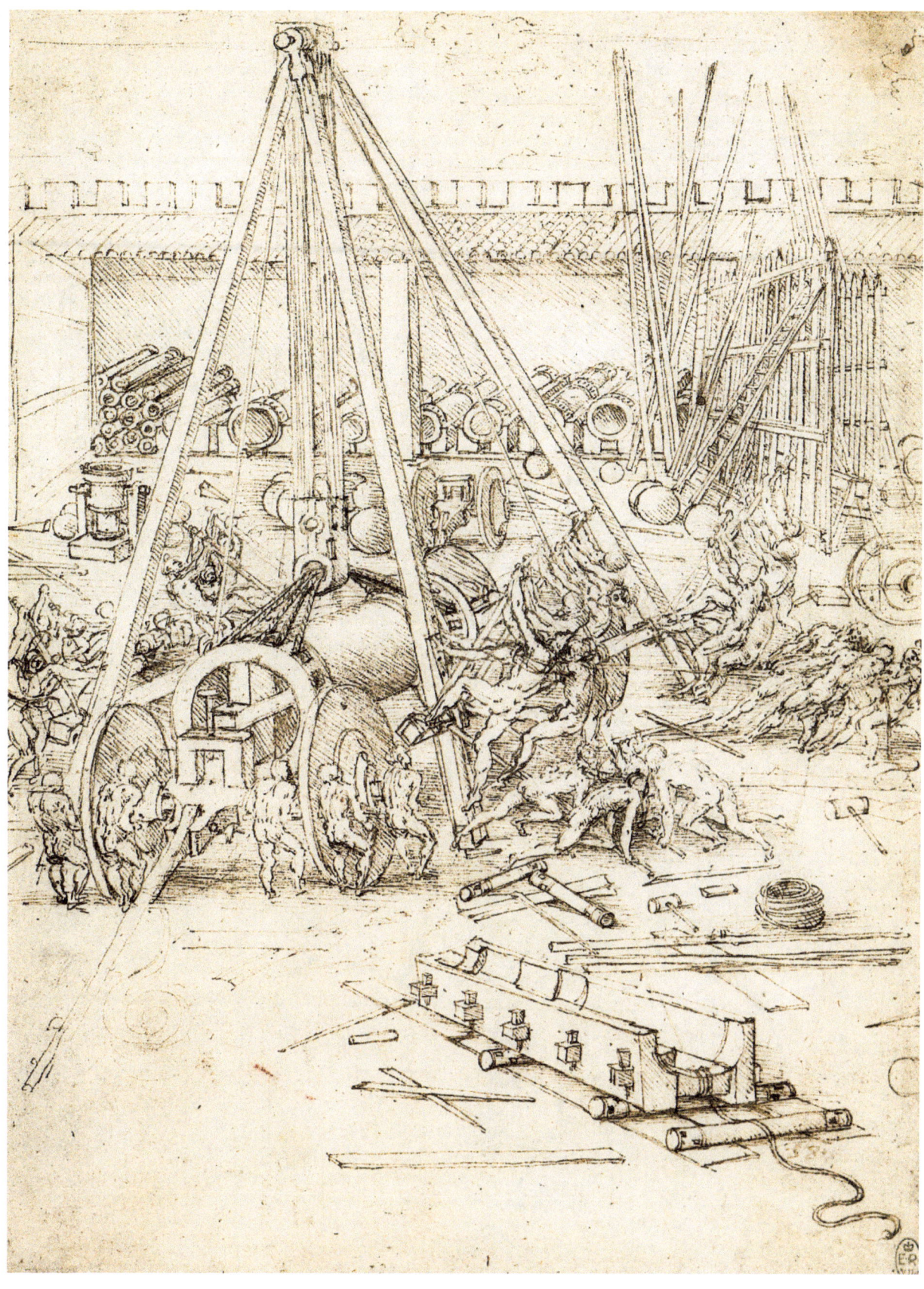

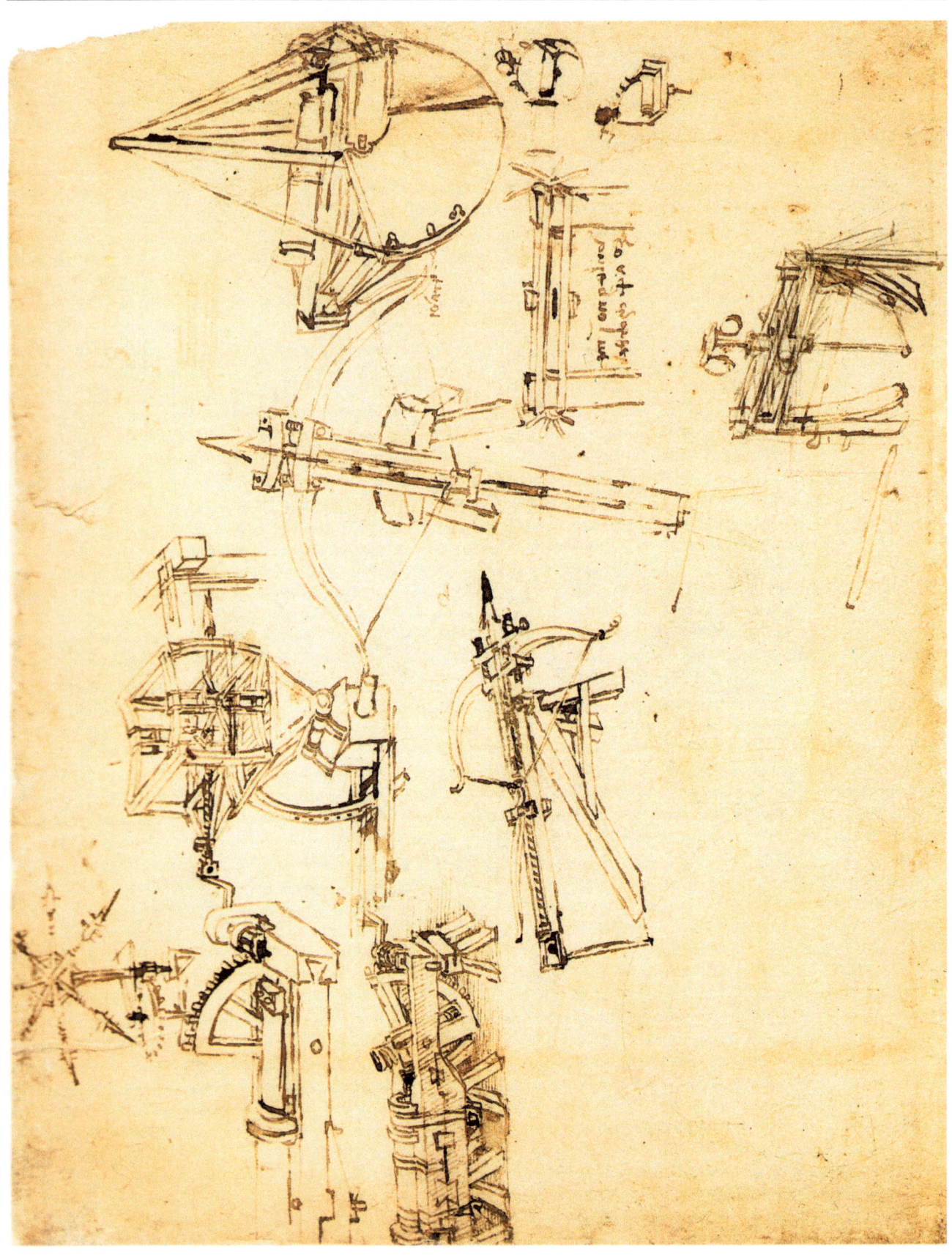

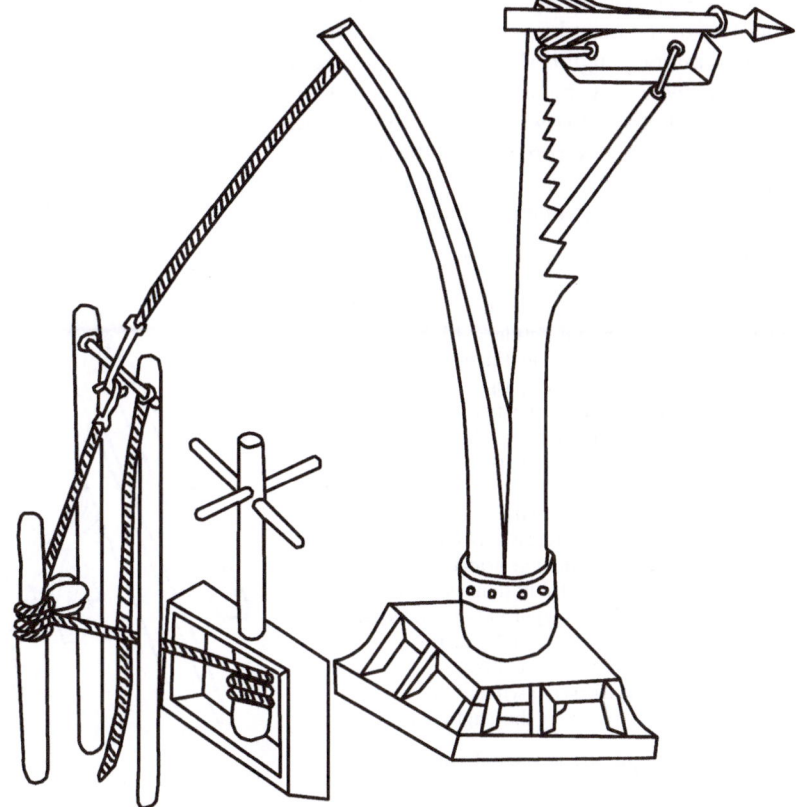

Bern Dibner (1897–1988), a well-known specialist on Leonardo's technical drawings, also mistranslates the text at the right of CA 149br. He writes:

> This crossbow spans 42 ells in its arms;
> that is where the rope is attached;
> and without its armour it is 1 ell and
> ⅔ in the thickest point, and ⅔ of an
> ell in the thinnest one. Its mounting is 14 ells.
> Its stem is 2 ells large and 40 long; and
> it charges 100 pounds of stone;
> and when it is on the way, the stem is
> lowered and the crossbow is aligned along the stem.[6]

A Roman ell is a large cubit, nearly equivalent to a *braccio mercantile milanesi* of
59.5 cm. This is a close estimate, although another option may be to consider the late
15th century Florentine *braccio* of 58.36 cm, if one were to accept 20th century scholars" opinions (as discussed below) that Leonardo made the drawing soon after travelling from Florence to Milan by April 1483. The Milanese *braccio* may however be more
accurate if the drawing dates to around 1487–1493. Dibner refers to the *armadura* as
"armour," which is incorrect. An *armadura* in this context is simply a crossbow armature, or the bow at the front of the crossbow stock. Leonardo's "*mo[n]tata b • 14*" refers
to an "elevation [or draw of] 14 *braccia*," or arm lengths, rather than Dibner's "mounting [of]
14 ells." Additionally, the meaning of the latter translation is unclear (e.g.: what is a
mounting of 14 ells?). Leonardo's "*porta lib[bre]* • 100 *di pietra*" translates as, "carries
100 pounds of stone," rather than "charges 100 pounds of stone."

---

[6] Bern Dibner in: Ladislao Reti (ed) (1974) *The unknown Leonardo*, Emil M. Bührer (designer), McGraw-
Hill, New York, p 177.

The examples of Liebel's and Dibner's translations show that two of the most knowledgeable scholars on early modern military engineering have misinterpreted the *Giant Crossbow* drawing. Still, the drawing is not easy to interpret. I offer below what appears to have been the first correct English interpretation of the statements on CA 149br:[7]

On the right side of CA 149br, Leonardo notes that:

1   Questa belestra · apre · nelle sue bracc[i]a[8]
2   cioè dove s'ap[p]icca · la corda · b · 42 · ·[9]
3   ed è nel più · grosso sanza l'armadura sua
4   · b · uno e 2 terzi, e nel più sottile ⅔ d[i] b
5   [h]a di mo[n]tata b · 14 · il suo tinierj[10]
6   è largho · b 2 · è lumgho 40 e por[-][11]
7   ta lib[bre] · 100 di pietra e quando è
8   in cam[m]ino il tenierj s'abbassa e la
9   balestra si diriz[z]a p[er] lo lumgo del tenierj

This crossbow opens at its arms,
that is where the rope is attached, 42 *braccia*,
and is at its thickest, without its armature,
1 and 2 thirds *braccia*, and at its thinnest, ⅔ of a *braccio*[12]
It has an elevation [or draw] of 14 *braccia*. Its carriage
is 2 *braccia* wide and 40 long and it carries
100 pounds of stone; and when it is
moving, the carriage lowers itself and the[13]
crossbow directs itself along the length of the carriage.[14]

At the bottom right of the page, Leonardo states:

10   Tirare de la corda
11   della · balestra

To pull the rope
of the crossbow.

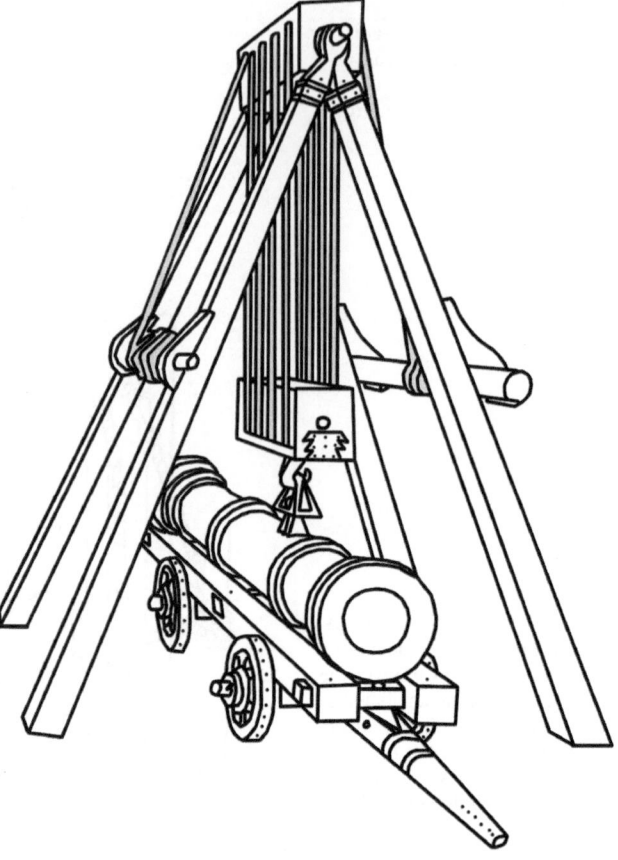

Fig. I.6. Illustration of the woodcut of a cannon with pulleys and carriage, the original illustration attributed to Matteo Pasti, in: Roberto Valturio, *De re militari*, 1472, Bodleian Library MS Douce 289, unfoliated, Oxford

[7] First published in: Landrus (2008) *The proportional consistency and geometry of Leeonardo's Giant Crossbow*, Leonardo 41(1):59. I am very grateful to Professors Martin Kemp and Dario Covi for their advice on the translation. The main source for this translation is: Giacomo Devoto, Gian Carlo Oli (eds) (1995) *Il dizionario della lingua italiana*, Le Monnier, Firenze.

[8] The line breaks and · marks in this transcription are in agreement with the line divisions (in reverse) and marks on CA 149br (Fig. I.1). Words lacking a letter in the original statement have that letter added in the present transcription in brackets.

[9] For the word "*braccia*" in this and subsequent instances, Leonardo used the symbol of an exaggerated form of b (the letter b, backwards).

[10] Leonardo's actual word for "montata" is "motata." There seems to be no such word as "motata" in Italian. Augusto Marinoni transcribes the word as "montata" in: Augusto Marinoni (transcr) (2000) *Leonardo da Vinci, Il Codice Atlantico della Biblioteca Ambrosiana di Milano*, 3 vols., (which comprise the 12 vol. set), Giunti, Florence, vol. I, p 201.

[11] In this line, "lungho" is Leonardo's spelling of "lungo," whereas this word does not have the 'h' in the ninth line. The division of the word "porta" between lines six and seven has left a reasonable space between the folio's text and the crossbowman to the left. This shows that Leonardo wrote the text after he drew the crossbowman and the crossbow.

[12] Literally, "più sottile" means "at its smallest," but it fits best in the present context as "thinnest" or "narrowest."

[13] Probably the best literal translation of "in cam[m]ino" is "en route," but within the present context, Leonardo simply refers to any instance when the crossbow is in motion. If the upper carriage were not lowered at the time that the machine is being moved or transported, the whole crossbow would tend to tip forward onto its front end.

[14] A literal interpretation of "si dirizza per lo lungo" would be "directs itself toward the length."

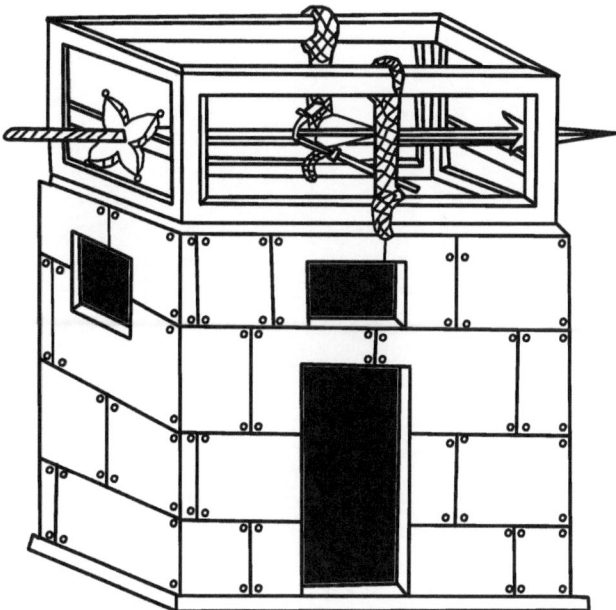

Fig. I.7. Illustration of the woodcut of a torsion spring springald, the original illustration attributed to Matteo Pasti, in: Roberto Valturio, *De re militari*, 1472, Bodleian Library MS Douce 289, unfoliated, Oxford

Under the centre diagram on the left of the page, he states:

12 Questo è il modo come • sta lo strumento • che va per la corda
13 el suo lasciare • nasce dal colpo di quel[l]o maz[z]o di sopra la noce

This is the way in which the instrument behaves. It works with the rope,[15]

its release commencing with the blow of this mallet above the nut.[16]

Under the bottom left diagram, he writes:

14 Questo fa il medesimo effetto de lo strumento di sopra •
15 salvo • che 'l suo lasciare si fa con la lieva ed è sanza strepi[t]o[17]

This makes the same effect of the instrument above, except that its release is done with the lever and it is without clamour.[18]

According to these statements, an interpretation of the giant crossbow's proposed dimensions may follow, as noted here, to the right of the transcription line numbers, and as supported in the rest of the present study. See also Appendix A.

1–2   The illustrated distance between each end of the armature, where "its rope is attached," is 42 *braccia* (24.51 m, or 80.4 ft).

3–4   The thickest section of the upper carriage—containing the screw, is 1⅔ *braccia* thick (97.5 cm, 38⅓ in).

4   The thinnest portion of the "carriage" is the lower carriage's thickness of ⅔ *braccio* (39 cm, or 15⅓ in), which is also the illustrated thickness of the top and bottom sections of the upper carriage.

5   The front end of the upper carriage can be elevated, while attached at the back end with an iron hinge, to a distance of 14 *braccia* (8.17 m, or 26.8 ft) from the ground.

5–6   The carriage width is 2 *braccia* (117 cm, or 46 inch, though the illustrated width would be 3 *braccia*, if it were consistent with other measurements on the drawing).

6   The carriage length is 40 *braccia* (23.34 m, or 76.6 ft, though only the length of the upper carriage section, which would measure approximately 37⅓ *braccia*, is illustrated).

6–7   The crossbow can carry (or shoot) 100-pound stones or baskets filled with "100 pounds of stone."

7–8   The carriage can lower its upper portion when in motion.

8–9   When the carriage's upper portion "lowers itself," the crossbow rests or "directs itself" along the length of the lower carriage or mount.

---

[15] This is not a strict literal translation of all words in the two lines, but serves as an interpretation that conveys the same meaning as Leonardo's statement.

[16] Leonardo writes "mazo," but likely refers to a "mazza": mallet.

[17] Leonardo's spelling of "strepito" is "strepido."

[18] Leonardo also refers to "strepido" or clamour on MS B f. 27v. See: Leonardo da Vinci (1999) *Manuscript A. The Manuscripts of Leonardo da Vinci in the Institut of France*, John Venerella (ed, trans, annot), Ente Raccolta Vinciana, Castello Sforzesco, Milan, p 80. C. Ravaisson-Mollien (ed) (1881–1891) *Manuscrit A, Les Manoscrits de Léonard de Vinci*, 6 vols., Maison Quantin, Paris, vol. I, folio 27 (verso).

◀ Fig. I.8.
◀ Fig. I.8.
Leonardo da Vinci, Preparatory
sketch for the *Giant Crossbow*,
Codex Atlanticus f. 147av
[prev. 52va], Biblioteca
Ambrosiana, Milan

10–11   The gearing that turns the central screw and thereby pulls the trigger mechanism, along with its rope, happens to be large gear, turned by a worm screw at
the centre of a capstan below the gear. Bars inserted at each end of the capstan
enable the rotation of the capstan by hand.

12–13   The first line refers to the screw and the side supports that hold the trigger
mechanism in place and allow it to work "with the rope," sliding along the upper part of the "carriage." The second line refers to the action of the trigger
mechanism. If the illustrated mallet strikes the metal knob at the top of the
trigger "instrument," the knob would push down a bar and the spring beneath
the bar. This would disengage the circular "nut" at the front of the trigger, which
would rotate forward and let the rope go.

14–15   This mechanism would have the same result as the one above it, except that it
works "with the lever and it is without clamour."

Explanations of these interpretations follow in Part II. Regarding the proportional
units that Leonardo used in the drawing, a summary of the estimated lengths is as
follows. These measurements refer to his third-part divisions of the crossbow with the
help of a compass and ruler. Such third-part divisions confirm the proportional relationships between the crossbow's dimensions as illustrated and as specified in the text. This
list therefore examines the drawing's proportions, more than its actual measurements.
There is no evidence to suggest that Leonardo carefully considered any measurements
in the drawing other than those stated in his written specifications on the folio. To
prove this, however, one needs to measure crossbow features on the original drawing
in order to check for specific measurements. Appendix A offers a list of my measurements, taken at the Biblioteca Ambrosiana in February 2004. Abbreviations in this list
include a Florentine *braccio* (48.23 cm), noted as "br". One can see on this list a proportional division of crossbow components into third portions, as noted on the "br"
column of lengths estimated in *braccia*. As for portions that conform to specific measurements, there is no evidence to suggest that Leonardo measured more than one or
two portions of the drawing in order to determine its possible scale in relation to the
crossbow's proposed life-size dimensions. As noted in the "est. scale" (estimated scale)
column in the middle of Appendix A, a comparison of the drawing's measurements and
his written specifications on the right of the drawing reveals significant differences
between the drawing's measurements and the proposed life-size dimensions. For example, the illustrated carriage width far exceeds (by 33%) that of the carriage width if
it were to represent two *braccia* on a drawing with an average illustrated *braccio* of
0.54 cm. Whereas the armature width and carriage length would measure 0.54 cm per
braccia, when comparing drawing measurements with Leonardo's specifications (drawing measurements ÷ 42, and 40), the carriage width would equal 0.82 cm per *braccio*.
More on the crossbow's proportions and measurements continues in Part II.

1

Proposal to Ludovico Sforza, CA 1082r [391ra]

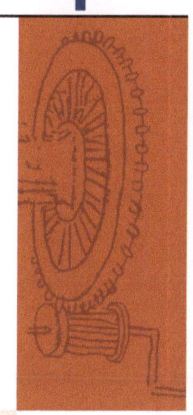

If we compare Leonardo's proposal to Ludovico Sforza with the *Giant Crossbow*, as has been the tradition, both items may belong to the period of 1483–1489. It should be noted, however that a perfectly reasonable date for the drawing would also be the in the early 1490s, according to Professor Carmen Bambach, particularly if we consider the precision of its execution and its likely inclusion in a treatise on military engineering.[1] Thus, though the letter proposes a machine like the *Giant Crossbow*, it is possible that the drawing post-dates the letter by a few years. The *Giant Crossbow* has been traditionally attributed to the 1485–1488 period, according to Calvi, Canestrini, Parsons, and Popham.[2] Pedretti dates it closer to c. 1485.[3] Popham refers to the relationship between this bow and Leonardo's notes in MS B on Valturius" *De re Militari*, which Popham dates to the mid-to-late 1480s.[4] He links these studies with Leonardo's letter to Ludovico Sforza (1452–1508) of Codex Atlanticus 1082r and with CA folios 149br, 85ar, 140ar, 141r, 145r, 147av, 147bv, 148ar, 148br, 150r, 151r, 152r, 159ar, 159br, 160br, 182br, 1048br, and 1070r (see Figs. I.1, 1.1–1.7).[5] Popham's 1481 date for the Sforza letter may be a year or two too early, for reasons discussed further below, though he correctly notes the relationship between the letter and the CA military drawings, even if the drawings post-date the letter by several years.[6] Calvi produced a definitive study of the famous Sforza letter, dating it to c. 1482.[7] Luca Beltrami and Pedretti also date the letter to 1482.[8] It would appear, however, that Leonardo had the letter written for him (as it is not in his handwriting) within a year or two after arriving in Milan in 1482–1483, and that he could have displayed the *Giant Crossbow* and military engineering treatise illustrations within the seven-year period following his arrival in 1483.

[1] As noted further above, this was discussed in person at the 2007 College Art Association meeting in New York.

[2] A. E. Popham's date is c. 1485–1488 in: Jonathan Cape, M. Kemp (ed, rev, intro) (1946/1994) *The drawings of Leonardo da Vinci*, London, pp 81–84, 162. For the 1485–1488 estimated date, see also: G. Calvi (1925) *I Manoscritti di Leonardo da Vinci dal punto di vista cronologico, storico e biografico*, Nicola Zanichelli, Bologna, p 107. G. Canestrini (1939) *Leonardo costruttore di macchine e veicoli*, Milan, pp 88–89. W. Parsons (1968) *Engineers and engineering in the Renaissance*, Harvard, Cambridge, Mass, p 45.

[3] Pedretti's date of c. 1485 is in: Carlo Pedretti (1978) *Leonardo da Vinci, Codex Atlanticus, A catalogue of its newly restored sheets*, Johnson Reprint Corp., New York, p 88.

[4] MS B, or Manuscript B, is the name posthumously given to this notebook of Leonardo which is now located at the Institute de France, Paris. Popham, p 83.

[5] Popham, pp 84–85 and 159–162. CA folios 1082r [391ra], 85ar [31ra], 140ar [50va], 141r [51ra], 145r [51vb], 147av [52va], 147bv [52vb], 148ar [53ra], 148br [53rb], 150r [54ra], 151r [54rb], 152r [54va], 159ar [57ra], 159br [57rb], 160br [57vb], 182br [64vb], 1048br [376rb], 1070 [387r].

[6] Popham, p 79.

[7] Calvi, *Manoscritti*, pp 65–70.

[8] L. Beltrami (1919) *Documenti e memorie riguardanti la vita e le opere di Leonardo da Vinci in ordine cronologico …*, Milan, no. 21. C. Pedretti (1977) *The literary works of Leonardo da Vinci, compiled and edited from the original manuscripts by Jean Paul Richter, commentary by Carlo Pedretti*, vol. II, University of California Press, Berkeley, p 295.

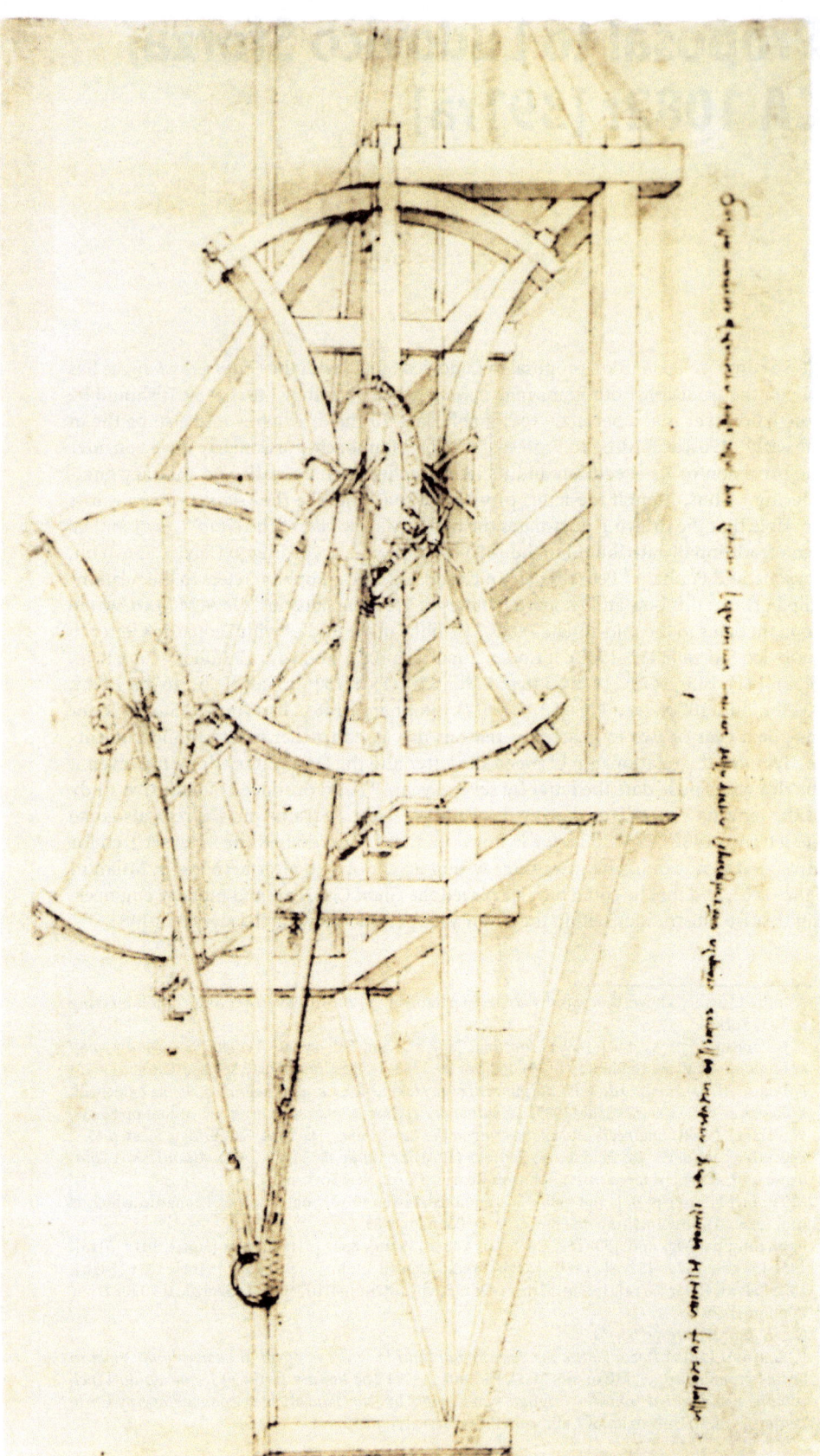

Fig. 1.1.
Leonardo da Vinci, A giant bent-frame springald, Codex Atlanticus f. 145r [prev. 51vb], Biblioteca Ambrosiana, Milan

He draughted this proposal to Ludovico:

My most illustrious Lord, having now sufficiently seen and considered the achievements of all those who count themselves masters and artificers of instruments of war, and having noted that the invention and performance of the said instruments is in no way different from that in common usage, I shall endeavour, while intending no discredit to anyone else, to bring myself to the attention of Your Excellency for the purpose of unfolding to you my secrets, and thereafter offering them at your complete disposal, and when the time is right bringing into effective operation all those things which are in part briefly listed below:

1. I have plans for very light, strong and easily portable bridges with which to pursue and, on some occasions, flee the enemy, and others, sturdy and indestructible either by fire or in battle, easy and convenient to lift and place in position. Also means of burning and destroying those of the enemy.
2. I know how, in the course of the siege of a terrain, to remove water from the moats and how to make an infinite number of bridges, mantlets and scaling ladders and other instruments necessary to such an enterprise.
3. Also, if one cannot, when besieging a terrain, proceed by bombardment either because of the height of the glacis or the strength of its situation and location, I have methods for destroying every fortress or other stronghold unless it has been founded upon rock or so forth.
4. I have also types of cannon, most convenient and easily portable, with which to hurl small stones almost like a hail-storm; and the smoke from the cannon will instil a great fear in the enemy on account of the grave damage and confusion.
9. Also, I have means of arriving at a designated spot through mines and secret winding passages constructed completely without noise, even if it should be necessary to pass underneath moats or any river.
5. Also, I will make covered vehicles, safe and unassailable, which will penetrate the enemy and their artillery, and there is no host of armed men so great that they would not break through it. And behind these the infantry will be able to follow, quite uninjured and unimpeded.
6. Also, should the need arise, I will make cannon, mortar and light ordinance of very beautiful and functional design that are quite out of the ordinary.
7. Where the use of cannon is impracticable, I will assemble catapults, mangonels, trebuchets and other instruments of wonderful efficiency not in general use. In short, as the variety of circumstances dictate, I will make an infinite number of items for attack and defence.
8. And should a sea battle be occasioned, I have examples of many instruments which are highly suitable either in attack or defence, and craft which will resist the fire of all the heaviest cannon and powder and smoke.
10. In time of peace I believe I can give as complete satisfaction as any other in the field of architecture, and the construction of both public and private buildings, and in conducting water from one place to another.

Also I can execute sculpture in marble, bronze, and clay. Likewise in painting, I can do everything possible as well as any other, whosoever he may be.

Moreover, work could be undertaken on the bronze horse which will be to the immortal glory and eternal honour of the auspicious memory of His Lordship your father, and of the illustrious house of Sforza.

And if any of the above-mentioned things seem impossible or impracticable to anyone, I am most readily disposed to demonstrate them in your park or in whatsoever place shall please Your Excellency, to whom I commend myself with all possible humility. [Codex Atlanticus 1082r/391ra][9]

His reason to list skills as a military engineer in nine out of the ten offers in this draught letter indicates that he was possibly aware of Ludovico's interest in hiring engineers around 1483 and in subsequent years. Additionally, the offer to demonstrate the machines in Ludovico's park (*parco*) may refer to Leonardo's personal awareness

---

[9] Translated by Martin Kemp and Margaret Walker in: *Leonardo on painting*, Yale University Press, New Haven, 1989, §612, pp 251–253. Augusto Marinoni's transcription is in: *Il Codice Atlantico*, Tomo III (vol. XII), pp 1937–1939. This was also transcribed and translated in: Jean Paul Richter (1883/1970) *The notebooks of Leonardo da Vinci*, Sampson, Low, Marston, Searle & Rivington, London, 1883; Dover ed. 1970, §1340, p 397. Leonardo had this proposal on CA 1082r written for him by a writer using a humanist script from left to right. This is a very different handwriting style from that of Leonardo's. He preferred to write in a mercantile script from right to left.

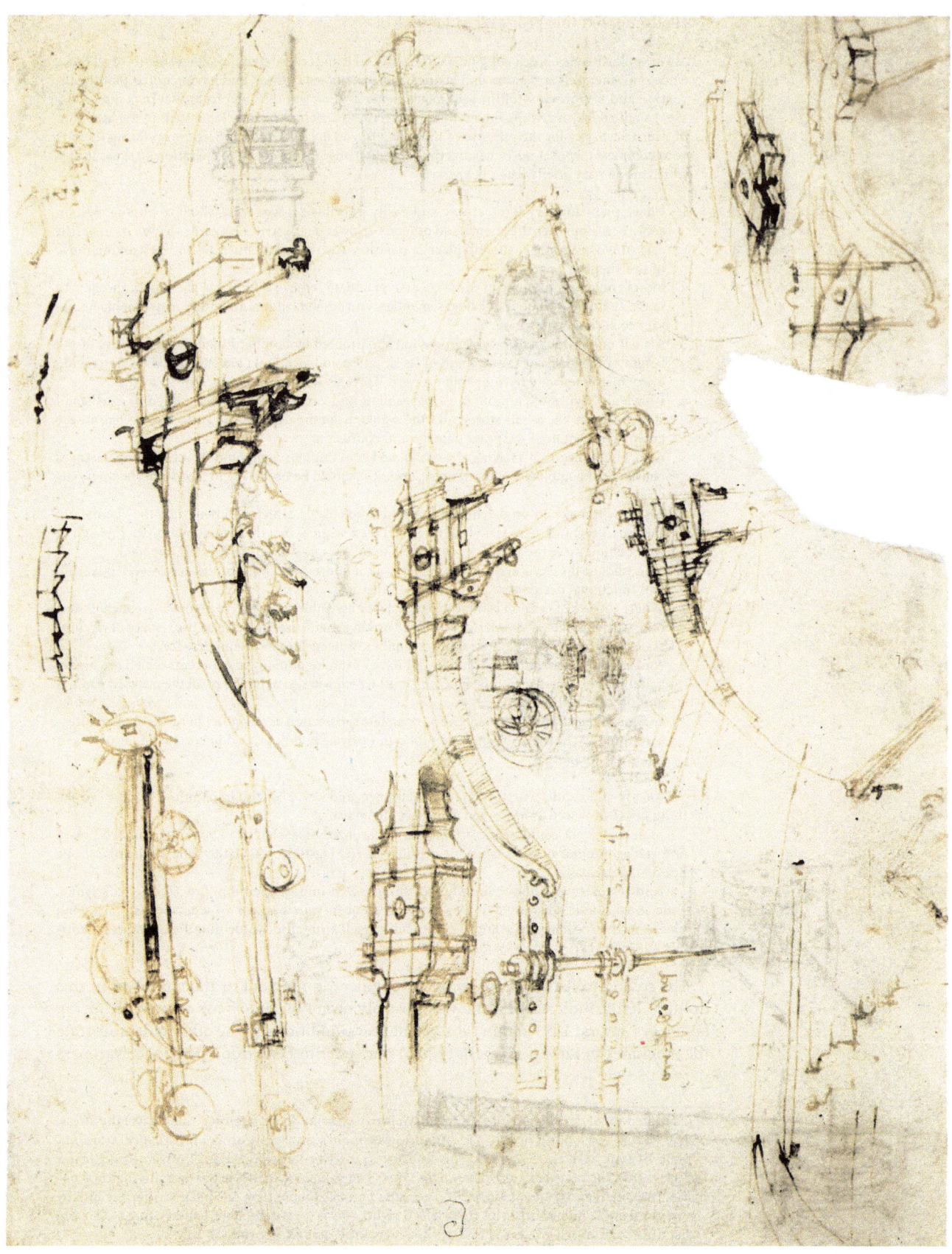

of the "park" of several acres inside the front section of the Sforza Castle. There is no evidence to suggest that he was personally more interested in engineering than in painting (though this could have been the case at various periods in his career). He was undoubtedly aware that the manufacture of arms in the heavily fortified city had, "given name, fame, and prosperity to Milan from as far back as the 13th century," to use the words of Ivor Hart.[10] A more obvious motivation for Leonardo's proposal, however, would have been the need for engineers at the time. Hart notes that, around 1483, Ambrogio Ferreri replaced the retiring Bartolommeo Gadio as the Sforzas' director of military engineering.[11] Ambrogio Ferrarri built fortifications for Galeazzo Maria Sforza in 1476 and there are other records of his work for the Sforzas from 1480 through 1499. Any promotion that he may have received in the early 1480s might have been expected for an engineer with his service record. Leonardo refers to "Messer Ambrosio Ferrere" on CA 887r [323rb] (possibly around 1493/1494) as a man of "some power" (*qualche commessione*) in the service of Ludovico.[12] In 1483, Gadio's post might have been temporarily vacant, encouraging Leonardo's written proposal and some of his studies of siege machinery. In any event, engineering assistance was generally in demand when Milan was at war with Venice in 1483–1484, costing the Sforza family nearly three quarters of their total budget.[13] Shortly after the Venetian Republic had taken Ferrara in the autumn of 1481, a League of states had joined forces against Venice. By February 1483, this League included Ferrara, Milan, Mantua, Florence, Naples, and the Papal States. As noted by Machiavelli, Ludovico Sforza made a decisive tactical manoeuvre in 1483; "Leaving the marquis of Ferrara to the defence of his own territories, he, with four thousand horse and two thousand foot—and joined by the duke of Calabria [Pope Sixtus' Legate to the League of states] with twelve thousand horse and five thousand foot—entered the territory of Bergamo, then Brescia, next that of Verona, and in defiance of the Venetians, plundered the whole country."[14] By August 1484, Ludovico had made the Venetians agree to a treaty that ended the war. There was a need at that time for cannon and accurate long-range siege engines, given the tactical benefits of these weapons. This is one reason to date the start of the Leonardo's initial interest in producing *De re militari* treatise project and drawings like the *Giant Crossbow* to around 1483–1484, rather than traditional date of 1482–1483.[15] As luck would have it, the timing of the draught letter also coincided with Ludovico's interest in honouring his father with an equestrian statue. Thus, Leonardo had an opportunity in the early 1480s to propose a full-time position that would involve the horse, royal portraits, festival designs, and engineering commissions, knowing the Sforza habit of keeping employees to a minimum by outsourcing individual projects as much as possible. Leonardo's full-time salaried position was not a reality until 1489, a *terminus ante quem* for the letter and possibly any related military engineering illustrations.

Locating Leonardo, c. 1480–1484

Although the proposal to Ludovico Sforza and initial military engineering drawings in Manuscript B and possibly those in the Codex Atlanticus were likely pre-

---

[10] Ivor Hart (1961) *The world of Leonardo da Vinci, man of science, engineer and dreamer of flight*, MacDonald, London, p 86.

[11] *Ibid.*, p 88.

[12] *Codice Atlantico*, Tomo III, p 1649.

[13] An early standard source for this is: Nino Valeri L'Italia (1949) *Nel'età dei principati dal 1343 al 1516*, vol. 5, Arnoldo Mondori, Verona, pp 611–615. The seventy percent drain on the Sforza budget for the war with Venice in 1483–1484 is noted most recently by Frank Zöllner in: F. Zöllner, with J. Nathan (2003) *Leonardo da Vinci, 1452–1519, the complete paintings and drawings*, Taschen, Köln, p 64.

[14] N. Machiavelli (1970) *The history of Florence*, Myron Gilmore (ed), Twayne Publishers, New York, book 8, chapter V, paragraph 8.

[15] See: Calvi, *Manoscritti*, pp 67–72. Pedretti, *Commentary*, vol. 2, p 295.

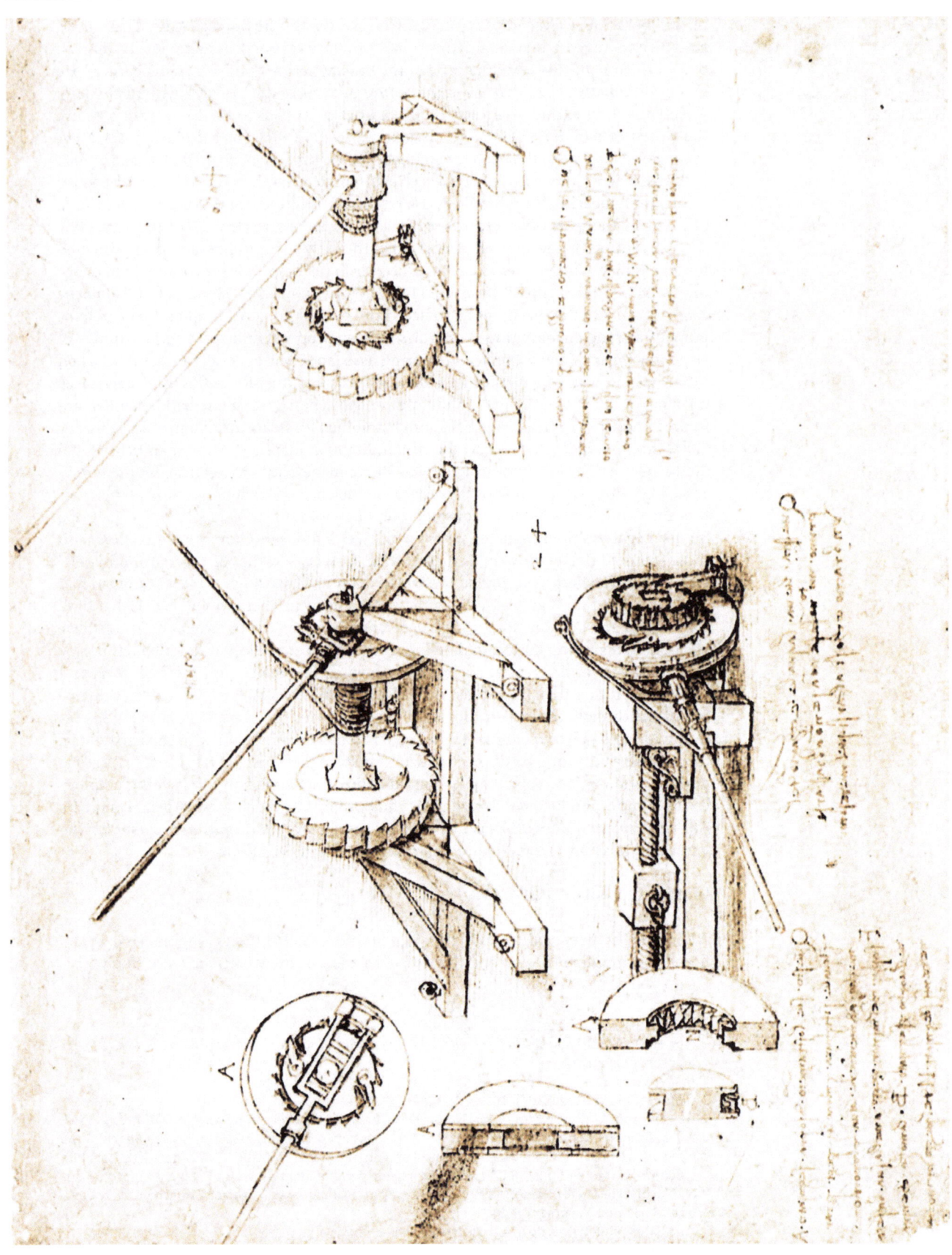

pared before 1489, the earliest dates for these studies are more difficult to determine. Kemp and Schofield believe that Leonardo had written his proposal to Ludovico after 1482, noting that Leonardo had probably gone to Milan late in 1482 as a painter and thereafter sought employment as an engineer.[16] In any event, it is likely that he was in the city around 25 April 1483, the contract date for a Milan-based confraternity's altarpiece now called the *Virgin of the Rocks*, the central painting for which is located at either the Louvre Museum, Paris, or at the National Gallery, London.[17] Thus, near the end of 1482, he probably expected to find work at least as a painter or sculptor in Milan, a city around two or three times the size of Florence.[18]

Another possibility is that Ludovico summoned Leonardo to Milan sometime between September 1481 and April 1483, and that Lorenzo de' Medici used the opportunity to send the young musician, Atalante Migliorotti, to present a *lira da braccio* to Ludovico. Various documents suggest this. First, Benedetto Dei records Leonardo and Atalante Migliorotti in Milan on the fifteenth of June 1480, though the exact date may be incorrect.[19] Second, Anonimo Gaddiano noted in 1540 that Leonardo was thirty years of age when he accompanied Atalante to Milan, which suggests a period for the visit of 1481–1483, when Leonardo was around twenty-nine to thirty-one years old.[20] Third, payment records for the *Adoration of the Magi* and other documents locate Leonardo in Florence from March 1481 through the end of September.[21] Fourth, Codex Atlanticus folio 888r-v [324r-v] contains the phrase, "head of the Duke," profile studies of old faces possibly for the Sforza monument, and early military engineering studies datable to around 1480–1482.[22] For reasons discussed below, these documents point to coinciding financial interests of the Medici, Ludovico Sforza's interest in making himself Duke of Milan, and Leonardo's possible interest in Milanese commissions between 1480 and 1482.

The chronicler, Benedetto Dei, made an account of Florentines in Milan on 15 June 1480, which included "Lionardo da Vinci dipintore, […], e Atalanta della viola." This is in the following document:

> Memoria da 15 di g[iug]nio 1480 di tutti e merchanti fiorentini che venuti sono a Milano al tempo di Benedetto Dei:
> (…) e Luigi Pulci, e Gian Perini, e Tenperano di messer Manno Tenperani, e Andrea Billincioni, e Dino Bettini, e Iachopo di Tanai de' Nerlli, e Franc° Ciarpellone, e ser Franc° Ghaddi, e Lucha del

---

[16] Professor Kemp discussed with me by e-mail on 29 May 2002 the likelihood of a 1483–1484 date for the letter. Schofield discusses the probability of a date anywhere between 1483 and 1489 for Leonardo's letter in: R. Schofield (1991) *Leonardo's Milanese architecture: career, sources and graphic techniques*, Achademia Leonardi Vinci 4:113–116.

[17] This contract is translated in Kemp and Walker (1989) §667, pp 258–270. A transcription by Janice Shell and Grazioso Sironi is in: *Zenale e Leonardo, tradizione e rinovamento della pittura lombara*, ex. cat., Museo Poldi Pezzoli, Milan, 1982, p 67. See also: Edoardo Villata (ed) (1999) *Leonardo da Vinci, I documenti e le testimonianze contemoranee*, Ente Raccolta Vinciana, Milano, Castello Sforzesco, no. 23, pp 19–34. And see: Luca Beltrami (1919) *Documenti e memorie riguardanti la vita e le opere di Leonardo da Vinci*, Milan, no. 24.

[18] Various estimates exist of the populations of both cities at the end of the 15th century. Most recently, Frank Zöllner states that Florence numbered 41 000 residents whereas Milan had a population of 125 000. Zöllner (2003) *Leonardo*, p 64. Galluzzi notes that, by 1482, Milan was already a fully developed metropolis of 200 000 inhabitants, more than twice the size of Florence. Galluzzi (1999) *The art of invention*, p 57. Hart states that, around 1483, Milan had a population of around 300 000; Hart, p 86.

[19] "Memoria da 15 di g[iug]nio 1480 di tutti e merchanti fiorentini che venuti sono a Milano al tempo di Benedetto Dei," *Codex Magliabechiano* II: 333, f. 51r, Biblioteca Nazionale, Florence.

[20] Anonimo Gaddiano's comments are in: *Codex Magliabechiano* XVII: 17, located in the Biblioteca Nazionale, Florence.

[21] See: Villata (1999) nos. 14–18, pp 12–14, from the Archivo di Stato di Firenze, Corporazioni religiose soppresse dal governo francese 140, 3, ff. 74r–81v.

[22] The military engineering studies are discussed further below. For the date of CA 888, see Calvi (1925) *Manoscritti*, p 59; and Carlo Pedretti (1978) *Catalogue*, pp 164–165.

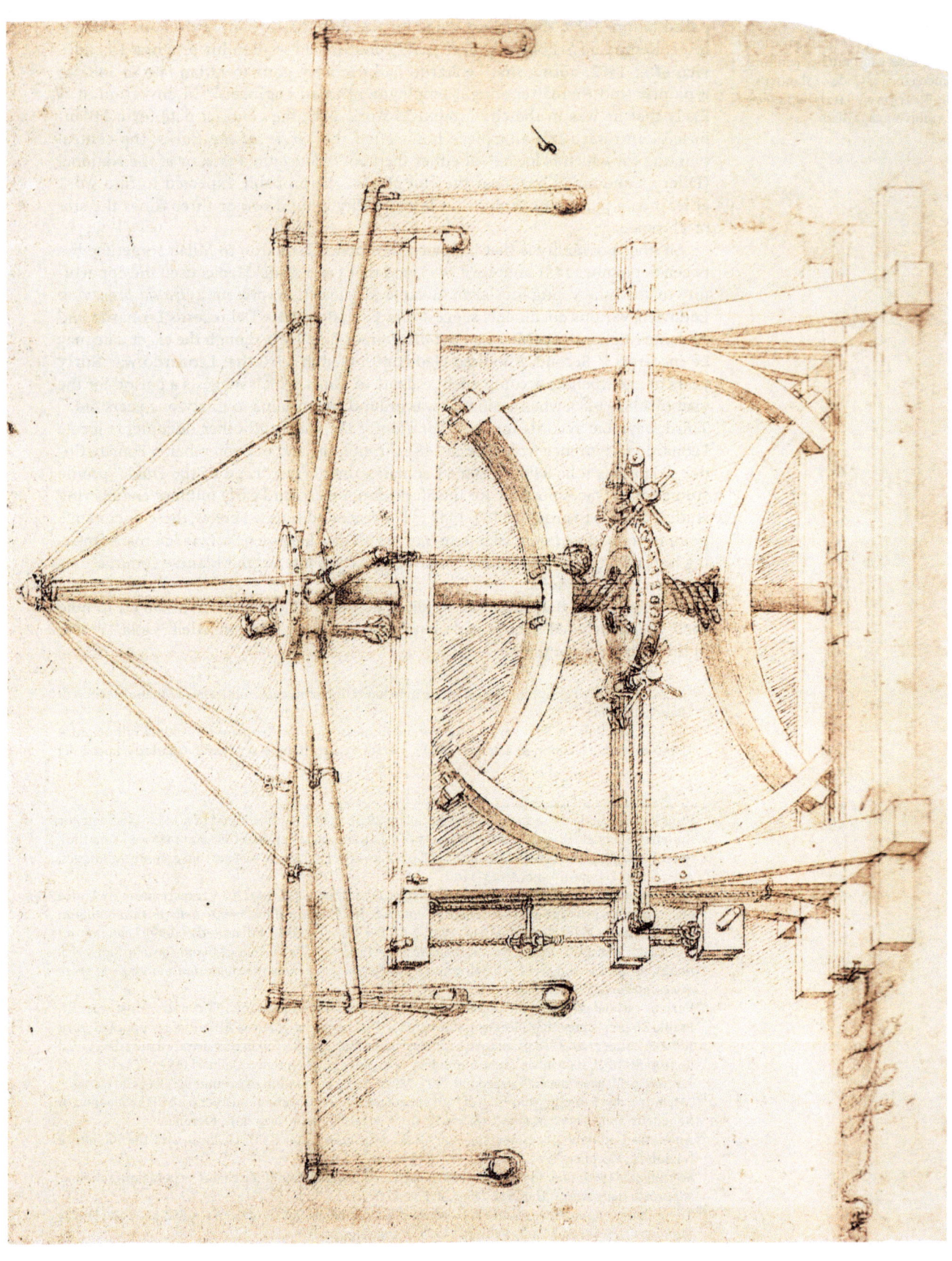

◀ **Fig. 1.4.**
Leonardo da Vinci, Illustration
of a large repeating sling, Codex
Atlanticus f. 159br [prev. 57br],
Biblioteca Ambrosiana, Milan

maestro Lucha, e Tomaso del Bene, e I° di Domenicho del Giochondo, e Agnolo Gherardini, e Ant°
Bartoli de' libri, e Bernardo di messer Benedetto d'Arezo, e Bernardo Paghanegli, e Ridolfo di Filippo
Ruciellai, e ser Nichollò Figiovanni cancelliero del onbasc[i]adore, e 'l conte Ghabriello Ginori, e
(…) Antinori, Chappone di Gino Chapponi, e Iachopo Iachopi, e Franc° di Meo delia Vacchia, e
Salvestro di Dino, e Tomaso Ghuidacci, e Franc° Berti, e Ruberto d'Ant° di Puccio, e Lionardo di
Filippo Bartoli, e Lionardo da Vinci dipintore, e Ridolfo Paghanegli di Pisa, e Atalanta delia viola, e
Franc° Cieho, e Andrea fratello di Malagigi de la v[i]vuola e (…) del maestro Mariotto, e Messer
(…) Lapacino prete(….)[23]

As stated in the title, much of the list consists of Florentine merchants (*merchanti
fiorentini*). This confirms the importance of Milan as a client of Florentine goods
and services, and thus the interests of the Medici in maintaining a good relationship with the wealthy city.[24] The last two-thirds of the list is quoted here, starting
with the famous Florentine poet, Luigi Pulci (b. 1432/d. 1484), author of comedic
epic *Il Morgente maggiore* (Venice 1481). Leonardo listed this title (as "*morgante*")
with a group of five books on Codex Trivulzianus folio 2r, around 1485–1490, next
to the drawing of an armoured boat attacking a castle's corner tower. Although
Lorenz Böninger doubts the reliability of Benedetto Dei's 1480 date of the list,
studies by Denise Budd and Raymond de Roover seem to confirm its accuracy.[25]
Budd offers a thorough analysis of the document's accuracy, reasonably correcting
Böninger's assessment that the date of the document refers to a general period.[26]
In any event, this document seems to place Leonardo in Milan sometime within a
year or two of 1480.

A document given much more attention in Leonardo scholarship, although with a
less accurate date, is Anonimo Gaddiano's reference to Leonardo in Codex Magliabechiano, XVII: 17:

> In his youth he was employed by Lorenzo de' Medici the Magnificent for whom he worked in the
> gardens of the square of San Marco in Florence. When he was thirty years of age, Lorenzo de' Medici
> the Magnificent sent him to present a lyre [*lira da braccio*] to the Duke of Milan, together with
> Atlante Miglioretti, because he played it exceptionally well. He then returned to Florence where he
> remained for longer [than he had been in Milan], but later, either from vexation or for some other
> reason, he left while working in the Council Chamber and returned to Milan where he remained for
> several years in the service of the Duke.[27]

If Leonardo had been "thirty years of age," we know from the documented date of
his birth—15 April 1452—that Anonimo may refer to the period between April 1482

---

[23] *Codex Magliabechiano* II: 333, f. 51r, transcribed by: Lorenz Böninger (1985) *Leonardo da Vinci
und Benedetto Dei in Mailand*, Mitteilungen des Kunsthistorisches Institutes in Florenz XXIX/2–3,
pp 386–387.

[24] See, for example: Benedetto Dei (1985) *La cronica dall'anno 1400 all'anno 1500*, R. Barducci (ed),
F. Papafava, Firenze, p 61; and Evelyn Welch (1995) *Art and authority in Renaissance Milan*, Yale
University Press, New Haven London, p 311, fn. 63.

[25] Böninger (1985) pp 386–387. Denise Budd (2002) *Leonardo da Vinci through Milan: Studies on the
documentary evidence*, doctoral dissertation, Columbia University, pp 148–172. Raymond de Roover
(1963) *The rise and decline of the Medici Bank, 1397–1494*, Harvard University Press, Cambridge,
MA, pp 16, 95, and 299.

[26] Budd (2002), pp 160, 148–172. Böninger (1985), p 387.

[27] Translation from: Andre Chastel (1961) *The genius of Leonardo da Vinci: Leonardo da Vinci
on art and the artist*, (originally in French), Ellen Callman (Eng. trans), Orion Press, New
York, p 5. *Codex Magliabechiano* XVII: 17, Biblioteca Nazionale, Florence. The original: "Stette
da giouane col Magnifico Lorenzo de Medicj, e dandolj prouisione per se il faceua lauorare
nel giardino sulla piaza di San Marcho dj Firenze. Et haueua 30 annj, che'l dal detto Magnifico Lorenzo fu mandato al duca di Milano insieme con Atalante Migliorottj a presentarlj
una lira, che unico era in sonare tale extruento. Torno dipoi in Firenze, doue stette piu tempo;
et dipoi o per indignatione che si fussj, o per altra causa, in mentre che lauoraua nella sala
del Consiglio de Signorj, si partj, e tornossene in Milano, doue al seruitio del duca stette piu
annj." Transcription from: Karl Frey (ed) (1892) *Il Codice Magliabechiano* c. XVII, 17. G. Grote,
Berlin, p 110.

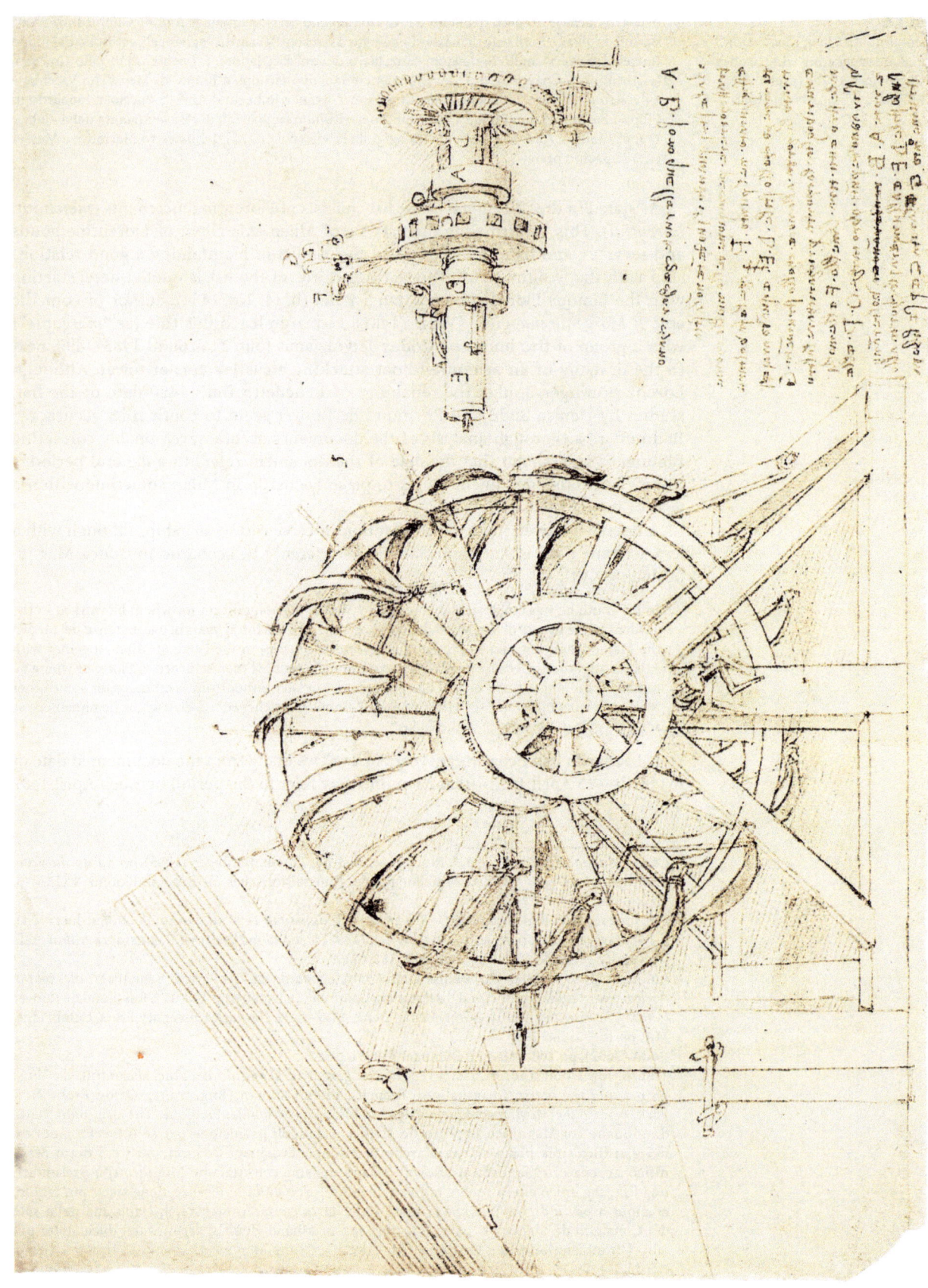

◄ Fig. 1.5.
Leonardo da Vinci, Giant
repeating crossbow, Codex
Atlanticus f. 182br [prev. 64vb],
Biblioteca Ambrosiana, Milan

and April 1483.[28] Nonetheless, a general identification of a person's age should not be taken literally. To refer to someone as twenty, thirty or forty years of age normally signified an approximate estimate for someone within one or two years of those ten-year periods. Hence, Anonimo Gaddiano's detailed account could locate Leonardo and Atalante in Milan any time between 1480 and 1482. This would agree with the date of Benedetto Dei's list. After the initial visit to Milan, Leonardo could have returned to Lorenzo de' Medici in Florence with a report on the successes of the ambassadorial exercise. Shortly thereafter Leonardo would have considered returning to Milan, possibly for a commission to paint the *Virgin of the Rocks* altarpiece now located in the Louvre.

Shortly after return from exile in September 1479 and the consequential control of Milan, Ludovico Sforza apparently looked for a sculptor to undertake the unfinished project for an equestrian monument of his father, Francesco (d. 1466).[29] Duke Galeazzo Maria Sforza had initiated the equestrian statue plan in 1473.[30] After July 1479, the Venetian Senate invited Verrocchio to submit a model for the bronze equestrian statue of captain Bartolomeo Colleoni of Bergamo (finished in 1488).[31] Ludovico was possibly aware of this and considered the political value of such a statue if he were to become Duke of Milan. Donatello's "Gattamelata" monument in Padua (1446–1453), the Marcus Aurelius in Rome (c. 180), and the Regisole in Pavia (c. 161) were famous symbols of their cities.[32] If Ludovico was in a position to offer the equestrian statue commission to Verrocchio, he was capable of offering this to Leonardo. According to Sabba Castiglione, Leonardo was at work on the "Sforza Horse" by 1483.[33] At that time, Leonardo was also at work on the Virgin of the Rocks and thus could not have committed much time to the horse, as perhaps indicated by the lack of visual evidence for this.

The Political Climate in Florence, c. 1480

Early in 1480, evidence suggests that Leonardo was actively looking for painting commissions. According to the records of Benedetto Dei and Anonimo Gaddiano, noted

---

[28] Regarding the date of birth see: *Notarile Antescosimiano P* 389 16192, f. 105v, 15 aprile 1452, Archivio di Stato, Florence: "1452, Nachue vn mio nipote, figliuolo di ser Piero mio figliuolo adì /15 d'apile In sabato a ore 3 di notte. Ebbe nome Lionardo...." Transcription from: *I documenti* 1999, no. 1, p 3. See also: E. Möller (1939) *Der Geburtstag des Lionardo*, 73. *Mostra di disegni,* manoscritti e documenti. Quinto Centenario delia Nascita di Leonardo da Vinci, Biblioteca Medicea Laurenziana, Firenze, 1952, no. 89. Renzo Cianchi (1952) *Vinci: Leonardo e la sua famiglia. Con appendice di documenti inediti*, Publ. no. 10, Museo delia scienze e della tecnica, Milano, pp 38–40. Raymond Stites (1970) *The sublimations of Leonardo da Vinci with a translation of the Codex Trivulzianus*, Smithsonian Institution, Washington, pp 6–7. James Beck (1988) *Leonardo's rapport with his father*, Antichita viva XXVIV(5–6), Oct.–Dec., p 10, no. 2. Pietro Marani (1999) *Leonardo una carriera dipintore*, F. Motta, Milano, p 342. Alessandro Vezzosi (ed) (2000) *Parleransi Ii omini ...*, Leonardo e l'Europa dal disegno delle idee alia profezia telematica, Exh. cat. Assisi, Napoli, San Benedetto del Tronto, e Milano, April–Dec, Relitalia, Perugia, p 40, with an illustration. Serge Bramly (1988/1991) *Leonardo, the artist and the man*, Michael Joseph, London, pp 37, 38, and 424, note 11.

[29] After the assassination of Galeazzo Maria Sforza in 1476, the Duchess Bona of Savoy exiled his brothers: Ludovico, Ottaviano, and Ascanio. She became regent to her son, Giangaleazzo. This is confirmed in Machiavelli's *History of Florence* VIII: 3.

[30] This is relatively well documented, and confirmed in: Kenneth Clark (1988) *Leonardo da Vinci*, Martin Kemp (intro), Penguin, London, p 139.

[31] Charles Seymour Jr. (1971) *The sculpture of Verrocchio*, New York Graphic Society Ltd., Greenwich, CT, p 17.

[32] The 'Regisole' commemorated the Roman Emperor Antonius Pius, AD 138–161. French soldiers destroyed the statue in 1796. Leonardo's study of the statue appears on Windsor folio 12317. Clark dates the drawing to around 1490. Clark (1969) *Catalogue*, p 22.

[33] Sabba Castiglione had lamented the horse's destruction in 1499, stating that Leonardo had worked on it for sixteen years. This would make the beginning of such work around 1483. Sabba Castiglione (1560) *Ricordi*, Venice, f. 57r. This is also noted in Clark (1939) *Leonardo*.

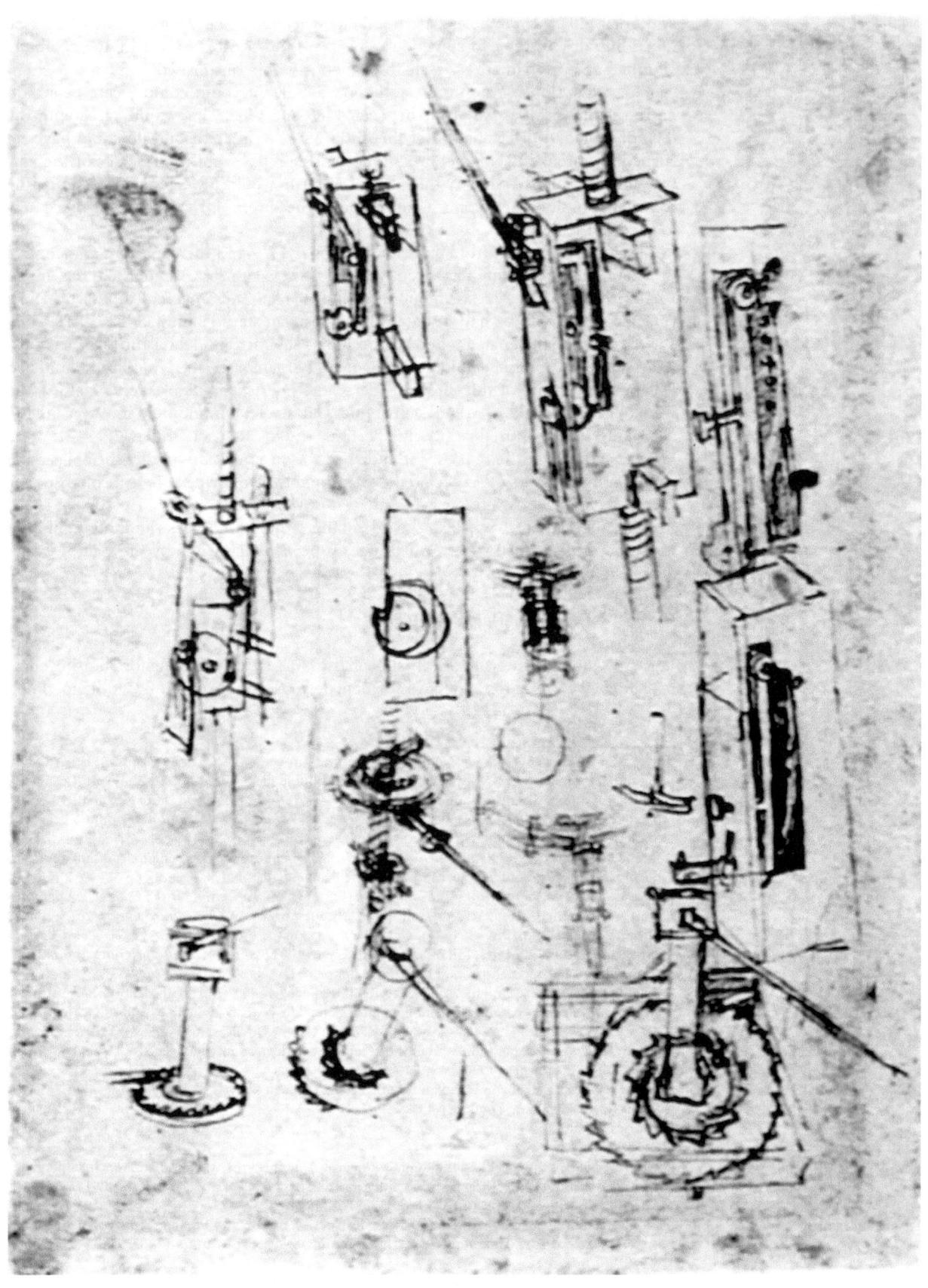

above, Leonardo was also apparently interested at this time in working outside of Florence. The discussion below emphasises the unstable political atmosphere in Florence around 1480. Also, the 1480–1482 military engineering studies and the possible Sforza monument studies on CA folio 888r-v, show Leonardo's early interest in the patronage of Ludovico Sforza. The last dated drawing from the first Florentine period offers additional proof of his interest in gaining new commissions, namely his drawing of "Bernardo di Bandino Baroncelli" on or just after 28 December 1479, located now at the Musée Bonnat, Bayonne.[34] Bernardo killed Lorenzo de' Medici's brother, Giuliano in the Duomo on 26 April 1478.[35] The detailed approach to Leonardo's drawing suggests its possible intention for the commission of a painted version of the subject. Denise Budd offers a compelling argument that Leonardo sought the commission to add Bernardo to the painting by Sandro Botticelli of eighty hanged men, condemned for involvement in the Pazzi conspiracy.[36] The men were depicted as hanged on the façade of the Bargello, or if exiled, depicted as hanged by the foot.[37] Botticelli oversaw the *pittura infamata* project, completed around July 1478.[38] Since Bernardo had fled to Constantinople, it is possible that he was painted as an exile, hanged by the foot. After Lorenzo de' Medici had Bernardo extradited back to Florence and hanged from the Bargello on 28 December 1479, at least two features were worth changing in Botticelli's painting: the addition of Bernardo's Turkish clothes—worn on order from Lorenzo, and an update to Lorenzo's inscription, under Bernardo's portrait: "an outlaw awaiting a crueller death."[39] Luca Landucci recorded the hanging of Bernardo on 28 December:

> 23 December. Bernardo Bandini de' Baroncegli was captured at Constantinople, the Grand Turk having given him up. He had fled from Florence when Giuliano de' Medici was murdered, believing that his life would be safe …
> 28 December. Bernardo Bandini was hanged at the windows of the *Palagio del Capitano*, he being the one who was said to have slain Giuliano de' Medici in the Conspiracy of the Pazzi. Certain arrangements had been made with the sultan that he should be given up.[40]

[34] The drawing is in: Popham (1946/1994), cat. 26, p 108. For the date, see: Luca Landucci (1883/1969) *Diario Fiorentino dal1450 al1516 di Luca Landucci continuato da un anonimofino al 1542,* Iodoco del Badia (ed), 1946, Studio Biblos, Firenze, p 33.

[35] This event is well known as the Pazzi conspiracy, an attempted coup to overthrow the Medici hegemony in Florence.

[36] Budd (2002), pp 145–153.

[37] By the mid 17th century, the painting had supposedly disintegrated from the façade. Herbert Horne (1908/1980) *Botticelli, painter of Florence,* John Pope-Hennessy (intro), Princeton University, Princeton, p 63.

[38] "Die xxj Julij 1478. / Item seruatis &c. deliberauerunt et stantiauerunt / Sandro boticelli pro eius labore in pingendo proditores florenos quadraginta largos, &c" in: Horne (transcr) (1908) *Partiti e Deliberazioni dei Signori Otto del 1478,* vol. 48, fol. 35, tergo, Archivio di Stato, Florence, Appendix II, Doc XVIII, p 305, see also: pp 63–64. Budd (2002), p 146.

[39] "Libro dei giustiziati", Biblioteca Nazionale, Florence, in: *Libro di varie notizie e memorie deUa venerabile Compagnia di Santa Maria della Croce al Tempio,* MS II, I, p 138, no. 435, dated 1479; Samuel Y. Edgerton Jr. (transcr) (1985) *Pictures and punishments: Art and criminal prosecution during the Florentine Renaissance,* Cornell University Press, Ithaca, p 106, fn. 31. Budd (2002) notes Lorenzo's orders that Bernardo wear his Turkish clothes, as described in the Libro dei giustiziati: "legato con catene e vistito come Turco.", p 151.

[40] Luca Landucci (1927) *A Florentine diary from 1450 to 1516 by Luca Landucci continued by an anonymous writer till 1542 with notes by Iodoco del Badia,* Alice de Rosen Jervis (trans), J. M. Dent & Sons, Ltd., London; E. P. Dutton & Co., New York, p 28. Landucci writes: "E a di 23 di dicenbre 1479, venne preso Bernardo Bandini de'Baroncegli di Gonstantinopoli, che 10 dette preso el Gran Turco; el quale s'era fuggito di Firenze quando fu morto Giuliano de'Medici, credendo essere sicuro delIa vita quivi … E a di 28 di dicenbre 1479, fu inpiccato, aile finestre del Capitano, Bernardo Bandini ch'era venuto preso di Gonstantinopoli, ch'era in quella congiura di messer Iacopo, e dissesi che fu lui quello che dette a Giuliano de' Medici. Ebbesi certi mezzi col Turco, che 10 concedette loro." Landucci, in: Badia (1883), p 33.

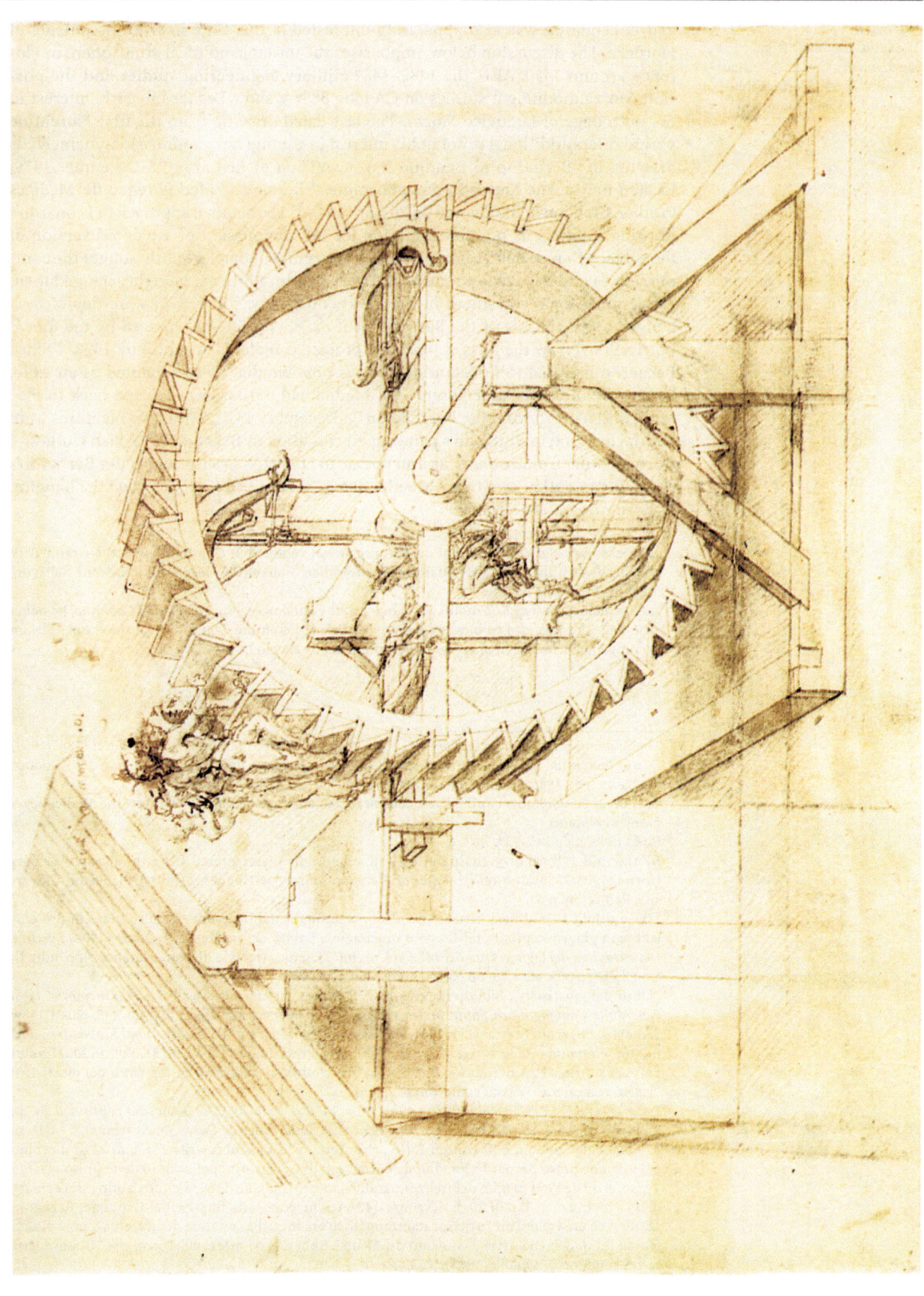

Leonardo's detailed study of Bernardo suggests that he was preparing to paint the portrait. He produced the following list:

A tan-coloured small cap
A doublet of black serge
A black jerkin lined
A blue coat lined
with fur of foxes' breasts
and the collar of the jerkin
covered with stippled velvet [of]
black and [red]
Bernardo di Bandino
Baroncelli
black hose.[41]

No record exists of Leonardo's possible involvement with the portrait series on the Bargello, although it makes sense that independent artists like him were constantly looking for commissions, especially from the Medici. Within a few months following his portrait of Bernardo in Turkish clothes, Benedetto Dei records Leonardo in Milan with Atalante.[42] As noted above, documents suggest that Leonardo travelled from Florence to Milan once or twice between 1480 and late 1482. Nonetheless, by April 1483, he had missed the Sistine Chapel's 1480–1481 fresco commissions that were awarded to Perugino, Botticelli, Ghirlandaio, Rosselli, Signorelli, Pinturicchio, Piero di Cosimo, and Bartolomeo della Gatta, he was to sign a contract in Milan to begin work on what would be the *Virgin of the Rocks*, and he was thus prepared to draught his letter of application for a military engineer position at the Sforza court. There was reason at this time to work on presentation drawings—such as preparatory sketches for the crossbows or even the *Giant Crossbow*—as supporting evidence for the letter of application. Also at this time, it is likely that Ludovico Sforza welcomed a proposal for the Sforza monument. The only problems with an early date for the *Giant Crossbow* is that it would predate Leonardo's other treatise illustrations begun around 1487–1489, and that the thorough precision of this drawing is remarkable by comparison to some earlier studies, including the 1470s Windsor "Lily" (W 12418r) and similar presentation quality studies begun with metal stylus incisions. In any event, when he signed the contract for the *Virgin of the Rocks* on the twenty-fifth of April, plans were undoubtedly underway to develop additional projects for the new studio in Milan. With future projects in mind, he developed between 1480 and 1489 written and illustrated evidence of his military engineering knowledge, while also offering to work as a sculptor, painter, and designer.

---

[41] Richter (1883) §664, p 345. His transcription of "soppannato di velluto appicchiettato nero e rrosso" is correct, though the translation is more accurately: "covered with stippled velvet [of] black and [red]," rather than "covered in black and white stippled velvet." Pedretti (1977), vol. I, p 380. Calvi (1925) *Manoscritti*, pp 31–34.
[42] *Codex Magliabechiano* II: 333, f. 51r, Böninger (transcr) (1985), pp 386–387.

2

Dating the Crossbow

If Leonardo had gained a commission at the Sforza Court for a series of siege engine designs—among other tasks—he could have begun working on many of these designs as early as 1483.[1] The following discussion outlines the documentary evidence of the drawings and writings associated with the development of the *Giant Crossbow* presentation drawing.

An appealing feature of the crossbow—at least to Leonardo—appears to have been its bat wing shaped armature. The early bat wing studies on MS B 74r (diagram, Fig. 2.1), 89v, and 100v are not specifically about the *Giant Crossbow*, but instead refer to the bat wing's structure. The following analysis offers proof of the relationships between these wing studies, other military studies, and the *Giant Crossbow* design. At issue is the extent to which these and many of the following military studies were part of the same project for Ludovico Sforza, as noted further above with reference to Leonardo's letter of c. 1484.

Fig. 2.1.
Illustration of Leonardo's flying machine wing on Manuscript B, MS B.N. It. 2037, f. 74r, Institut de France, Paris

---

[1] Paolo Galluzzi offers another approach to this problem of Leonardo's letter and Ludovico's possible commission, preferring to date Leonardo's military engineering drawings to the general period of 1482 to 1499: "The military-engineering drawings dating from Leonardo's first decade in Milan form a precise, graphic commentary on the letter (of 1482?) in which he offered his wide ranging technical expertise to Ludovico." Galluzzi (1999) *Art of invention*, p 58. The traditional date for Leonardo's letter is around 1485, as noted by Carlo Pedretti and Marco Cianchi in: C. Pedretti (1977) *Commentary*, vol. II, p 295. M. Cianchi (n.d., c. 1980) *Leonardo da Vinci's machines*, Becocci, Florence, p 18.

Manuscript B

An examination of the contexts for many of these military studies indicates that they are limited to three contiguous quires at one end of a group of ten quires, entitled "Manuscript B" by Giambattisa Venturi in 1796.[2] Each "quire" or booklet originally had five folded bifolia (two folios per sheet), making a total of ten folios (twenty sides of ten pages) per booklet. The last three quires have siege engine designs and other strategies for defence and attack by air, sea, land, and tunnelling. These relate to very similar drawings in the Codex Atlanticus, Windsor Royal Library, Turin Royal Library, the Accademia in Venice, the British Museum in London, and the École Nationale Supérieure des Beaux-Arts in Paris.[3] Compare for example the similar draughting styles and subjects in these drawings: CA 113v [40va], *Catapults and a Horse with three lances*, BM *Scythed chariot, armoured vehicle and halberd*; T 15583r, *Scythed chariots*; W 12653r, *Battle Chariot, a Soldier with Shield, and a Horseman with Three Lances*; and MS B Ashburnham f. A.2, *Halberds*.[4] Also, with regard to the surface treatment of these drawings that will be discussed further below, metal stylus lines on the Turin *Scythed Chariots* (15583r) resemble similar preparatory marks on the *Giant Crossbow* drawing and its neighbouring military engineering studies between CA 143r [51rc] and 166r [59vb].[5] These are early examples of Leonardo's use of a metal stylus with a straight edge. A good example of a freehand drawing with metal stylus is on page 99 of the Codex Trivulziano (previous 54r; see dia-

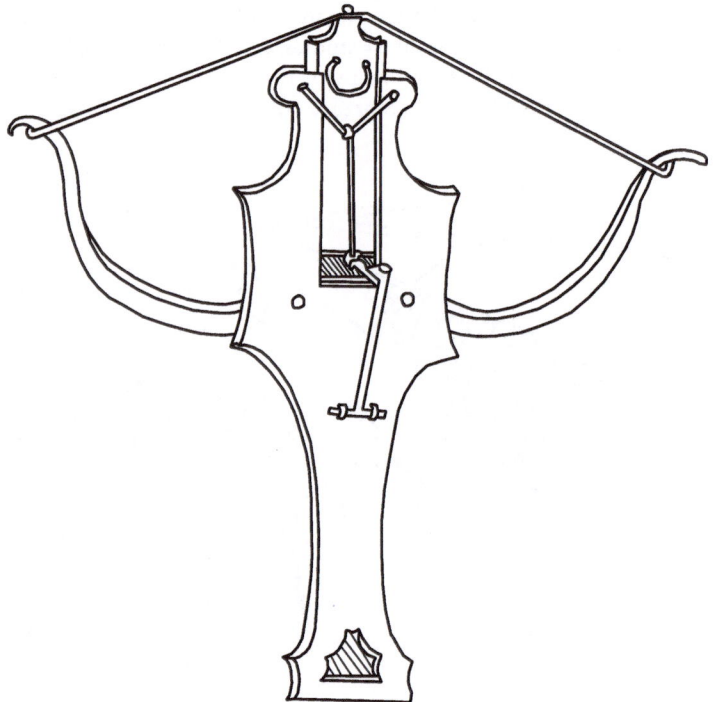

Fig. 2.2.
Illustration of Leonardo's drawing of the bottom side of a crossbow on Codex Trivulzianus N 2162, p. 99 [prev. f. 54r], Castello Sforzesco, Milan

---

[2] Nando De Toni (1974) *Giovanni Battista Venturi, ed, I manoscritti dell'Ambrosiana a Parigi nel 1797*, in: Frammenti Vinciani XXXI, as taken from: *Commentari dell' Ateneo di Brescia*, Geroldi, Brecia, p 3.

[3] Rather than restate the specific drawings at issue in the present study, a brief survey of some of these images may be seen among Johannes Nathan's set of drawings in Frank Zöllner and Johanne Nathan (2003) *Leonardo da Vinci, the complete paintings and drawings*, Taschen, Köln, pp 616–657. See also: A. E. Popham (1946/1994), pp 161–166, 297–320.

[4] *Ibid.*, for examples of the images. See also: Augusto Marinoni (1982) *Leonardo ingegnere militare*, (Exh. Cat.) Shell Italia, Milan.

[5] Metal stylus marks are partially visible in the Zöllner *Complete Paintings* image of the Turin *Scythed Chariots*. Although photographs in Codex Atlanticus facsimiles do not show most of the metal stylus marks on these drawings, I have seen the marks in person and I mention the findings further below.

gram, Fig. 2.2). This example, which post-dates CA 149br by several years, has metal sty-lus lines along the armature arms and at the un-inked stirrup at the top of the crossbow, all of which are just as visible on the original page.[6]

Though the three quires are at the end of a series of 100 posthumously numbered folios, it is unknown if they were filled in before, after, or during the time of the completion of the other quires. Often, Leonardo completed the right page and then the left page, as was the natural progression with his specular left-hand writing. So one can expect the quires to have a right-to-left, or left-to-right, or a combined sequence of progression. The latter progression is most common in Leonardo's notebooks. Perhaps the best reason for this combined approach is that he organised his notes according to subject matter.[7] For example, every quire in MS B contains military advice, although the second and third quires deal mainly with civil architecture, the fifth and sixth quires primarily examine military architecture, the third and fourth quires contain a majority of weapons, and the eighth, ninth and tenth quires are filled primarily with studies for flying machines, war galleys, and diving gear. Many studies, especially those of Valturio's *De re Militari*, as well as some that may be assoiated with Francesco di Giorgio's *Trattato di Architettura*, spill over to the margins of other folios. Initially, the quires were unstitched and part of a group of civil and military designs now located in the collections noted in the paragraph above. The original purpose of most quires in MS B was as a series on church architecture. Military studies overlap or surround these churches on a number of sheets.

Folios 84–87 are missing from ninth quire, though these probably contained the same proportional engineering subject matter as that of the ninth and tenth quires. The last quire was stolen by the Pisan mathematician, Brutus Icilio Timoleone (called Guglielmo Libri), around 1841–1844, who sold it shortly thereafter to Lord Ashburnham in England.[8] In the same way that Libri extracted the last quire from the codex, it would have been just as easy to get bifolia 84–87 and 85–86 from the centre of the ninth quire. These folios, the third folio of the first quire, and sixteen folios of MS E, have been lost since the time of the 1841–1844 theft.[9]

War with Venice

The purpose of this discussion of the contents of each quire of MS B is to explain something of the context in which Leonardo's early engineering studies developed. This and evidence of the rest of the present study show that these early military designs were for a treatise that would replace Valturio's *De re Militari*, as part of an attempt to gain the attention of Ludovico Sforza beginning as early as the time of Milan's siege of the Republic of Venice in 1483–1484. Even if most of Leonardo's military studies are of a later date than 1484, many of them were the product of his activities in 1480s Milan, during Ludovico's campaign to develop the military and the territories of Milan. Machiavelli's account of the war with Venice is as follows:

> Leaving the marquis of Ferrara to the defence of his own territories, he [Lodovico Sforza], with four thousand horse and two thousand foot, and joined by the Duke of Calabria with twelve thousand horse and five thousand foot, entered the territory of Bergamo, then Brescia, next that of Verona, and, in defiance of the Venetians, plundered the whole country; for it was with the greatest difficulty that Roberto and his forces could save the cities themselves. In the meantime, the marquis of

---

[6] Drs. Giovanna Colombo, Camilla Gavazzi and Ivanoc Riboli were very helpful with my arrangements to view this Codex. Fabio Saporetti, of Foto Saporetti in Milan, photographed the Codex.

[7] See: Claire Farago (1993) *Fractal geometry in the organization of Madrid MS II*, Academia Leonardi Vinci VI, pp 47–55, 12 plates.

[8] André Corbeau (1968) *Les soutractions de Libri*, in: *Les manoscrits de Léonard de Vinci: examen ctitique et historique de leurs elements externs*; Les manoscrits de Léonard de Vinci de la Bibliothèque nationale de Madrid: description critique et histoire, Centre regional de documentation pédagogique, Caen, pp 187–190. See also: Leonardo da Vinci (2002) *Manuscript E. The manuscripts of Leonardo da Vinci in the Institure de France*, John Venerella (ed, trans, annot), Ente Raccolta Vinciana, Castello Sforzesco, Milan, pp xiv–xv.

[9] *Ibid.* and *Manuscript E*, xiii–xv.

Ferrara had recovered a great part of his territories; for the duke of Lorraine, by whom he was attacked, having only at his command two thousand horse and one thousand foot, could not withstand him. Hence, during the whole of 1483, the affairs of the League were prosperous.

…

The position of Lodovico being known to the Venetians, they thought they could make it available for their own interests; and hoped, as they had often before done, to recover in peace all they had lost by war; and having secretly entered into a treaty with Lodovico, the terms were concluded in August, 1484.[10]

During 1482–1484, the Venetian naval fleet was the only likely opponent Ludovico might consider attacking by sea, distracting attention from Milan's borders. One of the likely reasons he fought the Venetians northwest of Ferrara was to keep the battlefront away from Milan. It is also possible that Leonardo was familiar with Ferrara particularly at the end of the 1470s and early 1480s, while he studied local fortifications, mills, locks and dams, recording his notes on sheets now in the Codex Atlanticus and Manuscript B. If this were the case, he would have been interested in the passionate intensity and directness of Cosmè Tura's paintings—especially the 1469 *Annunciation* altarpiece—in Ferrara, which may have influenced Leonardo's Louvre version of the *Virgin of the Rocks*, commissioned on 25 April 1483.[11] As it happens, his designs for flying machines and naval weapons would have been most helpful in defence of Ferrara's Po Delta in 1483, allowing quick escapes, assaults, and manoeuvrability around the waterways. Leonardo's written and illustrated proposals to Ludovico were alternatively useful as a means of honouring the Sforzas' unfinished business of retaking Genoa and Corsica, not to mention the completion of an equestrian statue of Francesco Sforza. When Ludovico took control of Milan in 1479–1480, he automatically took over the unfulfilled duties of Bona of Savoy to retake control of Genoa, and of Galeazzo Maria to build an equestrian statue of Francesco.[12]

---

[10] Niccolo Machiavelli (1962) *Istorie fiorentine*, Milan, translated by Laura F. Banfield and Harvey C. Mansfield Jr. as *Florentine histories*, Princeton, 1988, book 8, chapter 5.

[11] Cosmè Tura's *Annunciation*, now in the Museo del Duomo in Ferrara, was possibly on display in the Duomo in 1483. The painting measures 349 × 305 cm, the right half of which resembles the format and some of the contents of the Louvre's *Virgin of the Rocks* (1483–1486), measuring 199 × 122 cm. The year of 1483 marks a turning point in Leonardo's approach to the emotions and directness of the subject matter in his paintings. Cosmè Tura (b? c. 1430; d. 1495) worked for the Ferrara court of the Este family between 1458 and the mid-1480s. Of course the frescos produced by Cosmé Tura and Francseco del Cossa in Ferrara's Schifanoia Palace (complete by 1470) were also remarkably unlike anything Leonardo had seen in Florence, with an exceptional attention to the emotional states of the figures and the naturalism of the varied materials of their surroundings. Although much of this effect is now lost with the ageing of the frescoes, a glimpse of this captivating approach survives in the *Merchants' Alterpiece* of Francesco del Cossa, commissioned in Bologna in 1474, and now located in the Pinacoteca Nazionale, Bologna.

[12] Bona Sforza lost control of Genoa around August 1478 and Galeazzo Maria's initial plans for the monument to his father date back to 1473. Pope Sixtus IV died in 1484, which could have diverted Ludovico Sforza's attention back to Genoa after the 1484 treaty with Venice. No evidence exists, however, of this interest in an invasion. Instead, Milanese troops were practically welcomed into Genoa in 1487, when the Genoese Cardinal Paolo Grefoso was prepared to give the city to the Duke of Milan. This conveniently removed the Florentine forces that were beginning to occupy the area. The Florentines, as allies of Milan, retreated to Sarzana where they were in the process of taking Genoa's city from the Casa San Giogio. After Ludovico's easy takeover of Genoa, Milan could help with the construction of ships that would help the Genoese deal with pirates of the Ligurian Sea. Hence, some of Leonardo's ship and military engineering designs might relate to Ludovico's support for the Genoese during his rule, from 1487 to 1499. Additionally, the only record of Leonardo's visit to Genoa was for his inspection of a ruined harbour in 1498. One could also argue that Leonardo submitted some of his ship designs to the Venetian Senate in 1500. Nonetheless, the earliest set of Leonardo's military engineering studies could not post-date 1485. The best sources on the complexities of late 15th century Genoese politics are: Agostino Giustiniani (1557) *Annali della repubblica di Geneva*, Genoa; and for the period, 1488–1514, Bartolomeo Senarega (pre-1557) *De Rebus Genuensibus*, in: Emilio Pandiani (ed) (1932) *Rierum Italicarum Scriptures*, n.s., 24, pt. 8, Bologna. See also: Valeri (1949), pp 611–615, 667–678. Steven A. Epstein (1996) *Genoa and the Genoese, 958–1528*, University of North Carolina, Chapel Hill. Denys Hay and John Law (1989) *Italy in the age of the Renaissance, 1380–1530*, Longman, New York, pp 162–164, 244–247. Macchiavelli (1962) *Istorie fiorentine*, book 8, chapters 2–7. Regarding the Sforza Horse, see: Kenneth Clark (1988) *Leonardo da Vinci*, Martin Kemp (intro), Penguin, London, p 139.

Leonardo's interest in leaving Florence after June 1480 was perhaps partially to escape the rapidly increasing dangers there of the Council of Seventy, and to seek the patronage of a court with a relatively stable future, unlike that of the conceivably endangered Medici.[13] Leonardo was no doubt aware of Genoa's support of Pope Sixtus IV's Crusade to retake Otranto from the Turks (a serious threat to the rest of Italy) during 1480–1481, and Ludovico's possible interest during 1480–1481 in regaining Milan's former (1464–1478) control of Genoa and Corsica. Nonetheless, Leonardo would not expect Ludovico to take Genoa during or immediately after the Pope's Crusade; and by 1483, Ludovico was helping the troops of Sixtus IV/the Duke of Calabria and Ercole d'Este in Ferrara. Possibly prepared for the Ferrarese war, the last three quires of MS B offer detailed advice on war galleys, diving gear, flying machines, and defensive strategies, often mixed together on the same folios. On these and related folios, Leonardo also made anatomical studies of bats and birds, examined their strengths and proportions, compared strong and light-weight materials, studied the dynamics of air-flow and air currents, weighed the static capabilities of different pulley arrangements, designed models of jointed wings for testing, and engineered specialised traction and torsion mechanisms. Manuscript B folios 73v-76r and 79r-v contain innovative designs for an ornithopter, a human operated machine usually shaped like an aircraft that propels itself with the flap of its wings. MS B 74r illustrates the proposed function of the ornithopter wings, with the explanation, "Method for how the wing becomes perforated when it rises upward, and how it becomes all unified when it lowers."[14] On most pages immediately adjacent to this study, from 75v through 90v, flying machine and naval warfare studies are side by side. Although the naval warfare studies appear below pre-existing flying machine studies on some pages, they have most likely spilled over from similar sheets in the ninth quire on naval warfare, from 91v through 99v. Thus, problems of attack and defence by air, sea, and land occupy especially the last three quires of MS B. These preparations for a battle with Venetians at sea therefore appear to be part of a complete programme relative to Leonardo's written proposal to Ludovico Sforza around 1483/1484 and the Codex Atlanticus siege engine presentations and related studies in other collections.

A Programme for Siege and Defence Strategies

Because the present study proposes for the first time in Leonardo scholarship this link between preparatory studies for the *Giant Crossbow* and Leonardo's early programme for air, sea and land siege and defence strategies, additional proof of this programme is as follows. At issue are the historical and thematic contexts of this early military engineering programme. The following discussion outlines the various links between the subject matter, draughting styles, and historical aspects of the early programme, locating a preparatory programme within a 1482–1490 context that would lead to the siege engine series that includes the *Giant Crossbow*. This is proof of the thoroughly studied, though often impractical, approaches of Leonardo's early mechanical solutions for giant siege weapons, flying machines, navel strategies, cannon, incendiary missiles, and bridges. Some of his earliest approaches to these six subjects are as follows.

---

[13] The complex circumstances related to the Council of Seventy in 1480, and especially regarding developments during 1471–1494 are addressed in: Nicolai Rubenstein (1966) *The Government of Florence under the Medici (1434 to 1494)*, Clarendon, Oxford, pp 161–210. Rapid constitutional changes between March and July 1480 show that the Medici, the Signoria, the *Balìa*, and the *Cento*—four of the primary political forces of Florence—were rapidly losing their previous powers to govern the city. As Rubenstein notes, "Nor is it difficult to believe the Este ambassador that there was much popular discontent with recent developments. In his dispatch [to Ercole d'Este] of 3 July, he quotes a citizen who had asked Tommaso Soderini, a member of the *Balìa* and of the Seventy, why he no longer said, as he used to do, 'you will run into danger', *periculareti*; to which Tommaso replied: 'because we are danger', *perchè siamo periculati*" (p. 201, Archivio di Stato, Modena, Est., Firenze, 2).

[14] Leonardo da Vinci (2003) *Manuscript B, The Manuscripts of Leonardo da Vinci in the Institut de France*, John Venerella (ed, trans, annot), Ente Raccolta Vinciana, Castello Sforzesco, Milan, p 120.

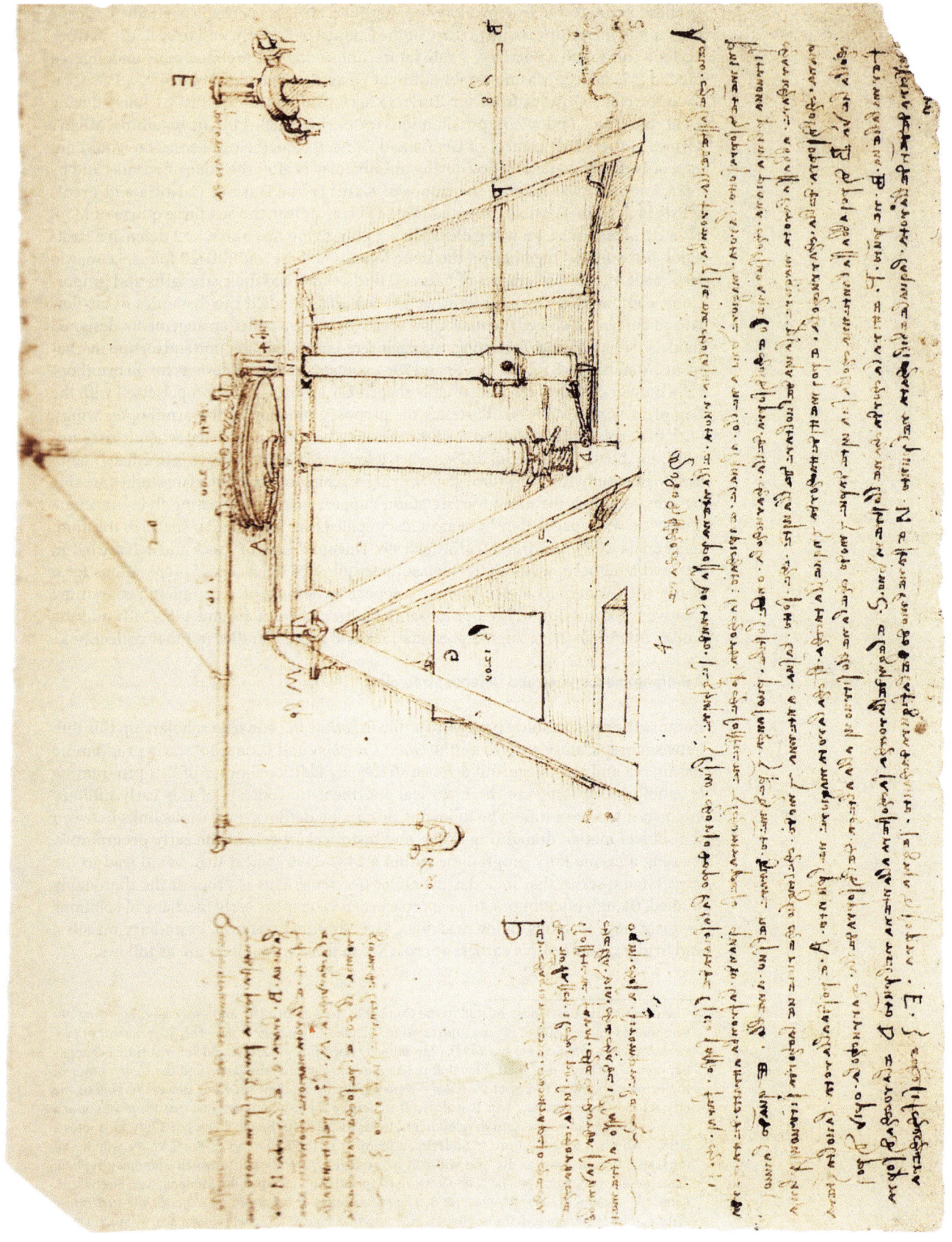

◀ **Fig. 2.3.**
Leonardo da Vinci, Large sling
engine, Codex Atlanticus f. 182ar
[prev. 64va], Biblioteca
Ambrosiana, Milan

First, the *Giant Crossbow* is part of a series of giant siege weapons in the Codex Atlanticus and at the Windsor Royal Library. The giant arsenal includes the vertical springald on CA 181r (Fig. I.2), bent frame springald on CA 145r (Fig. 1.1), giant sling on CA 182ar (Fig. 2.3) repeating multiple sling on CA 159br (Fig. 1.4), multiple cannon device on a boat in W 12632r, the famous cannon factory of W 12647r (Fig. I.3), and repeating crossbow machines on CA 182br and 1070r (Figs. 1.5 and 1.7).

Second, some of the flying machine studies may date to the period of 1480 to 1487. Examples of this group include: the wing study on UF 9r [UF 447Er] (c. 1480), the statement on the back of the sheet (UF 9v [UF 447Ev]) that, "this is the way that birds descend;" the parachute and flying machine studies on CA 860r [313va] (c. 1480–1485), 1058v [381va] (c. 1480–1485), and 1051r [377rb] (c. 1483–1487); bat wing studies on CA 848r [309bv], and MS B 74r (diagram, Fig. 2.1); and flying machine studies on CA 747r [276br] and MS B 73v-75r, 76r-80r, and 87v-90v. Uffizi 9v is reproduced by A. E. Popham, plate 50, who dates it to around the time of the *Adoration*, c. 1480, and offers the transcription: "questo e il modo del chalare degli uccelli".[15] Carlo Pedretti dates CA 848r to c. 1487–1490.[16] He also dates CA 747r to c. 1487 "or even c. 1485."[17] He bases these dates on the drawings' relationships to the MS B folios and later flying machine studies. There is sufficient evidence to place some of the folios of MS B, as well as CA 848r, within the 1483–1485 context of his work. In 1978, Pedretti dated CA 860r to c. 1480, CA 1051r [377rb] to c. 1483–1485, and CA 1058v to c. 1485.[18] In 1999, he re-dated CA 1058v to c. 1480, stating that, "Leonardo began his studies on flight much earlier than has been commonly believed, as early as the time of the *Adoration of the Magi*, in 1481, and thus before going to Milan."[19] Pedretti originally based the CA 1058v date on a comparison with MS B 88v and 90v, a valid comparison of the sketches on the three sheets of the weight applied by the human body to a flying machine lever that forces the motion of a wing.[20] Compound pulleys on the vertical ornithopters refer to the simple principle of multiplying the force of human power enough to move a giant wing. One can see that flapping devices were only a serious consideration between 1480 (UF 9r [UF 447Er], CA 1058v [381va]) and 1484 (vertical and horizontal ornithopters of MS B 73v-90v and CA 846v-848r). After 1484, Leonardo abandoned the heavy vertical ornithopter structure in favour of glider designs, some of which include horizontal ornithopter sections.

Third, various studies from 1480–1483 address problems of military assault and defence in bodies of water. With the exception of flight over water, as noted above, studies of naval warfare are as follows: diving gear and flotation devices for "walking on water" on CA 26r [7r], 909v [333v], and 1069r [386rb]; the "boat used to sink ships" on MS B 11r; methods for building a submarine on CA 881r [320vb], the culverin (long cannon) carrying vessels with anti-recoil devices on CA 1052ar [378ra], 172r [61ar], 149ar [53va], 581r [217br], and 693r [257cr]; the ship designs on Windsor 12650r-v and CA 876va [319v]; and the webbed glove and flotation devices on MS B 81v and CA 748r [276va]. Calvi and Pedretti date CA 26r [7r] to 1480–1482.[21] Pedretti dates 1069r and its fragment (lower left) of a "flying machine" to 1480–1482.[22] Pedretti also states that 909v [333v], "has become famous with the publications of Müller-Walde, Baratta, and Solmi, and is often associated with a defence programme proposed by Leonardo to

---

[15] Popham, p 25.

[16] Pedretti (1978), *Catalogue*, Pt II, pp 144.

[17] Pedretti (1978), *Catalogue*, Pt II, p 97.

[18] Pedretti (1978), *Catalogue*, Pt II, pp 257, 149 and 262.

[19] C.x Pedretti (1999) *Leonardo, The Machines*, Giunti, Florence, p 78.

[20] Pedretti (1978) *Catalogue*, Pt II, p 262.

[21] Calvi (1925/1982) *Manoscritti*, 1982 ed.: pp 65–67, 1925 ed.: pp 28–30. Pedretti (1978) *Catalogue*, Pt I, p 33.

[22] Pedretti (1978) *Catalogue*, pp 268–269. The study referred to by Pedretti as the "mechanism of a flying machine" (p 269) also appears on 1059v [381vb], which he dates to 1480–1482. Pedretti (1978) *Catalogue*, p 262. Without the central screw, this mechanism on 1069r and 1059v resembles upper portions of the vertical ornithopter on MS B 80r and 90r.

the Venetian Senate in 1500 (see note to f. 638dv [234vc]), and this in spite of the evidence provided by Calvi (1925) of a date at the time of MS B or even earlier [than
1487]."[23] Although Calvi suggests that the latest "probable" date of CA 909v could be
1487, with reference to Sforza strategies against pirates of the Ligurian Sea, he and
Pedretti (as noted above) also consider an earlier date. Calvi initially referred to the
drawing's possible relationship to the War of Ferrara, stating:

> I have reasons to doubt and I ask myself if the folios cited and the projects for underwater attack
> were drawn not in favour of the Venetians in 1500 but against them in an earlier period, around
> 1483–1484, when the duke of Milan, together with Sisto IV, the King of Naples and the duke of
> Ferrara were engaged in the conflict against Venice; or, perhaps even more likely (as one might
> think which is why I want to come and explain the issue concerning one of the folios) at the time
> around 1487 when the Milanese government, after the annexation of Liguria, had to deal with mari-
> time defences against the Corsairs.[24]

Furthermore, evidence suggesting a date for CA 909v closer to 1483 than to 1487 is
the note at the bottom of 909v, stating that: "You need to take an impression of one of
the three iron screws of the Opera of Santa Liberata."[25] This and other notes on the
page Leonardo had written to himself about the type of iron screw of Brunelleschi's
design that he had seen in the Florence Cathedral workshop. The workshop was part
of the administration of the Florence Cathedral, often referred to after as the "opera di
santa liberata." (Leonardo's actual word in this case is "liberata" instead of "reparata".
Although the church of Santa Reparata previously stood in the place of the Florence
Cathedral, there had been a prohibition by the Republic of Florence since 1412 against
the discussion of any link between the church and the mythical patron saint, Santa
Reparata.)[26] Given the likely strained relationship between Milan and Florence immediately after Milan blocked the Florentine advance on Genoa in 1487, Ludovico and
Leonardo probably wanted to avoid the Medici that year. In 1482–1483, Leonardo's
underwater attack strategy would have been useful to the Ferrarese, with whom Milan
and Florence were allies. If Leonardo had accompanied Ludovico to Ferrara, then he
was closer to Florence, where he could consider having a copy made of Brunelleschi's
iron screw for the Ferrarese.[27] MS B 11r and CA 881r [320vb] may not be earlier than
1485, as Pedretti notes, but they appear to be part of the same programme noted im-

---

[23] P. Müller-Walde (1899) *Beiträge zur Kenntnis des Leonardo da Vinci*, pp 60–80. M. Baratta (1905)
*Curiosità Vinciane: perchè Leonardo da Vinci scriveva a rovescio, Leonardo da Vinci enigmofilo,
Leonardo da Vinci nella invenzione dei palombari e degli apparecchi di salvataggio marittimo*, Fratelli
Bocca, Torino, pp 111–134. E. Solmi and S. Solmi (1976) *Scritti vinciani: Le fonti dei manoscritti di
Leonardo da Vinci e altri studi*. (orig. 1908), La nuova Italia, Firenze, pp 169–179. Pedretti (1978)
Catalogue, Pt II, pp 174–175. See also: Paolo Galluzzi (1999) *The Art of invention: Leonardo and Re-
naissance engineers*, The Science Museum, Exhibition Road, London, 15 October 1999–24 April 2000,
Istituto e Museo di Storia della Scienza, Florence, pp 48 and 54.

[24] This is my general translation of: "Io mi permetto di dubitarne e mi chiedo se non potrebbero invece
i fogli citati e il progetto di attacco subacqueo esser stati disegnati non a favore dei veneziani nel
1500, ma o contro di essi, in un'epoca assai anteriore, intorno al 1483–1484, quando il duca di Milano,
collegato con Sisto IV, col re di Napoli, col duca di Ferrara, si trovava impegnato nella lotta contro
Venezia; o, forse più probabilmente, (come si potrebbe pensare per ciò ch'io verrò tosto a dire intorno
ad uno dei fogli che interessano questa questione) in un tempo non lontano dal 1487, quando il
governo Milanese dovette, per l'annessione della I; Liguria, occuparsi di difese marinare, anche conrro
corsair," as stated in Calvi (1982) *Manoscritti*, p 72.

[25] Leonardo states: "vuoisi <'n>prò<n>tare vna delle 3 vidj / dj fero dellopera dj santa liberata", transc.
Pedretti (1978) *Catalogue*, Pt II, pp 175.

[26] The story and consequences of Florence's embarrassment, when Naples sent a plaster cast of the
mythical Santa Reparata's right arm—instead of what could have been arm bones, is recounted by
Franklin Toker (1999) in: *Amid rubble and myth: Excavating beneath Florence's Cathedral*, Humanities 20(2) (March/April):14–18. This is also posted at: http://www.neh.gov/news/humanities/1999-
03/toker.html (as of June 2005).

[27] See also: Carlo Pedretti (1981/1985) *Leonardo architetto*, Electa Editrice, Milan (Eng. ed. 1985), pp 8–9.

mediately above, which could place them two years earlier.[28] Alternatively, if Ludovico Sforza had had an interest in regaining control of Corsica in the early 1480s, Leonardo's naval strategies could have proven useful in that endeavour. After 1466, and especially between 1479 and 1484, Corsica tried to remove itself from the suzerainty of the dukes of Milan, finally succeeding by 1484. As for the other naval drawings, Pedretti compares the "style, contents, and technique" of CA 881r with that of CA 1052ar [378ra] and 172r [61ar], dating them to around 1485.[29] Pedretti notes that the boat on 172r appears on folios 149ar [53va] (Fig. I.4), 581r [217br], and 693r [257cr].[30] CA 149a was previously glued to the *Giant Crossbow* drawing, 149br, presumably with reference to the great crossbows on 149ar. Pedretti confirms that the style of 581r "points to the time of MS B, but it would be slightly earlier, c. 1485," and that the ductus and style of 693r "appear close to those of the notes and drawings in MS B."[31] Kenneth Clark compares W 12650r-v with CA 876v, and MS B, giving them similar dates of 1487–1490 though he notes that W 12650r appears to be earlier, c. 1485–1487.[32] MS B 81v and CA 748r [276va] demonstrate a life preserver, webbed fins (MS B), and a bridge supported by boats (on 748r) that could be useful at sea, or when crossing Ferrara's Po waterways. As noted by Augusto Marinoni, the human figure in these drawings can be found on CA 26r [7r], a drawing of around 1480–1482, according to Calvi and Pedretti (as noted above).[33]

Fourth, innovative cannon studies on MS B, CA and Windsor folios bear direct relationships to the military engineering studies of c. 1480–1485. The cannon with elevating arcs and machine guns on CA 32rb [9rb], 157r [56va], 76vb [26vb], 94r [34va], 114v [40vb], and 157v [56vb] are from 1480–1482, according to Pedretti and Calvi, given the direct relationships between these studies—and especially CA 32rb—to that of CA 47v [14va].[34] The latter drawing has the same cannon as 32rb and 76vb, as well as a sketch of "mills which are used on the Po River," as stated by Leonardo.[35] Calvi and Pedretti see this Po River reference as related to the handwriting style of Leonardo's last days in Florence, though the reference could also note his civil and military engineering interests while in transit to or from Ferrara or Milan between 1481 and 1484. The cannon on CA folios above also closely resemble the style of long barrelled "eighteen-pounder" or medieval dart (see 32rb) cannon on Windsor 12652r and MS B 24v, 31r, 32r, 50v, and 54r. Clark dates W 12652r to the period of MS B, but also gives a number of reasons to date this drawing to 1483–1485, based on the sketches related to the *Virgin of the Rocks*, and the mortar related to CA 59bv [18rb], the parent sheet of the 1483–1484 W 12725r.[36] The most obvious link, however, between the 1480–1482 CA 47v and MS B is the cannon barrel screw attachment visible on that sheet and on MS B 24v, 31r, and 32r. Thus, the date for MS B could be c. 1480–1490, beginning with military and civil engineering studies along the Po River and ending with Leonardo's copies of Francesco di Giorgio Martini's *Trattato di architettura* (written c. 1470).

Fifth, mortars noted above, as well as fragmenting bombs and incendiary missiles date to around 1483–1487. With regard to the *Giant Crossbow*, the most noteworthy missiles are the incendiary darts on W 12651r and MS B 50v. As stated on the latter drawing, "this is a dart to be released by a great crossbow" (questo e uno dardo daessere tracto da uno gran balestra). Examples from the same period of studies include the explosive missiles and balls on CA 31r [9ra], 33r [9va], MS B 30v, 31v, 37r, 45v, 55v, and

---

[28] Pedretti (1978) *Catalogue*, Pt II, pp 160–161.

[29] Pedretti (1978) *Catalogue*, Pt II, p 258 and Pt I, p 95.

[30] Pedretti (1978) *Catalogue*, Pt I, p 87–88.

[31] Pedretti (1978) *Catalogue*, Pt II, pp 27 and 76.

[32] Clark (1969) *Catalogue*, pp 141–142.

[33] *Codice Atlantico*, 2000, v. III, pp 1448–1449.

[34] Pedretti (1978) *Catalogue*, Pt I pp 35, 42, 56, 66–67, 74, 90. Calvi (1925/1982) *Manoscritti*, 1982 ed.: pp 50–60, 1925 ed.: pp 30–41. See also: Galluzzi (1999) *Art of invention*, pp 50 and 56.

[35] *Ibid.*

[36] Clark (1969) *Catalogue*, pp 142–144.

80v. Pedretti attributes CA 31r and 33r to c. 1485, and drawings at the other sides of those sheets (33r [9rb], 34r [9vb]) to c. 1480–1482.[37] Examining the styles of drawings on both sides of the sheets, one could argue that their similar styles and adaptations of 14th century cannon technology locate them within the same period, such as that of military activity in Ferrara between 1482 and 1484. Clark notes that Leonardo possibly designed the weapons on MS B 43-50 for Ludovico Sforza's armourer, Gentile de' Borri.[38]

Sixth, Leonardo produced several temporary bridge designs between 1480 and 1485, especially in the form of a wooden trestle that would be relatively quick and easy to build on site. These would have proven most useful when defending Ferrara's Po Delta from Venetians in the early 1480s. Bridges attributed to the earliest period by Pedretti include those on CA 888v [324v] (1480–1482), CA 55r [16va] (1483–1485), CA 902br [329rb] (1483–1485).[39] The first of these sheets, CA 888v, Pedretti dates to 1480–1482, and it could be of a time closer to the *Giant Crossbow*. The basket/string combination and the wood lamination studies at the top of the sheet point to similar designs and the sketchy style on the CA 147av and 147bv giant crossbow studies. The last two drawings, CA 55r and 902bv MS B 23R, have the same bridge design. A tree branch hook designed to secure these bridges to the shore was cut from 902br, the companion sheet of 902bv. The two sheets were of the same sheet or the same group before the 17th century division of sheets, and the tiny bridge fragment cut from 902br is now at Windsor, W 12439r.[40] Most importantly, however, this direct link between 902br and 902bv, 888v, and 55r confirms that the drawings were part of the same set of military engineering studies associated with the crossbow engine trebuchet and the *De re Militari* style springald (*dardo e frammola*) on 902br. As noted further above, Leonardo's studies in MS B and CA of a springald very similar to that in Valturio's *De re Militari* were seemingly part of an effort to copy, if not emulate, much of that book shortly after its second printing in 1483 (in Verona). Bridge designs Pedretti believes to be of a later period— c. 1485–1487—include those on sheets CA 57v-69ar (Fig. 2.4), 71r [23ra], and 71v [23va].[41] With this group, one could attach the bridge designs on MS B 23r, of the same period and style, but with different designs. The bridge structure on 57v–69ar immediately preceding the giant crossbow studies above and below it. The same form of bridge appears on CA 71r, below the drawing of a mortar. The verso of the same sheet (71v) contains an arched bridge design, surrounded by the cannon discussed further above. Clark dates the fragment W 12469r—previously cut from folio 71v—to c. 1485.[42] The same arched bridge design on 71v appears on CA 69ar [22ra]. The date of this sheet is at issue because of the small sketch of the interlocking bridge beams at the bottom right corner, as this drawing bears a direct relationship with a similar design at the bottom of CA 888v, the date for which is c. 1480–1482. Pedretti's entry for 888v in the CA Catalogue states:

> Various studies of military engineering and architecture (weight lifting devices, temporary bridges, and a device to set a wooden truss to a desired bend). This seems a little later than the recto, to which see note. The ductus points to a date about 1482–1483 and may be taken to suggest that certain sections of MS B are in fact earlier than 1487. For the system of temporary bridges compare f. 55 recto, to which see note for related sheets. The device to bend a wooden truss appears in other early drawings, e.g. on ff. 47 verso, 90 recto, 91 verso, and 946 recto. Compare also the light sketch on the much later f. 886 verso. A variation of the same device to prepare trusses for bridges is on f. 917iii [c] recto, to which see note.[43]

---

[37] Pedretti (1978) *Catalogue*, Pt 1, p 35–36.

[38] Clark (1969) *Catalogue*, p 142.

[39] Pedretti (1978) *Catalogue*, pp 165, 45, 171.

[40] Clark (1969) *Catalogue*, p 70.

[41] Pedretti (1978) *Catalogue*, pp 47, 53, 54.

[42] Clark (1969) *Catalogue*, p 77.

[43] Pedretti (1978) *Catalogue*, p 165. This is the note regarding CA 55r [16v]: "Style and handwriting indicate the early years of Leonardo's first Milanese period. See: The preliminary sketches on f. 902 recto, a sheet related to 888 verso, c. 1483. See also the slightly later drawings on f. 37 recto, c. 1487. For other types of such military bridges, see: f. 855 recto, which also dates from about the same time" p 45.

Fig. 2.4. ▶
Leonardo da Vinci, Preparatory sketch for the *Giant Crossbow*, among other engineering studies, Codex Atlanticus f. 57v [17ra], shown as originally attached to 69ar [prev. 22ra] on the right, Biblioteca Ambrosiana, Milan

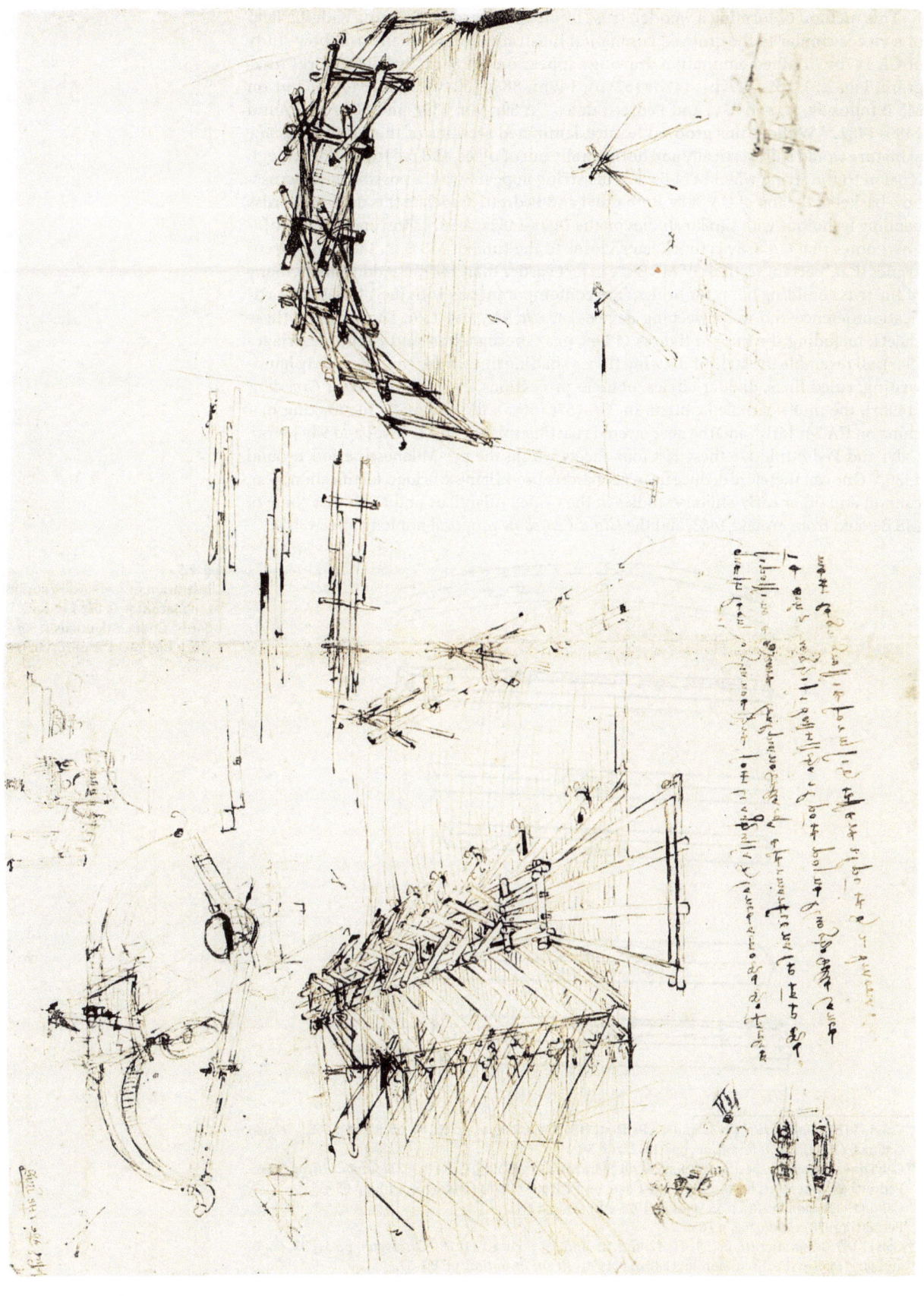

This method of forming a wooden truss by attaching grooved boards with the help of a vice is similar to the grooved lamination illustration on the giant crossbow study of CA 147bv. Toothed lamination drawings appear on CA 89r [32va], 90r [33ra] (diagram, Fig. 2.5), 139r [49vb], 147bv [52vb], 149br, 888v [324v], 946r [344va], and on MS B folios 5v, 6r, and 27v; and Pedretti dates CA 89r, 90r, 139r, and 888v to around 1480–1482.[44] Without this grooved feature, laminated sections of the *Giant Crossbow* armature would shift vertically and horizontally out of place, and fail to provide enough tension to the string when bent back. This string appears with a possible giant crossbow basket at the top of CA 888v. Both Calvi and Pedretti associate this drawing's truss bending technique with similar studies on the 1480–1482 CA 90r [33ra] and 91v [33vb].[45] Calvi notes that CA 888v is much later, closer to the time of MS B [c. 1487].[46] Pedretti argues that, "certain sections of MS B are in fact earlier than 1487."[47] Additionally, many of the truss building notes for bridges are contemporaneous with the 1480–1482 fortification defence and wall-sweeping devices on 89r, 94r, and 139r. Drawings on these sheets, including the human figures (139r), gear mechanisms, and cannon carriages (94r), all resemble the style of drawing (loose shading lines, bulky muscles, early handwriting, ruled lines, divider circles, oblique projections, etc.) for the *Giant Crossbow* (149br), the multi-barrelled cannon on CA 157r [56va], the automatic file-making machine on CA 24r [6rb], and the air-powered roasting spit on CA 21r [5va] and 94v [34ra]. Calvi and Pedretti locate these last four sheets within the pre-Milanese period, around 1480.[48] One can therefore deduce from Leonardo's pre-Milanese bridge, fortification, gear, cannon and other early military studies in the Codex Atlanticus and MS B that some of MS B could from around 1483, and the *Giant Crossbow* proposal not long afterward.

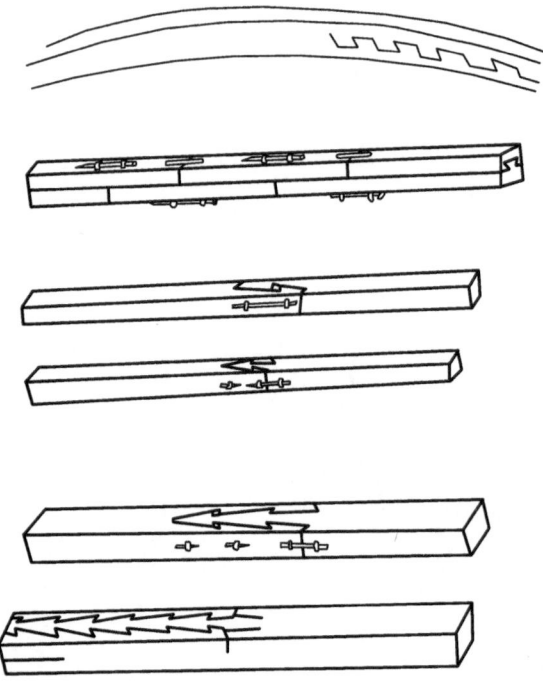

Fig. 2.5.
Illustration of Leonardo's studies of the techniques used to join boards, Codex Atlanticus f. 90r [33ra], Biblioteca Ambrosiana, Milan

---

[44] Calvi (1925) *Manoscritti*, pp 33 and 42. Pedretti (1978) *Catalogue*, pp 62, 63, 65, 84, and 165. See also: Galluzzi (1999) *Art of invention*, pp 51, 52, and 56.

[45] Calvi does not offer a specific date for CA 90r [33ra] and 91v [33vb]; Calvi (1982) *Manoscritti*, pp 55, 65. Pedretti gives reasons for the 1480–1482 date on Pedretti (1978) *Catalogue*, Pt I, pp 65–66.

[46] Calvi (1982) *Manoscritti*, p 55 (p 41 in 1925 ed.).

[47] Pedretti (1978) *Catalogue*, p 165.

[48] Calvi (1925) *Manoscritti*, pp 38, 41, 42, figs. 25 and 19. Pedretti (1978) *Catalogue*, pp 31, 32, 66, 67 and 90. See also the discussion in: Galluzzi (1999) *Art of invention*, pp 50–57.

3

A Real Project

The first indication that Leonardo may have intended to design a working giant crossbow—rather than just an impressive drawing of the bow—appears in the form of numerous drawings of portions of this giant crossbow. Were he to have produced just a drawing of a fanciful object, there would have been no need for the extensive existing studies of the giant bow. Additionally, for whom was the *Giant Crossbow* presentation drawing intended? If evidence suggests that it was for Ludovico Sforza (as noted above), it is doubtful that this particular duke had much patience for fanciful drawings of siege weapons. (Siege weaponry and imaginative illustrations were completely different interests for the powerful military commander. The practical technical problems of siege warfare were deadly enough to discourage an imaginary approach to military toys within the Court of Milan.) A number of studies related to the *Giant Crossbow* were produced between the periods when Milan was embroiled in an expensive war with Venice and the period when Leonardo started working for Ludovico in 1489 and expanded his treatise project. Beginning in 1483, Ludovico probably viewed—and may have commissioned—Leonardo's designs for a range of weapons, from handheld spear throwers and crossbows to cannon and siege machinery. Dozens of these early drawings survive in the form of personal studies and formal presentations. The following discussion examines a portion of this group, in addition to later drawings relating to the *Giant Crossbow* drawing.

Studies and Measurements

Numerous examples of studies dated from 1478 through to 1508 show Leonardo's interests in all aspects of crossbow design, construction, and usage, some of which are listed in Appendix B. So many of these drawings are related in style or subject that they could date to the 1483–1489, close to the time of the *Giant Crossbow* drawing. Some drawings in this group, and which follow, may only have an indirect relationship to CA 149br. Nonetheless, an appropriate study of the dates of these drawings would expand the present book unnecessarily. For this reason, unless otherwise noted, most of the examples in the following group accompany their respective dates as generally accepted.[1]

---

[1] Pedretti (1978) *Catalogue*, p 42ff. Following the chronological transliteration system of Leonardo's script by G. Calvi (*Manoscritti* 1925), many of Pedretti's dates for drawings around 1490–1495 are based on the presence of a certain form of the capital Q, which looks like this: ⟨Q⟩ . Among Leonardo's later drawings the tail of his Q would sweep up into text. Also, the pre-1500 forms of the lowercase l would tend to have this backwards loop: ℓ. Drawings after 1500 had only a hook at the top of the l, not a loop. Even if this were consistently applicable as a method for dating drawings, problems remain for dating engineering studies that lack enough text on which to make a judgement. Where there is very little alternative evidence, there is a tendency among scholars to date more formal, less hurried writing, to an earlier period. Pedretti made use of other forms of convincing evidence to date the drawings, but problems remain about the dates for some of the engineering studies. Pedretti's notes on Leonardo's handwriting may be found in his *Literary Works … Commentary*, 1977, pp 91–92.

The dates indicate the possible historical relationships between numerous examples and the *Giant Crossbow* drawing.

Before discussing in the next sections Leonardo's illustrated and written dimensions, it is worth noting that any designs he might have produced for the Sforza Court could have been measured in Milanese *braccia* (arm lengths) of 59.5 cm each, rather than in Florentine *braccia* of 58.4 cm.[2] He could have been aware of the measuring standards of the Sforza Court, even if no record exists of the *braccio mercantile milanesi* or the *braccio a panno fiorentino* in his notebooks. Throughout his notebooks, however, he uses for weight and financial estimates the Milanese *oncia* (ounce) and *dinaro* as much as, if not more than, the Florentine *soldo*.[3] A minor difference in measurement would not seem to be of critical importance to the construction of an exceptionally large crossbow forty *braccia* in length, but a patron, builder, or carpenter might have considered this important, measuring a Milanese forty *braccia* bow at almost a foot and a half (40 cm) longer than a Florentine forty *braccia* bow. Also, as will be discussed further below, measurements for the trigger mechanisms at the bottom left of the *Giant Crossbow* drawing had to be accurately measured to the exact Milanese *once* ($\frac{1}{12}$ *braccio* or 4.9 cm) in order to fit in their positions at the centre of the crossbow. Even if Leonardo did not depend heavily on rigorous measurements for the design of the bow, as will be discussed, he would not have expected a Milanese client to read the measurements in his drawings in Florentine *braccia*. With regard to the present study, however, the Milanese metrological standard is most important because of its twelve-part division of a *braccio*.[4] A Milanese *braccio* equals twelve *once* and twelve *once* equal 144 *punti*.[5] The *Giant Crossbow*'s portions happen to be in factors of twelve, indicating that the Milanese standard of measure might have influenced the proportions of Leonardo's design. Numerical factors or divisors in the design are predominantly third and fourth portions of the "divident" of twelve.[6] Less prominent in this design are the two-part or tenth-part divisors of the Florentine units, whereby that *braccio* is two *palmi*, or twenty *soldi*, or sixty *quattrini*, or 240 *denari*.[7] For reasons noted above, most of the following measurements will be in Milanese *braccia* and *once*, according to their sizes in modern metres and centimetres.

---

[2] With reference to the most consistently recorded measurements of the late 15th century, the present study uses the early standards discussed by Angelo Martini (1883/1976) in his *Manuale di metrologia*, E. Loescher, Turin, 1883 and Editrice Edizioni Romane d'Arte, Rome, 1976. Measurement standards in 15th century Italy were exceptionally diverse. Indeed, this lack of consistent measurement standards encouraged 15th century artists to depend less on exact measurements and more on proportions, simple geometry and third portions for their designs. Grant O'Brien proves that there was an obvious preference for simple geometry and third portion in the design of early modern keyboard instruments. His papers on this problem are at the web site of the Russell Collection of Early Keyboard Instruments in Edinburgh: <http://www.music.ed.ac.uk/russell> (online, January 2007). His tables of standard measurements offer a thorough view of the many differences in early modern measurements. Perhaps the most useful early source on Milanese metrological standards is Giovanni Croci's (1860) *Dizionario universale dei pesi e delle misure in uso presso gli antichi e moderni con ragguaglio ai pesi e misure del sistema metrico*, Milan.

[3] He used, for example, the Milanese mercantile standard for canal digging projects during the early 1490s, noting on CA 46rb, around 1490–1495, that: "if the trench were 16 *braccia* wide and 4 deep, coming to 4 *lire* for the work, 4 Milan *dinari* the square *braccio*." Richter (1977) *Literary Works*, §1001, p 228. Pedretti (1977) *Commentary*, vol. II, p 174.

[4] Whereas the Florentine *braccio* was equal to twenty *soldi* of 2.9 cm each, the Milanese *braccio* measured twelve *once* of 4.95 cm each. Various authors note these values, including especially Martini (1883), p 783, and Croci (1860), p 38. Their records confirm that these measurements predate the measurement standards legislation of 1782.

[5] Croci, p 38: 1 *braccio* = 12 *once* = 144 *punti* = 1 728 *atomi* = 20 736 *momenti*.

[6] One can divide the 'divident' of twelve by the 'factors' (divisors) three and four. Other factors of twelve include two and six.

[7] *Ibid.*; also, O'Brien's database of Italian measurements offers detailed evidence of the Florentine preference for half and ten-part divisions: <http://dionysos.music.ed.ac.uk/russell/metrology/table.html> (online, May 2003).

Leonardo's Sources

The "great crossbow" is a large-scale crossbow that has a thicker armature and is several times stronger than the standard hand-held crossbow.[8] To the tradition of proportional improvements in crossbow design, Leonardo made a significant contribution, albeit in his own unpublished treatise illustrations and notes. Hence, he appears to have made no significant contribution to the history of crossbow technology. And very little survives of early crossbow and siege engine technology, such that it is difficult to compare Leonardo's studies with the designs he inherited. The following history of the giant crossbow will help to illustrate some of the differences between his "great crossbow" designs and those of his predecessors. It is possible that he had seen illustrations of some of these examples, had heard about them, or had read about them.

The first arrowheads date to as early as 50000 BCE. The first painting of the use of a bow and arrows is in Tassili, Egypt, from around 7500 BCE.[9] Until the mid 1980s, a commonly accepted reliable record of the first handheld crossbow (manuballista) was that noted in the battle of Ma-Ling in China in 341 BCE.[10] Since then, Yang Hong (1985) and Zhu Fenghan (1995) have discussed archaeological evidence of bone, shell and stone trigger mechanisms for Chinese crossbows from as early as 2000 BCE.[11] But early literary sources, such as Sun Tzu's *Art of War* (5th century BCE), attribute the invention of the crossbow to Mr. Ch'in in the 7th century BCE. The earliest bronze trigger mechanisms for crossbows, from around 600 BCE, were found at a grave burial in Qufu, China.[12] This is the approximate date of the first known complete trigger mechanism of a mechanical crossbow found in Shandong, as discussed by Stephen Selby (2000).[13]

---

[8] The most useful secondary sources on the great crossbow include the following: D. Baatz (1978a) *Recent finds of ancient artillery*, Britannia IX:1–17; D. Baatz (1978) *Das Torsiongeschütz von Hatra*, Antike Welt 4:50–57; D. Baatz (1980) *Ein Katapult der Legio IV Macedonia aus Cremonia*, Römische Mitteilungen 87:283–299; D. Baatz (1994) *Bauten und Katapulte des römischen Heeres*, Stuttgart; M. C. Bishop and J. C. N. Coulston (1993) *Roman military equipment*, London; C. Boube-Piccot (1994) *Les bronzes antiques du Maroc*, IV, L'equipement militaire et l'armement, Paris; R. C. Clephan (1903) *Notes on Roman and Medieval military engines etc*, Archaeologia Aeliana 24:69–114; F. Cumont (1936), *The excavations at Dura-Europos preliminary report of sixth season of work*, Yale, New Haven; K. Devries (2002) *A cumulative bibliography of Medieval military history and technology*, Brill, Leiden; H. Diels (1924) *Antike Technik*, Leipzig; C. Gravett, R. Hook and C. Hook (illus) (1990) *Medieval siege warfare*, Osprey, Oxford; P. Holder (1987) *Roman artillery I*, Military Illustrated 2:31–37; J. G. Landels (1980) *Engineering in the ancient world*, Chatto and Windus, London; F. Lepper and S. S. Frere (1988) *Trajan's column*, Alan Sutton, Gloucester; J. Liebel (1998) *Springalds and great crossbows*, Royal Armouries, Leeds; E. W. Marsden (1969) *Greek and Roman artillery: Historical development*, Oxford U., Oxford; E. W. Marsden (1971) *Greek and Roman artillery: Technical treatises*, Oxford U., Oxford; W. F. Patterson (1990) *A guide to the crossbow*, Society of Archer-Antiquaries; R. Payne-Gallwey (1903) *The crossbow (with ... an appendix on the catapult, balista and Turkish bow)*, Longmans, Green and Co., London; D. Nicolle (1999) *Arms and armour of the crusading era 1050–1350: Islam, Eastern Europe and Asia*, Greenhill, London; D. Nicolle, S. Thompson (illus) (2002) *Medieval siege weapons (1), Western Europe AD 585–1385*, Osprey, Oxford; I. A. Richmond (1936) *Trajan's army on Trajan's column*, Papers of the British School at Rome xiii:1–40; R. Schneider (1906) *Heron's Cheiroballistra*, Römische Mittelungen 21:142–168; E. Schramm (1904/1906) *Zu der Rekonstruktion griechisch-römischer Geschütze*, Jahrbuch der Gesellschaft für lothringische Geschichte und Altertumskunde 1904 xvi:142–160, 1906 xviii:276–283; M. Thévenot (ed) (1693) *Veterum mathematicorum ... opera Graece et Latine ... nunc primum edita*, Paris; C. Wescher (ed) (1867) *La Poliorcetique des Grecs*, Paris.

[9] <http://www.centenaryarchers.gil.com.au/history.htm> (online, January 2008).

[10] Paterson, p 25.

[11] Yang Hong (1985) *Zhongguo Gu Bingqi Luncong* (Collected essays on ancient Chinese weapons), Cultural Relics Press, Beijing. Zhu Fenghan (1995) *Gudai Zhongguo Qingtongqi* (Ancient Chinese bronzes), Nankai University Press, Tianjin.

[12] Fenghan (1995).

[13] Stephen Selby (2000) *Chinese archery*, University Press, Hong Kong.

Bronze arrowheads measuring 1.7×0.2 inch found with the Shandong mechanisms indicate that the bows were likely built to shoot light arrows of about two ounces, starting at 225 miles per hour, for a distance of 100–400 yards. According to this ratio of arrowhead weight to bow design, the bow was built for fast loading, light arrows, with optimal accuracy within 150 yards. These bronze arrowheads could pierce armour if shot at short range. Increased speed of the launched arrow meant greater accuracy, mainly because of the reduced need to lift the bow's aim above the target in order to send an arrow high enough for it to descend to the appropriate distance.[14] Most self-bows (made with one piece of wood) only have a draw tension of 37–42 pounds and shoot an arrow 65–200 yards, starting at 150 miles per hour. Body armour can normally withstand these assaults.

The Greek *katapeltees* (shield piercing dart throwers) that were in the forms of the *gastraphetees* and *oxybelees* and were famous for shooting the "sharp-pointed missiles" that helped defeat the Carthaginian navy in western Sicily in 399 BCE. Dionysius of Syracuse had the darts "catapulted" from shore.[15] Although much less sophisticated than its Chinese shotgun-style predecessor, the *gastraphetees* could shoot three-pound darts nearly 300 yards, or small arrows even further. So named "belly shooter" (*gastra phetes*), this straight-spring, non-torsion bow could launch a dart while resting on the lower portion of the rib cage or at the waist. It reached fifteen feet in length. The armature (bow or lath portion that pulls the string taught at the front end) was normally wrapped in animal sinew, such as from a bull's neck or a dear's hamstring, for added elasticity and strength. Depending on the expected trajectory, the bow could be spanned (string and bow pulled) with two hands back to the desired position along the ratchet bar and then the crossbow lifted to the appropriate height, loaded and shot.

Using a similar ratchet bar and sinew treated armature on a larger scale, the Greek *oxybeles* of 399–353 BCE could shoot six ounce, twenty-six inch bolts over 400 yards. The addition of a base, windlass, and ratchet bar or ratchet wheel, made possible the span of a much stronger armature and thereby the launch of heavy darts much further with greater accuracy. More than two crossbowmen were needed to operate the bow at peak efficiency. After swift placement and set-up, the bow could be aimed, loaded and shot at a rate of three or four darts per minute. Soldiers could turn the chassis and then tilt it by placing the central supporting arm of the rear portion into different notched positions underneath. Although it was designed for darts, it could also shoot stones—with the proper sling attached. This would be necessary when expensive arrows were in short supply.

Roman troops might have used this bow, but they chose to develop another Greek invention: the "U-spring" or torsion screw equipped ballista. Greek names for these dart and stone launching ballistae were: *euthytonon*, of 350 BCE to 100 CE, and *palintonon*, of 334 BCE to 60 BCE, respectively. The Latin term for "crossbow" is a *ballistra* or *cheiroballistra*, whereas the stone thrower is a *ballista*. In some cases, any of these names refer to either machine. Both machines used a torsion screw engine, consisting of two twisted-rope springs at the front, to which rods were attached. When these rods were pulled back and then released, they would swing forward to normal position and snap taut the attached sling, thereby launching the arrow, dart, bolt, or stone. Hemp, sinew, hair, or any combination of the three, when treated for elasticity with an herbal unguent like spikenard, would make the twisted ropes for each spring. As long as the

---

[14] Also, the metal trigger cradle would drop forward to release the string from a vertical position to a horizontal position, which meant that the string would not be able to bounce off of the trigger edge, as it could from a rotating bone nut that falls into a round crevice when released. Called the Chinese lock, this trigger cradle was never commonly adopted in the West.

[15] Diodorus Siculus (1976) Library of history XIV:41, 3–6, Loeb Classical Library, vol. VI. C. H. Oldfather (trans), Harvard, Cambridge, MA

springs and the positions of their swinging arms were identical—in opposite directions—and the rest of the structure generally sound, the ballista would shoot straight. Proportional adjustments in the differential of torsion screw tensions were managed with large washers at the top of each screw, making this weapon more accurate than its predecessors. Also these adjustable torsion screws were much easier and quicker to mass-produce than their torsion spring predecessors, which required various stages of herbal treatments, drying and ageing.

The earliest visual evidence of these torsion screws appears on a balustrade relief at Pergamum and the tombstone of Vedennius, both dating from around 170 BCE. But these particular U-spring engines are Scorpios, modified with curved rods, improved adjustable washers (that tighten the rope-screws), and much lighter frames. Vitruvius referred to this dart-shooting machine around 60 BCE as the *scorpionum catapultarum*, or scorpio, which was well known in the Renaissance.[16]

Perge (Turkey) has the best example of the earliest known fortifications, from around 200 BCE, made for the installation of ballistae. Eric Marsden's analysis of this architecture shows that three ballistae per window could defend the area within 400 yards with the ability to attack any point within eighty percent of that yard instantly with at least six ballistae.[17] Proportional calculations obviously determined the optimal balance between the build-up of defensive masonry and the width of open windows in order to attack any spot within the first 200 yards with at least nine ballistae. Around 1485, Leonardo considered similar calculations for his studies of the distribution of canon fire over fortifications, as in *Codex Atlanticus* 59bv [18rb] and MS B 74r, and *Windsor* 12275r and 12337v. Actually, his 42 × 40 *braccia* (80 ft, 24 m) *Giant Crossbow* would have fit on the top floor of a corner tower of the Sforza Castle.

Possibly the most effective crossbow prior to Leonardo is the *cheiroballistra* (Fig. 1.6), a highly improved scorpio. The new torsion spring engine had adjustable, shielded springs and was much lighter and thus more manoeuvreable than the scorpio. So important were the *ballistrae* that six of them figure prominently in two scenes (40 and 66) of the Column of Trajan. No depictions appear on this column of straight-spring ballistae or U-spring scorpios. Roman successes in foreign sieges partially resulted from the benefits of these light and powerful *cheiroballistrae*. Caesar used them in Africa in 46 BCE and Germanicus used them in Germany in 15 CE.[18] Their improvements over the scorpio and other ballistae included the light metal frame, waterproof metal cylinders over the rope-screw, modular sections for portability and repair, and the upper arched strut allowing a broader angle of vision and a precise central aim. Thus it travelled well, withstood difficult weather conditions as well as oncoming missiles, swiftly took aim, and had easily replaceable modular parts. Although its shooting capacity was less than that of the *oxybeles* and ballista, its tactical benefits meant that it outperformed its predecessors in portability, set-up, manoeuvrability, accuracy and maintenance. As an essential improvement for dealing with moving targets, the arch of the thin front crossbar enabled one to see at the same time a broad field of vision and the specific target. N. Cooper and R. Cooper built a full-scale *cheiroballeistra* based on Marsden's translation and found that each portion of the machine was proportionately designed for its function. "One man could just about carry the *cheiroballistra* itself, a base and ammunition, for a short distance without too much hardship," states Marsden.[19]

---

[16] Vitruvius (1934/1999) *On architecture*, X:10–16, Loeb Classical Library 280, F. Granger (trans), Harvard, Cambridge, MA, (1999 ed.).

[17] Marsden (1969) pp 116–163.

[18] Ceasar, *African War*, 29. Tacitus, *Annals*, II: 20.

[19] Marsden (1971) *Greek and Roman artillery: Technical treatises*, p 233.

Similar light siege engines were used by Roman-led armies in Europe and the Middle East from around 43 CE, for which substantial archaeological evidence exists in the form of square wounds in skulls and the remains of Roman 14 and 26 inch iron bolts, averaging weights of six ounces or more.[20] Some of the five-inch bolt-heads had metal cages for the attachment of flammable material.[21] A piece of 1st century CE Roman jewellery, now called the *Cupid Gem*, displays the earliest representation of the spanning of a *cheiroballistra*.

Prior to the Romanian excavation of a *cheiroballistra* in 1974, reconstructions of this machine were based incorrectly on what could be gleaned from the textual evidence.[22] Parts from this excavation had proven that certain 11th century Byzantine diagrams were consistent with Roman texts on the *cheiroballistra*.[23] Without the ability to use these diagrams with the earlier texts, previous reconstructions were inaccurate. Following the texts of Philon's *Belopoeica* (catapult manual) (49–78) of the 3rd century BCE, and Heron's *Belopoeica* (71–119) of the 1st century CE, Victor Prou produced the first reconstruction in 1877.[24] He mistakenly assumed that it had hand-shaped bronze springs. No trace of the original *cheiroballistra* exists between the 3rd century CE and 1995. At this latter date, Len Morgan built an appropriate reconstruction of the mark III *ballistra* known from 340 BCE.[25] Nineteenth century European studies of this machine followed 3rd century BCE texts by Philon (c. 212 BCE) and Heron (c. 65 CE), and Vitruvius (50 BCE, *De Architectura* X: X–XVI). These authors copied 3rd century BCE editions of Ctesibius' *Commentaries* (now lost) and Biton's *Construction of War Machines* (43–68), adding their own amendments, such as measurements. No evidence exists of the successful use of post 3rd century CE diagrammatic examples of the *cheiroballistra* until this material was compared to archaeological evidence in 1995.

Leonardo and his contemporaries were well aware of Vitruvius and an as yet undefined portion of Heron's and Ctesibius' work on machines governed by water, steam, air and weights.[26] Around 65–75 CE, Heron added to his treatise the sections on mechanics from Ctesibius, Philon, Vitruvius, and others. But none of the original diagrams survive from these sources. Had they been available in the 15th century, there is no evidence of their identification with *cheiroballistrae* or similar machines (with the exception of the possible source for the 14th century French espringal, as discussed further below). Visual sources for Leonardo of these pre-3rd century CE designs may have been the recently excavated Column of Trajan (re-erected by Paul III in 1540), sarcophagi, tombstones (Marsden pl. 1), gems and other antiquities.[27] No record exists of his contact with these objects or with Rome prior to 1485, the latest likely date for his *Giant Crossbow*, although he had access to collections of antiquities in Florence and Milan.

If Leonardo had access to early illustrations of the *cheiroballistra* (Fig. 3.1), he was probably limited to examples similar to the Byzantine copies that survive today in Paris, Rome, and Vienna:

[20] *Ibid.*, plate 5. J. D. Vicente, M. P. Punter, B. Ezquerra (1997) *La catapulta tardo-republicana y otro equipamiento militar de 'La Caridad' (Caminreal, Teruel)*, (Spain), Journal of Roman Military Equipment Studies 8:167–199.

[21] S. James (1983) *Archaeological evidence for Roman incendiary projectiles*, Saalburg Jahrbuch 39:142–143.

[22] N. Gudea and D. Baatz (1974) *Teile Spätrömischer Ballisten aus Gornea und Orsova (Rumanien)*, Saalburg Jahrbuch xxxi:50–72.

[23] Alan Wilkins (1995) *Reconstructing the cheiroballistra*, Journal of Roman Military Equipment Studies 6:5–60.

[24] V. Prou (1877) *La Chirobaliste d'Heron d'Alexandrie*, Notices et Extraits des manuscrits de la Bibliothèque nationale et autres bibliothèques 26, Paris, pp 1–319.

[25] *Ibid.*, pp 5–7.

[26] Noted on CA 96va and 219va. Martin Kemp (1981) *Leonardo da Vinci, The marvellous works of nature and man*, Harvard, Cambridge, MA, p 170.

[27] Marsden (1969) *Greek and Roman artillery: Historical development*, plates 1, 2 and 9–13.

Fig. 3.1.
Illustration of a cheiroballistra

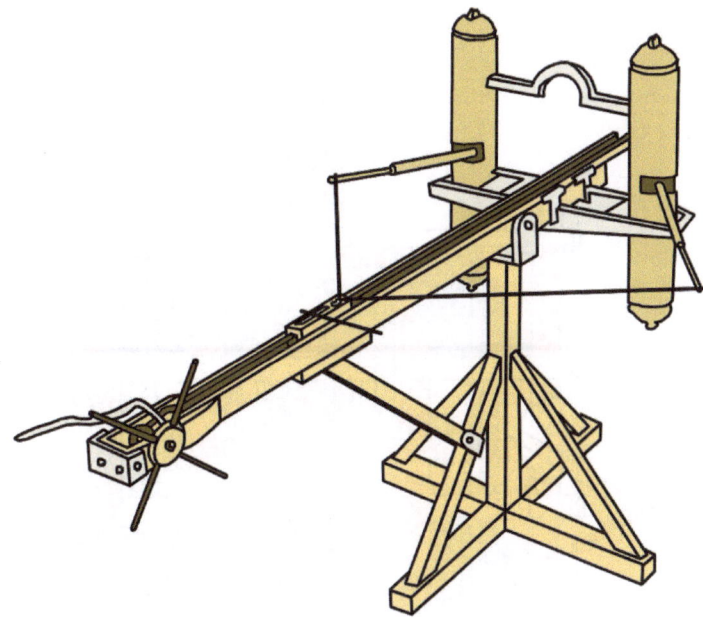

Par 607	Codex Parisinus graec. 607
	Bibliothéque Nationale, Paris
	11th C. military treatises
Par 2442	Codex Parisinus graec. 2442[28]
	Bibliothéque Nationale, Paris
	11th C. military treatise (Fig. 3.2)
Vat 1164	Codex Vaticanus 1164
	Biblioteca Apostolica Vaticana, Rome
	Late 10th/early 11th century military treatise
Wien 120	Fragmenta Vindobonensia 120
	Österreichische Nationalbibliothek, Vienna

Par 607 and *Wien 120* contain copies of diagrams and instructions attributed to Biton and Heron, whereas *Par 2442* and *Vat 1164* contain similar material attributed to these authors as well as to Philon.[29] The best copy, from which the other manuscripts seem to be copies twice or thrice removed, is the *Par 607*.

Par 607 was at one time part of the 50 000 volume library at Buda (of present-day Budapest) that was founded by King Matthias Corvinus (reigned 1457 to 1490).[30] Although removed before Ottoman Turks destroyed the library in 1526, the MS was later at the Monastery of Vatopedi on Mt. Athos. Suleiman the Magnificent looted the library and might have found and stored the military *Par 607* in a library at the Ottoman occupied Mt. Athos. Nonetheless, given the close ties between Corvinus and the Milanese Court during the 1470s and 1480s, the likelihood of Leonardo's commission to paint a Madonna for Corvinas in 1485, Corvinas' habit of commissioning and collecting thousands manuscripts, and the threat of the Ottoman invasion of Hungary, Corvinas could have transferred military manuscripts to Milan or vice-versa.[31]

---

[28] Magdalen College, Oxford, has a sixteenth century copy of a portion of this MS, as seen at: <http:// image.ox.ac.uk/show?collection=magdalen&manuscript=msgr14>, MS Gr.14, see especially folio 8.

[29] Wescher, pp v ff. Marsden (1971) *Technical treatises*, pp 7–14.

[30] Wescher, p xv. J. M. Moore (1965) *The manuscript tradition of Polybius*, Cambridge, p 134.

[31] Regarding Leonardo's possible commission, see: Kemp, p 210. Only 22 codices are know to exist from the Bibliotheca Corviniana and there are only 215 Corvinas extant in 49 libraries worldwide <http:/ /www.frankfurt.matav.hu/angol/konyvkiad.htm>.

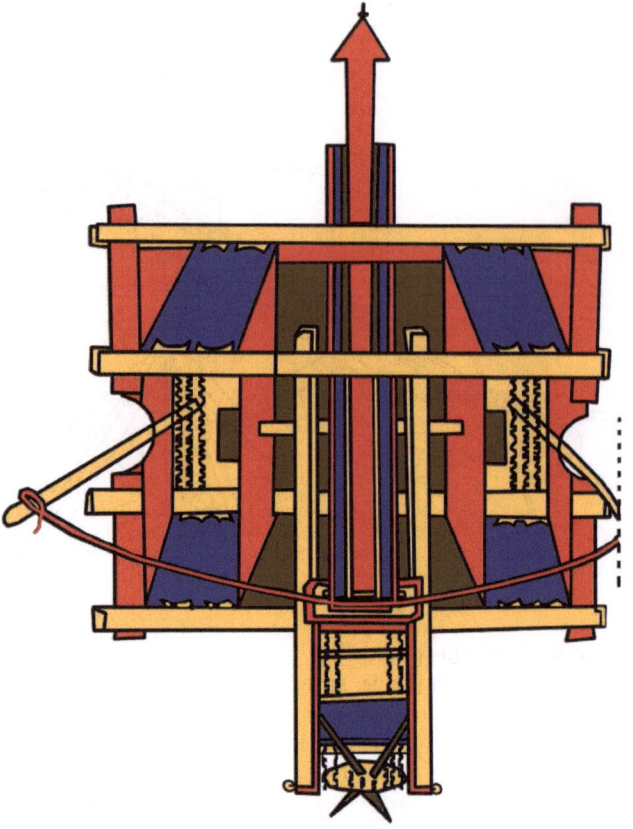

Fig. 3.2.
Illustration of an 11th century Byzantine diagram of a Roman style cheiroballistra, possibly copied from Heron's *Belopoiika*; MS Codex Parisinus graec. 2442, f. 76v, Bibliothéque Nationale, Paris

The problem with these diagrams, however, is that they would require a military engineer and a Greek translator to decipher the poorly copied plans. Eleventh century monks or artists copied the combined depictions of rear and overhead views of ballista portions in foreshortened composites. Medieval copyists of Greek and Roman illustrations often had this difficulty with perspective.[32] No direct evidence exists of Leonardo's possible interest in these particular diagrams, but there was obviously considerable contemporary interest in these manuscripts in the humanist circles in which he moved.

Incorrect interpretations of ballistrae reproductions likely contributed to their decline after the 3rd century CE. The four Byzantine manuscripts listed above copy 3rd century CE Roman manuscripts written in Greek. At this time, no further evidence exists of the *cheiroballistra*. Ammianus Marcellinus (c. 363) and Vegetius (c. 385) were the last to write about this machine.

Also at this time, these authors refer to the *onager* (wild ass), a new stone-throwing engine that is much easier to make, to use, and to affect more devastating results.[33] Without explaining this engine in detail, it should be noted that *onagari* are the ancestors of the popular massive object-throwing *mangonel* and trebuchet catapults. All three of these machines were commonly used in European and Eastern sieges from the 4th century until the more versatile and powerful canon had gained sufficient popularity in the 16th century. From the early 14th century there is evidence of the trebuchet at the walls of Siena, as painted by Simone Martini, and in a manuscript drawing of the Scottish siege of Carlisle.

---

[32] See: A. Stückelberger (1994) *Bild und Wort: Das illustrierte Fachbuch in der antiken Naturwissenschaft, Medizin und Technik*, Zabern, Mainz. Also: D. Baatz's recent site on catapults <http://home.t-online.de/home/d.baatz/catapult.htm>.

[33] Ammianus Marcellinus XXIII:4.4–7. Vegetius II:25.

Villard de Honnecourt produced an early diagram of a trebuchet base-frame around 1250. This is one of the first Western schematic drawings of a siege engine detail. It accompanied another illustration, now lost, of the supporting frame, bean-sling and counterweight. The text at the bottom states:

> If you wish to build the strong engine called the trebuchet pay close attention. Here is the base as it rests on the ground. In front are the two windlasses and the double rope by which the pole is hauled down, as you may see on the other page [now lost]. The weight that must be hauled back is very great, for the counterpoise is very heavy being a hopper full of earth. This is fully two fathoms long, eight feet wide and twelve feet deep. Remember that before the missile is discharged, it must rest on the front stanchion.[34]

Even if the *fulminalis* were never built, it is still possible that someone knowledgeable about military engineering originally designed it. Phrygian caps of the soldiers reveal the late Roman use of Persian soldiers and technology. Swirling pattern decorations along the side indicate that this is meant to be a beautiful object. Octagonal latticework on each side of the *fulminalis* (Fig. 3.3) may be part of a ratchet mechanism holding each windlass rotation, in addition to protecting the soldiers and offering handlebars with which to lift and steer the rear portion. For this machine, a straight non-torsion spring is used, such as a sinew treated wooden pole, protected inside the front chassis. Possibly influenced by drawings similar to the Paris manuscripts 607 folio 56v and 2442 folio 76v (Fig. 3.2), the rectangular frame houses windlass portions at the rear and trigger portions at the front. As seen at the front end of the *fulminalis* illustration, one lever holds down a trip wire and the other lever is poised to lift the wire. Much of what triggers the machine hides below the upper chassis and no available written description exists of the trip wire mechanism. Attached to the hidden windlass one can see a four-part compound pulley at the far end, the other end of which is attached to a hooked form holding back the arrow's platform frame. As the windlass pulls the arrow's platform frame, the only tension (T) required to pull the weight (W) of this frame—thanks to the four-part compound pulley—is a fourth of the original weight (such that $T = W/4$). With the addition of a second four-part compound pulley, from the rear of the entire central frame to the metal crossbar, the windlass pulls a tension only one quarter of the central frame's weight. Thus, from windlass to pulley, from pulley to arrow platform frame, from arrow platform frame to central platform frame (which rests on boards of the lower frame), from central platform frame to pulley, from pulley to *cheiroballistra* style metal crossbar, is the pull strength of nearly 800 pounds for every 100 pounds of tension applied to the windlass. Thus the windlass can be wound with one or two soldiers pulling 100 pounds of tension.

Fig. 3.3.
Illustration of an anonymous painting of a ballista fulminalis, *De rebus bellicus*, Bodleian Library, Oxford

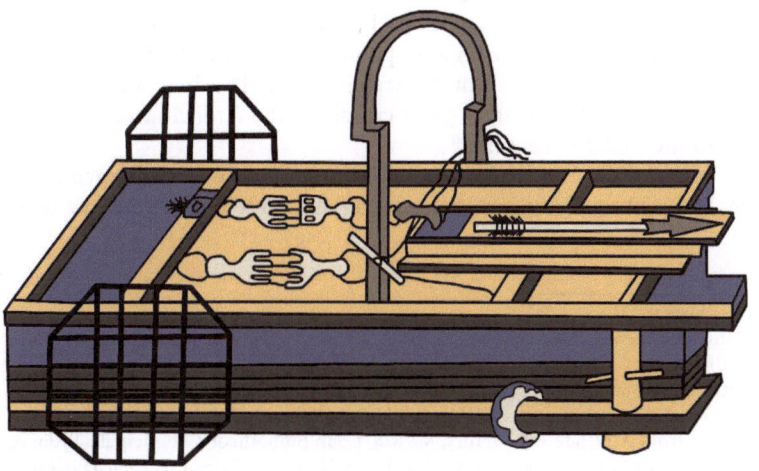

[34] T. Bowie (1959) *The sketchbook of Villard de Honnecourt*, Indiana U., Bloomington.

The best technological breakthrough is obviously the compound pulley distribution of weight, although this converted the traditional string launcher into a massive sliding upper chassis, attached to straight spring with pulleys. Its structure is dangerous and the upper board would have quickly deteriorated, but it is not as difficult to build as its carefully balanced torsion screw predecessors. Tensions of the left and right portions of its engine did not require balancing with the same degree of precision, craftsmanship and calculation. Underneath the top platform, to each side of the metal crossbar could be attached (presumably) a sinew treated wooden pole, much like that shown with the *ballista quadrirotis*. The entire upper portion, including both internal frames and metal crossbar, are designed to snap forward, shooting an arrow at four to eight times the power of a non-pulley torsion screw or non-torsion straight spring engine. Presumably, a gunner managed two ropes with two handles. One handle, when pulled forward, held the rope holding down the bottom end of the hook at the arrow's platform frame. This rope is seen reaching around the metal crossbar at the far end, down under the arrow's platform frame, up at the front of the general launching platform frame, and back to the metal crossbar. If the other handle were pulled, it would lift the hook at the arrow platform frame. This rope reaches around the metal crossbar, across the launching platform, under the hook, and to a mid-height portion of the crossbar illustrated with a white rod, to which both ropes were attached. Such features allow for the spanning and launching of large spears by as few as three soldiers, sending missiles conceivably four times the distance of previous crossbows.

In theory this might work; but in practise it would be ridiculously difficult to manage. Hence, no evidence exists of the construction of a *ballista fulminalis*. Compared to the simpler 4th century *onagari* and smaller crossbows, the *fulminalis* would be too expensive to build and maintain. More interesting for the present study is the extant to which the available visual information about *cheiroballistrae* was used to design a siege engine proposal. This methodology parallels to some extant Leonardo's approach to designing his *Giant Crossbow* around 1485.

By the late 9th century, "catapult" *ballistae* that shot bolts and stones were well known in Europe. Around 885 CE, Abbo Cernuus notes that Paris used a hundred of these "catapults" on its walls in order to impale invading Vikings with *ballista* bolts. Ebolus—Abbot of Saint-Germain des Prés and a principle defender—supposedly yelled at the enemy, "carry the slain to the kitchen!" noting that several men were skewered with a single shot.[35] But no illustrated evidence exists of these large crossbows prior to the 14th century. A French depiction of the 37 BCE Siege of Jerusalem from around 995 only shows hand held crossbows and battering rams. (Late medieval illustrators added the most recent siege engines to depictions of ancient sieges.) Other European illustrations from the twelfth and thirteen centuries note the extensive use of these small bows as well as trebuchets.[36]

The Portuguese town of Santarem used a large-scale crossbow in 1184 to kill the Muwahhid Caliph Abu Yaqub with a 5.5 pound bolt.[37] They had previously obtained this weapon, a *qaws al-lawab*, from the Arab invaders. This bow was probably used in Spain at that time, although their great crossbow, the *ballista de torno*, is not mentioned until the mid-13th century.[38]

In the early 13th century, three kinds of great crossbow were known in France, as noted by the scribes of Philip II Augustus in 1204–1206: the crossbow with stirrup

---

[35] Payne-Gallwey, p 271.

[36] See: Beatus' *Commentaries on the apocolypse*, from Gerona c. 1100 and from the Arab *lu'ab*, MS J. II.j, f. 190r, Bibliotheca Nazionale, Turin; Peter of Eboli's *Chronicle* of the "Siege of Naples," early 13th century, MS Cod. 120/II, f. 15a, Burgerbibliothek, Bern; *Maciejowski Bible*, "Saul's army destroying Nahash," mid 13th century, f. 23v, Pierpont Morgan Library, New York; William of Tyre's *History of outremer*, from Acre, late 13th century, MS 828, f. 33r, Bibliotheque Municipale, Lyon; Annales de Genes, MS Lat. 10136, f. 107r, 141v, 142r, Bibliotheque Nationale, Paris; *City Charter of Carslisle*, 1316, Cumbria Records Office, Carlisle.

[37] Nicolle (2002) p 33.

[38] *Ibid.*

(*balistam ad estrif*), two-foot crossbow (*balistam ad duos pedes*), and windlass crossbow (*balistam ad tornum*).[39] The two-foot crossbow, so-named for shooting two-foot bolts, had an armature of up to six feet in length, often made of Portuguese or French yew, treated with goat horn and sinew. This is similar to the bow in Conrad Kyeser's 1405 *Bellifortis*, fols. 76r–79r, illustrated with its spanning mechanism: a compound pulley attached to a post, with a metal claw to pull the bowstring.

A bow stave (bare wooden front portion) of yew, found in the moat of Berkhamstead Castle, happens to be the best studied evidence of an early 13th century (c. 1216) large-scale wooden-frame or wall-mounted crossbow.[40] Measuring four feet in length, the thick stave required a spanning mechanism for a draw weight of 154 pounds at a span length of only twelve inches.[41] This could shoot bolts weighing up to a pound, according to the discovery of six-ounce bolt heads dating back to the 13th century. The short span length contributed to a quick loading procedure and ease of use along some of the narrow passageways atop castle walls.

But the Berkhamstead type of armature was just as likely to have been part of a winding screw mechanism, as seen in Walter de Milemete's treatise of 1326, fol. 69r.[42] The most accurate early representations of this machine happen to be among Leonardo's manuscripts CA 149ar, Uffizi 446ev, CA 147bv and CA 147av. He also used the central screw spanning mechanism for his *Giant Crossbow*. Large crossbows like that in Milemete's treatise were most likely installed on fortifications at Florence and Milan in the 15th century. Taccola produced a drawing of this design in 1449, showing a "composite" armature treated with sinew, horn, and/or other material.[43] This crossbow uses a medieval tensioning bench to span the bow. A tensioning bench is the bowyer's (crossbow carpenter's) tool for shaping, ageing, and stringing the crossbow. Possibly the earliest illustration of this tensioning bench bow happens to be the mid 14th century example of the Vienna MS 3069, folio 22v, with what seems to be an iron crank handle and screw. These traditionally used armatures six feet in length, made of yew and treated with horn and sinew. They could store almost a ton (~1 764 lbs) of energy and shoot metal tipped 26 inch bolts a maximum distance of 593 feet (198 yds), and with reasonable accuracy, half that distance.[44]

The Milemete treatise records a turning point in siege technology, with the first Western illustration of a cannon.[45] A late 14th century manuscript at the Bodleian Library illustrates the use of a small cannon, trebuchet, and crossbow.[46] Cannon were not popular in European siege warfare in until the late 15th century.

The French built highly effective box format springald (field-launcher or *lanciacampi*) crossbows in the early 14th century. Springalds were easier to transport than the massive stone throwing machines and were best suited for defensive emplacements above fortifications. Made of beech, elm or oak, they were often around five feet wide, five feet tall, and ranged in length between six and twelve feet. In 1339, the French sent to the recently captured Castle Cornet on Guernsey:

> …eighty small crossbows, nine large two-feet crossbows [held down and spanned with two feet], three heavy crossbows spanned with windlasses, one frame-mounted windlass crossbow and two *espringals garnis de ij braes cordes*. In other words, they already had their twisted skeins of horsehair attached. The ammunition for these *espringals* consisted of four barrels of heavy quarrels of bolts feathered with latten, a type of brass.[47]

[39] E. Audoin (1913) *Essai sur l'armée royale au temps de Philippe Auguste*, Champion, Paris, pp 187–197.

[40] R. C. Brown (1967) *Observations on the Berkhamstead Bow*, Journal of the Society of Archer Antiquaries 10:12–17.

[41] *Ibid.*

[42] Walter de Milemete, de Nobilitatibus, Sapientiis, et Prudentiis Regum, Christ Church MS no. 92.

[43] Mariano Taccola (1449) *De machinis* (aka. *De rebus militaribus*), MS Codex Latinus Monacensis 28800, Bayerische Staatsbibliothek.

[44] Liebel, pp 61–68.

[45] *Ibid.*, Milemete, f. 70v.

[46] MS Bodl. 264, f. 255r, Bodleian Library, Oxford.

[47] Nicolle (2002) p 47.

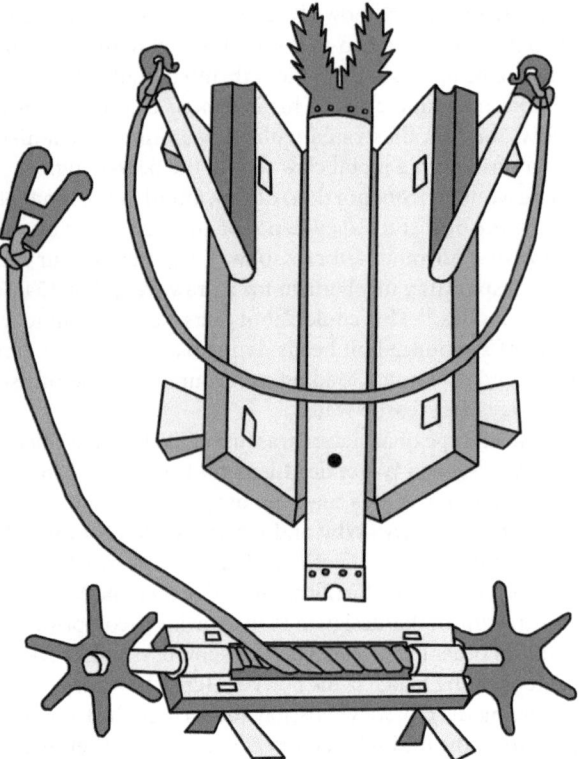

Fig. 3.4.
Illustration of a woodcut of the
components of a springald style
battering ram in Conrad Kyeser's
Bellifortis, Universitätsbibliothek,
MS 63, f. 80v, Universitätsbibliothek, Göttingen

Although generally more powerful than the great crossbow, the springald was not
as versatile, since it could not be moved quickly during a siege, nor would it operate
properly in wet conditions at sea. For this reason, the great crossbow, or "*balista de
torno vel pesarola*," was favoured by the Venetians, who required in 1255 that ships
carry fifty quarrels of ammunition per *balista*.[48] By 1500, Venetian galleys were equipped
with cannon.

The springald was probably based on surviving information about the Roman
cheiroballistra such as antique remnants and 11th century Byzantine diagram copies
that eventually found their way to the Paris Bibliothéque Nationale as the Codex
Parisinus graec. 607, and especially folio 76v of Codex Parisinus graec. 2442. The earliest known illustrations of the French/German box format springald machine include
the springald-style battering ram found in Conrad Kyeser's *Bellifortis* of 1405 (Fig. 3.4)
and the bolt-shooting springald drawing attributed to Matteo Pasti in Roberto Valturio's
De re militari of 1451 (diagram, Fig. I.7).[49] To the trained eye, folio 76v, of Parisinus
graec. 2442 (Fig. 3.2), looks almost like the *cheiroballistra* that it illustrates, though it
could have been misinterpreted as a springald design, viewed from the rear, with a
box-shaped chassis and torsion rope springs stretching from top to bottom. The
springald was also more efficient. Whereas the 4th century *ballista fulminalis* concept

---

[48] *Ibid.*, p 35.

[49] A note glued to the inside of a Bodleian Library copy of this Verona edition (Douce 289) identifies
Matteo de'Pasti (f. 1441–1467) as the original illustrator of the ninety-six woodcuts. He had the
added assignment, on behalf of Sigismund Malatesta (Duke of Rimini and General of the Papal
Forces), to give a copy of his manuscript to the Sultan Mohamet II in 1451. The Bodleian Library's
Douce 289 is the only fully coloured copy of the three 1472 editions in the library's collection.
Matteo supposedly did not reach the Sultan, as he was arrested in Crete by Venetian authorities,
who sent him to Venice to explain why he was carrying military information to the Turks. Although the manuscript contained only medieval technology, the Venetians were at least initially
convinced that the Turks could make use of it.

of a *cheiroballistra* (Fig. 3.3) stored an estimated 800 pounds of tension, the springald stored nearly two tons (3 969 lbs) of draw tension.[50] Another influence on the springald could have been the Arab *qaws al-ziyar* "skein bow" (with torsion springs, a larger version of the *qaws al-lawab*). Royal scribes for King Louis wrote about the "*balistrarium silvestrarum vel spingardarum,*" a kind of Arab torsion spring *qaws al-ziyar* that was used with devastating effect against Crusaders in Egypt in 1250.[51] Although no one has identified a visual example of the *qaws al-ziyar*, its capability resembles that of the springald by virtue of the 5.5 pound, 26 inch bolts that it could shoot. Unlike the *cheiroballistra*, the springald required fewer parts, consisted primarily of wood and twisted horsehair, shot heavier bolts, and was easier and cheaper to make and operate.[52] Another technical benefit of the springald was the addition of the nut trigger mechanism: a circular catch that rotates forward, releasing the bowstring without any resistance (Fig. 3.5a). Leonardo added this type of trigger to his *Giant Crossbow*, first making preparatory designs on folio CA 1048br (Fig. 1.6). He would have seen this kind of trigger as a standard feature of the great crossbow, as illustrated on folio 34v of the Löffelholz manuscript (Nuremberg, c. 1505) (Fig. 3.5b), and as he illustrated on CA 149ar [53va] (Fig. I.4).

In addition to accurate great crossbow drawings, Leonardo offers the most accurate late 15th century drawing of the springald, at the top centre of CA 149ar. He makes a significant modification to this machine for which there are no existing examples: the

Fig. 3.5a.
Detail of the trigger mechanism on Codex Atlanticus f. 149br, with superimposed arrows that show directions of the components

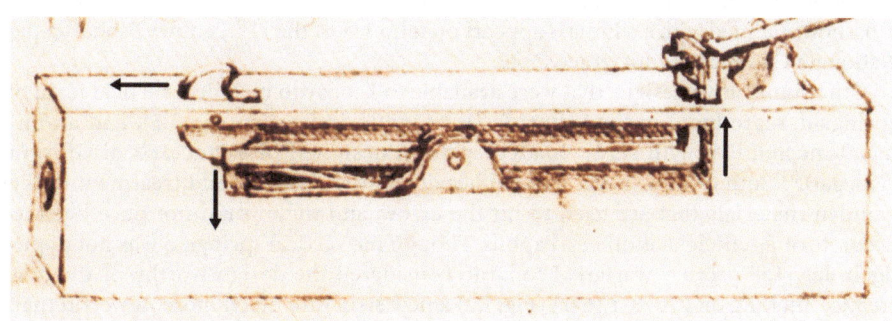

Fig. 3.5b.
Illustration of a great crossbow and incendiary missile in the Löffelholz manuscript, MS lat. 599 f. 34v, Bayerische Staatsbibliothek, Munich

---

[50] Liebel, pp 61–62.
[51] Nicolle (2002) p 35.
[52] See: Liebel, pp 2–22.

frame is allowed to bend, adding considerable tension to the twisted rope. He was possibly influenced by late 15^(th) century examples (manuscript and/or actual) of German steel spring catapults (powered by tensed steel rods). Next to CA 149ar, he illustrates a plan for fastening the corners of the flexible frame. Exploring the limitations of this bent frame engine, he produced the presentation drawing on CA 145r [51vb] (Fig. 1.1) of a springald powered almost exclusively by its bent frame (rather than mainly by the torsion screw). This particular design would appear to be his invention, especially because the torsion screw's arms are too long and the winding mechanism at the end of the worm screw's pole is too small for the model to work properly. Springalds had nearly vanished from siege warfare almost a century before Leonardo's design. Nonetheless he found ways to reinvent this machine—so much so that CA 145r is hardly recognisable as a springald. Although Carlo Pedretti dates this drawing and similar CA sheets 140ar [50va] and 140br [50vb] to around 1500, these drawings should be recognised for characteristics in design, ingenuity, and treatment nearly identical to Leonardo's studies at the time of the *Giant Crossbow*.[53]

Italians also used metal armatures on smaller bows during the 15^(th) century, of the kinds demonstrated in Renaissance paintings of the *Martyrdom of St. Sebastian* by Signorelli, Perugino, Francesco del Cossa, and Piero and Antonio del Pollaiuolo. Larger versions of these were spanned with a windlass mechanism, instead of the belt hook in the Pollaiuolo painting, and had the added benefit of withstanding wet weather and lasting longer than the wooden bows. The earliest illustration of a metal armature crossbow spanned with a windlass appears on folio 13v of the 11^(th) century Bibliothéque Nationale Codex Parisinus graec. 2442.

Other innovative designs that were available to Leonardo include a vertical form of springald, reproduced in the Valturio's *De re militari* (diagram, Fig. I.5), and an unusual mechanism with screw-loaded springs, designed by Francesco di Giorgio (Fig. 3.6).[54] Both of these items depend heavily on the elasticity and treatment of the wooden materials that are used to hit the arrow, and they would not have been as accurate or as efficient as other weapons. Though the vertical springald was not popular in late 15^(th) century warfare, Leonardo considered the design worthy of the new treatise, making on CA 181r [64r] (Fig. I.2) a new structure much more powerful than its predecessor. This giant machine could have fired nearly ten arrows in an instant, with force derived from three beams of wood, instead of the traditional single beam. To cope with the large size of the structure, he added a winch and gear system at the bottom, along with a large ratchet (presumably metal) trigger mechanism at the top.

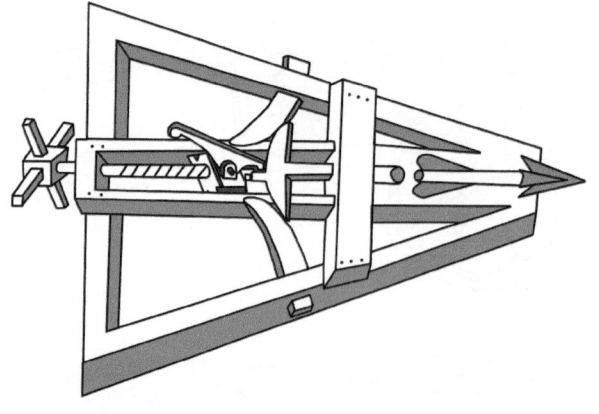

Fig. 3.6.
Illustration of Francesco di Giorgio Martini drawing of a dart smacking device, MS Salluzziano 148, f. 61v (detail), Biblioteca Reale, Turin

---

[53] Pedretti (1978) *Catalogue*, pp 85–86.

[54] Springald mechanism illustrated possibly by Matteo Pasti for Robertus Valturius, *De re militari*, Verona, 1472; Bodleian Library MS Douce 289. Screw-loaded dart smacking device illustrated by Francesco di Giorgio Martini in MS Salluzziano 148, Biblioteca Reale, Turin, f. 61v (detail).

At a time when cannon were becoming the most efficient form of siege artillery, Leonardo was not content to work simply with cannon designs, of which we have a thorough repertoire from him. He exhibited an extensive understanding of the visual history of crossbow and springald designs in order to produce studies of siege engines no longer popular in late 15th century Northern Italy. He had found different kinds of siege engine designs and—as with his cannon inventions—reinvented those machines on a giant scale unfamiliar to his contemporaries and with new mechanical features fundamental to the altered scale of the new structure. To design machines at a greater scale, he used proportional calculations to determine visual and practical proportions of the appropriate types, sizes, and strengths of necessary materials. This inventive process, with regard to the visual construction his *Giant Crossbow*, is the subject of the Part II.

The present section's approach to documentary evidence relating to the *Giant Crossbow* locates the drawing within its historical context and offers evidence that Leonardo wanted the design to be taken seriously. The documents give a glimpse of the broad range of his approaches to proportional and mechanical solutions for the giant crossbow project. These solutions include methods for determining the design layout, page margins, machine design, human proportions, rope tension, length of draw, missile weight, missile trajectory, qualities of wood, metal alloys, trigger design, wooden and metal spring capabilities, load bearing structures, screw thread technology, etc. Although these are practical considerations for the giant crossbow design, existing evidence related to this design is entirely theoretical and does not necessarily suggest that the *Giant Crossbow* is a realistic project. No evidence exists of the successful construction of this siege engine. What the documents suggest, however, is that Leonardo developed this drawing as a reasonable mechanical diagram for a treatise, illustrated for Ludovico Sforza or one of his commanders. The following study will use the extensive preparatory studies noted above to show that the *Giant Crossbow* is an unusually resolved design for a treatise illustration, and as such, it serves as an exemplary instance of proportional design, particularly in terms of the precision arrangement of what appear to be constructible modular components of this dimetric scale model.

Part II Making It Work: Functionality of the *Giant Crossbow* Design

Publications that mention Leonardo's *Giant Crossbow* on *Codex Atlanticus* (CA) folio 149br [53vb] (Fig. I.1) praise its compelling artistic design, though they doubt its mechanical capabilities.[1] As a partial result of this abbreviated approach, there is no adequate study of the drawing's mechanical design. Nonetheless, this drawing is an unusually thorough example of Leonardo's mechanical and graphic design skills. The present study will show the ways in which his skills as an engineer and as an artist enabled him to design the *Giant Crossbow* presentation drawing. At issue is the theoretical and proportional consistency between the crossbow's sections in the drawing and the possible corresponding life-size sections to which the drawing may refer. For example, proportional divisions of this bow into modules of third portions are not immediately obvious features of its remarkably complex design. The illustrated accuracy of these proportional modules are nonetheless rendered in dimetric orthographic projection and according to a specific scale at which one can examine them as functional components. Hence, the capability of this design to address theoretically various practical problems with some accuracy calls into question the bow's possible reliability if built to scale. It would seem that one could build the bow according to its design on paper, but it is not possible to determine if the bow would function properly other then by devoting extensive resources to building it. In any event, even if the drawing serves no other purpose than to advertise an idea, in this form it succeeds as a virtuoso example of the design techniques used by 15th century artist/engineers. At issue in the following study is the extent to which proportional techniques formed the basis of Leonardo's theoretical approach to machine design. The features of the bow will be surveyed as well as the proportional techniques used in its design.

A thorough study of Leonardo's approaches to practical problems of giant crossbow construction is not yet possible partially because there is no reliable evidence of a construction that has been particularly faithful to his *Giant Crossbow* proposal. Introduced in the present study are a number of these practical issues mainly as a way to link his theories to practical problems if one were to attempt a faithful construction of the machine. This includes a very brief mention of the nearly full-scale model built by ITN Factual Television in the summer of 2002 (Figs. II.1–II.6), the

---

[1] See: G. Canestrini (1939) *Leonardo costruttore di machine e veicoli*, Milan. A. Uccelli (ed) (1940) *I libri di meccanica di Leonardo da Vinci*, Milan. Margaret Cooper (1965) *The inventions of Leonardo da Vinci*, Macmillan, New York. G. De Toni (1969) *Studio di meccanica*, in: "Leonardo da Vinci," atti del *Simposio Internazionale di Storia della Scienza*, Florence. P. Rossi (1971) *I filosofi e le machine (1400–1700)*, Milan. B. Gille (1972) *Leonardo e gli ingegneri del Rinascimento*, Milan. Charles Gibbs-Smith (1978) *The inventions of Leonardo da Vinci*, Phaidon, London. Marco Cianchi (nd) *Leonardo da Vinci's Machines*, Becocci, Milan. P. Galluzzi (1996) *Art of invention: Leonardo and Renaissance Engineers*, Giunti, Florence. Pedretti (1999) *Leonardo, the Machines*, Giunti, Florence. L. Tursini (1951) *Leonardo e l'arte militare*, Rivista d'ingegneria 10 (Oct.).

Fig. II.1.
Author's photograph of a reconstruction of Leonardo's *Giant Crossbow* by ITN Factual Television, a view of the filming on Salisbury Plain in September 2002

Fig. II.2.
Author's photograph of a reconstruction of Leonardo's *Giant Crossbow* by ITN Factual Television, now located at the Royal Armouries, Fort Nelson, Hampshire, UK

Fig. II.3.
Author's photograph of a reconstruction of Leonardo's *Giant Crossbow* by ITN Factual Television, now located at the Royal Armouries, Fort Nelson, Hampshire, UK

construction of which was meant to follow, albeit in modified form, the theoretical designs of Leonardo's drawings.[2] ITN made an expensive investment in the construction of his crossbow, although the final design differed so much from his original proposal that a discussion of its practical problems would have virtually no basis in the original design. An armature design similar to that of Leonardo's, though shorter and made of metal, was used for the movie *Lord of the Rings: The Two Towers*.[3] It remains to be seen, however, if the *Giant Crossbow* is a practical design. What can be said of this is that Leonardo offers on CA 149br and preparatory drawings a legible presentation of his thoughtful approaches to this mechanical problem. A court engineer or a duke would have appreciated the detail and precision of this design.

Thirty-five preparatory and related sketches useful for building a giant crossbow prove Leonardo's commitment to designing the proposed model exhibited in his *Giant Crossbow* illustration of c. 1483–1492. Previous scholarship considered the *Giant Crossbow* drawing an impractical design, meant to impress rather than work. As will be shown, each portion of this model is carefully proportioned to work in cooperation with the overall design, although there are considerable inaccuracies. Proportional inaccuracies include a crossbowman who would measure approximately 12 feet in height and the disproportionately large back end of the crossbow's upper carriage (body) in the drawing, which would be ⅓ *braccio* (arm-length) too large by comparison to its thickness on the side and the thickness Leonardo notes in the text on the right. The design is most likely a preparatory study for a formal presentation intended for a treatise dedicated to Ludovico Sforza, as part of Leonardo's offer to "build unusual bombards, mortars and light ordinance with beautiful and useful shapes." He likely knew that Ludovico was one of the commanders most experienced in military technology in Europe, and that an expensive siege weapon proposal had to be as practical as it was impressive. Leonardo studied various proportional calculations associated with the function of a crossbow. The design of various components of the bow required a knowledge of statics, the branch of mechanics that governs the proportions of force necessary to manipulate stationary load bearing systems like screws, wheels, cams, ratchets, springs, balance arms, pivots, pulleys, etc. In order to understand the structure necessary to fire "100 pounds of stone" a reasonable distance, Leonardo also made various calculations about the bow's dynamics: proportional studies of the levels of force necessary to move certain objects an optimal distance.

This study will also highlight the various compromises and disconnections between the theoretical and practical uses of proportional guides in machine design and construction. Considered in this study is a range of dynamical models, from proportional principles in optics, to those in music, in fluid motion, and in ballistics. While new theoretical approaches to the dynamics of optics were relatively in decline during the 15th century, there had been an increase in manuscript and printed studies of the dynamics of ballistics. A discussion follows of Leonardo's inheritance of this increasing

---

[2] I produced an essay and report regarding the possible construction and history of the *Giant Crossbow*, and provided additional research assistance for the documentary script for ITN Factual in April 2002, stating that it is illustrated in such a way that it could have been constructed, based on Leonardo's numerous sketches relative to the CA 149br presentation drawing. Martin Kemp and Paolo Galluzzi agreed with the finding and offered suggestions for this and subsequent research. Having an interest in building a full-scale version of the *Giant Crossbow* for a documentary, ITN, encouraged by this finding, hired a team that built a nearly full-scale model. After providing research assistance for ITN during the summer of 2002, I had no further involvement with the construction of the crossbow. I was included in the beginning of the film, as part of the workshop discussion about the crossbow, and later attended the crossbow testing on the Salisbury Plain in September. The documentary, "Leonardo's Dream Machines," aired on 10 and 17 February 2003, on Channel Four in the UK. PBS purchased the documentary in 2005 and converting it to a *PBS Home Video* DVD.

[3] This is the second of three movies, based on J. R. R. Tolkein's *Lord of the Rings*. The siege ballista used to launch grappeling hooks in the Battle of the Hornburg (aka. the Battle of Helm's Deep) appears to be a small version of Leonardo's *Giant Crossbow*. The Weta Workshop built it.

Fig. II.4.
Author's photograph of a reconstruction of Leonardo's *Giant Crossbow* by ITN Factual Television, now located at the Royal Armouries, Fort Nelson, Hampshire, UK

Fig. II.5.
Author's photograph of a reconstruction of Leonardo's *Giant Crossbow* by ITN Factual Television, now located at the Royal Armouries, Fort Nelson, Hampshire, UK

Fig. II.6.
Author's photograph of a reconstruction of Leonardo's *Giant Crossbow* by ITN Factual Television, now located at the Royal Armouries, Fort Nelson, Hampshire, UK

preoccupation with these medieval impetus theories and the proportional methods with which he tried to interpret them. Codex Atlanticus and Codex Madrid folios with seemingly random calculations are found in this study to propose arithmetical solutions to dynamical problems. Recorded in these manuscripts are his notes on the principles of practical dynamics, with questions and calculations on how to solve these problems. That his approaches to these problems were primarily proportional suggests that he depended on medieval dynamical concepts before 1495. Inaccuracies in dynamics, statics and mathematics were reasons for designing a *Giant Crossbow* that would not work properly, although the proportional principles for its construction conform appropriately to medieval theory. Proof of this will be discussed with regard to a nearly full-scale model recently built and tested for an ITN Factual documentary. Alternatives to that construction will be considered.

Evidence of Leonardo's *Giant Crossbow* project exposes the practical limitations and benefits of his theoretical understanding of proportions in load bearing structures, in ballistics, in the representations of human form, in metal alloys, in aesthetic features, in screw thread technology, in metal spring design, in the different strengths of wood, and in the treatment of rope, etc. His notes on the crossbow offer a rare glimpse of the extent to which proportion theories formed the basis of nearly every aspect of machine design, involving work on human proportions, geometric pictorial composition, statics, dynamics, and arithmetical and geometrical proportions.

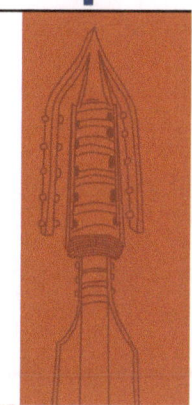

4

The Armature's Design and Proposed Construction

The armature is the portion of the *Giant Crossbow* in which Leonardo seems to have shown the most interest, for which the most preliminary studies exist, and on which he starts his handwritten notes on CA 149br, where he estimated the carriage's proportional length and its placement on the sheet.[1] Leonardo refers to the armature or laths across the front end of the *Giant Crossbow* as the *armadura*, or armature. He notes this at the end of the third line to the right of CA 149br (Fig. I.1). A great crossbow armature normally consists of one or more laths perpendicular to the body of a crossbow, attached to the body's front end, and responsible for applying tension to a string attached at both ends of its lath or laths. This arm or bow flexes back with the pull of the centre of the string, giving a bolt or stone placed at that centre its force and motion when the string is released and the armature springs forward with the string. Around 1494/1495, Leonardo produced a detailed study of the mechanics of a great crossbow with this kind of simple or compound lath armature on CA 142r [51rb] (diagram, Fig. 4.1). The design of the *Giant Crossbow* armature, however, differs considerably from what one would have seen on great crossbows in the 15th century. The present section of this study will outline some of the possible reasons for the unusual design of this armature.

The giant armature on CA 149br is the crossbow's most decorative feature and thus the most relevant to Leonardo's proposal to make "ordinance of very beautiful and functional design":

7. … I will make cannon, mortar and light ordinance of very beautiful and functional design that are quite out of the ordinary.
8. Where the use of cannon is impracticable, I will assemble catapults, mangonels, trebuchets and other instruments of wonderful efficiency not in general use. (CA 1082r [391ra])[2]

The armature is not a single wooden lath or a simple set of equal-length laminated laths, which was the customary body for "compound" great crossbow armatures, treated with sinew, horn, and an herbal unguent.[3] Instead, it is a complex design of modular laths that conform to an overall appearance of narrow bat wings. Curves at

---

[1] These preliminary studies would include two armature designs on CA 147av, eight sketches of an armature on CA 147bv, two armatures on CA 57v and possibly various indirectly related studies of armatures.

[2] Translated by Kemp and Walker, §612, p 252.

[3] Preparation of the armatures of medieval compound crossbows, or "great crossbows", sometimes involved the application of an herbal unguent like spikenard in order to preserve the elasticity of sinew from a bull's neck or a dear's hamstring. The 'two-foot crossbow,' so-named for the length of bolt that it shot, had an armature of up to six feet in length, most often made of yew, and treated with goat horn and sinew. As an example, three kinds of great crossbow are documented in France in the 13th century, noted by the scribes of Philip II Augustus in 1204–1206: the crossbow with stirrup (*balistam ad estrif*), two-foot crossbow (*balistam ad duos pedes*), and windlass crossbow (*balistam ad tornum*). E. Audoin (1913) *Essai sur l'armée royale au temps de Philippe Auguste*, Champion, Paris, pp 187–197.

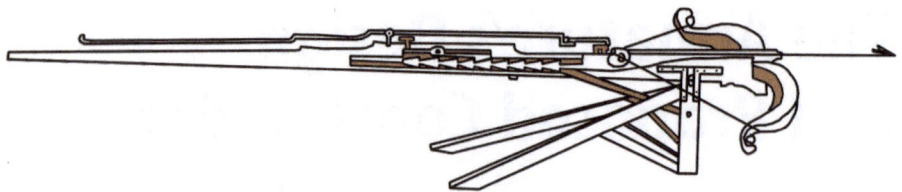

Fig. 4.1.
Illustration of Leonardo's
drawing of a great crossbow,
Codex Atlanticus 142r [51rb],
Biblioteca Ambrosiana, Milan

the centre front and rear of the armature resemble the front and rear curves of most bat wings. Before a discussion of the physical format of these modular laths, an analysis of the possible reasons for its bat wing design immediately follows.

The narrow bat wing design (▶) is more obvious when looking at the crossbow from directly above, as in the diagram of Fig. 4.2. Leonardo later used this form for his flying machine designs, such as the examples on CA 70br [22vb] and on CA 846v [309av]. The earliest record of this wing design is among the studies on Manuscript B folio 74r, from the mid 1480s (Fig. 2.1). He designed on CA 149br a large bat-like machine, using a structure that he would have considered a design proven by nature herself. At around the time when he wrote his letter to Ludovico, or shortly thereafter, he produced on MS B 74r his earliest known detailed drawing of the bat wing's structure. This was drawn at the time when he was copying Valturio's *De re Militari* onto the pages of MS B, producing his first known designs for flying machines in MS B (ff. 73v–80v, 88r–90r) soon after which he designed the *Giant Crossbow* (the date of which discussed in the previous section of the present study).

A number of Leonardo's earliest studies offer evidence of his reasons for its massive size and bat wing form. An immediate impression of the *Giant Crossbow* drawing suggests that it is a formal presentation drawing, intended as a treatise illustration, and therefore to impress a wealthy patron, such as Ludovico Sforza. Unlike CA 147bv [52vb] and 147av [52va], 149br contains complete drawings, and has formal written explanations of its diagrams (written in Leonardo's reverse script). He likely chose to write *alla mancina* because his left hand could produce his best looking handwriting. Without reading the text, one can judge the relative size of the machine by observing its proportion to that of the crossbowman on its upper carriage. In the 15th century, great crossbows were considered extra siege weapons, for use in places for which there were not enough cannon, trebuchets, or mangonels for a similar purpose. If a great crossbow design were to impress a wealthy patron who owned many cannon, existing crossbows and presumably a number of trebuchets and mangonels, it had to offer what the rest of the Sforzas' siege arsenal could not offer. Statements on CA 149br claim that the machine can quietly launch a 100-pound stone. The seventh line of text at the right of the sheet mentions the stone, and the last line on the left refers to the quiet release of the trigger. Although the *Giant Crossbow*'s practical possibilities are discussed further below, it is worth noting here that its design gives an immediate impression of its proposed capabilities to anyone who understands siege engines. Obviously, the drawing represents a machine capable of the direct thrust of a heavy missile with a relatively quiet trigger. Unlike the expensive bronze cannon, it is mostly composed of wood, it does not require gunpowder, expensive missiles or lead coated stones, and it can launch missiles without much of a sound. Rather than toss a missile less accurately into the air with a trebuchet or mangonel, the *Giant Crossbow* would make a direct hit at short distances or shoot objects more accurately along a lower trajectory. As for its visual appeance as that of a giant bat, this too would be a significant pshychological and memorable advantage at a siege.

Leonardo's notes suggest that he seriously considered the bat-like structure of the armature. Several times in his notebooks—from the mid 1480s through as late as 1515— he made notes about the bat and its wings.[4] Relevant to the present study is his study

---

[4] See folios of MS B (mid 1480s) 74r, 89v, 100v, MS H 12r and 14r (c. 1493–1494), MS K 3r (c. 1503–1505), MS Sul Volo (on the Flight of Birds) 16 [15]r (c. 1505), MS F 41v (c. 1508), Windsor 19087r (c. 1513), MS G 63v (c. 1510–1515). The dates given in this instance, with the exception of MS B, are from Kemp (1981) p 21–22. The group of MS B folios from around 60r through 90r seem to be from the mid 1480s.

Fig. 4.2.
Illustration the appearance of the
Giant Crossbow's basic features
as they might look from a viewpoint directly above

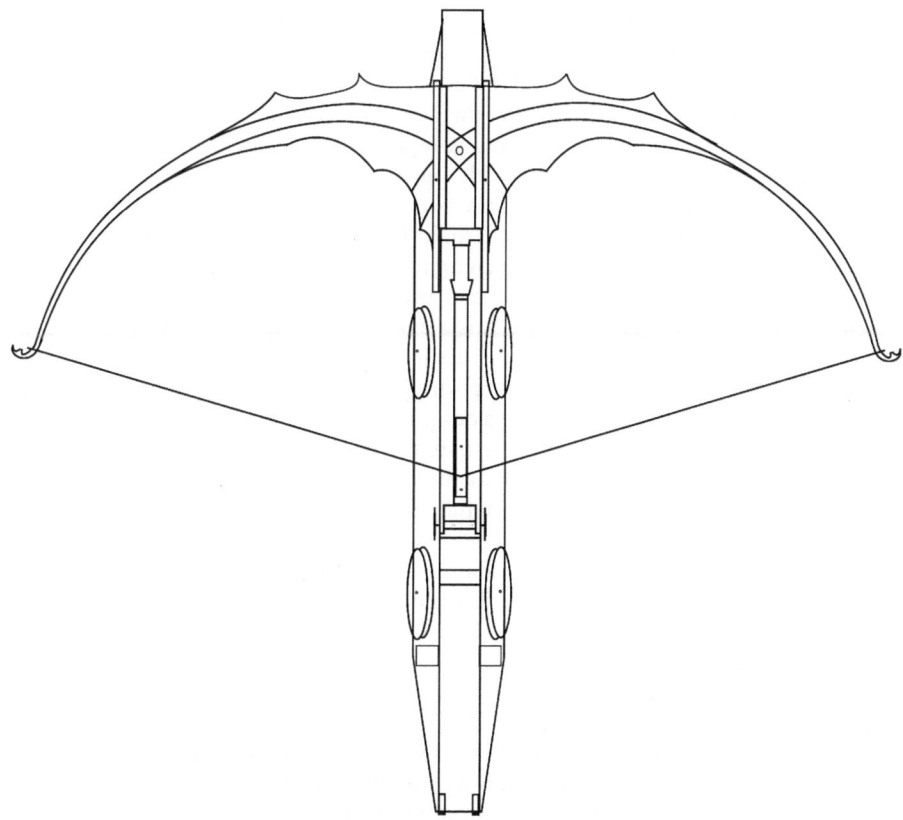

on MS B 89v of the structural capabilities of the wings of bats and birds, from the mid 1480s. At the top of the page, above the drawing of a bat with outspread wings, he states, "This weighs two ounces; it opens one half *braccio*."[5] To the right of the bat drawing, above the drawing of an eagle with outspread wings, he states, "This weighs 240 ounces; it opens three *braccia*."[6] Immediately below these two examples, he writes:

> I say that if the bat weighs two ounces and opens out to one-half *braccio*, that the eagle, proportionally, ought to pen out to no less than sixty *braccia*, though we see by experience that it does not exceed the width of three *braccia*. And it might appear to many who had not seen, nor had ever seen such animals, that one of these two ought not to be able to fly, considering that if the bat has his weight well proportioned to the width of its wings, then the eagle would have too little. And that if the eagle, with his, were well equipped, then the other would have too much, and that they would seem disproportionate, and not useful, for its needs. And we have seen the one and the other to be borne with maximum dexterity by their wings—and especially the bat, which, with its rapid turning and swerving, vanquishes the rapid veering and escaping of the flies, bluebottles, and other similar, small animals.
>
> Let the weight of a coin be applied to the top of a reed. You will see it bend to the ground. Take a thousand of these reeds and bind them together tightly, and fix them at the bottom, and even them off at the top, and load them. You will see that where, according to the first reason, it ought to sustain but about three and one-half pounds, that it will sustain more than forty.[7]

Suggestions of Leonardo's interest in the bat as a possible model for a harmful, efficient siege machine appear in later manuscript notes. In his bestiary of around

---

[5] See: Ravaisson-Mollien (1883) *Manuscrits B & D*, f. 89v. Translation by Venerella (2002) *MS B*, p 140.

[6] *Ibid.*

[7] *Ibid.* Venerella, pp 140–141. Bluebottles are also known as blow-flies. Leonardo refers to them as "*mosconi*." The common "coin" in the second paragraph is a "*dinaro*" or dinar. See: Edward McCurdy (1959) *The notebooks of Leonardo da Vinci*, George Braziller, New York, (1st ed 1939), pp 438–439.

1493–1494 he defines the bat as the "vice which cannot endure where virtue is…." as it "loses its sight more where the light has more radiance."[8] Around 1505, he instructs in his "Treatise on Flight" (MS *Sul Volo*) that, for the construction of a flying machine, "your bird should have no other model than the bat because its membranes serve as an armour or rather as a means of binding together the pieces of its armour, that is the framework of the wings."[9] Again with regard to this flying machine, he suggests around 1508 that one should "dissect the bat, study it carefully, and on this model construct the machine."[10] Thus, he refers to the bat as that which avoids virtue (light), which has armoured wings, and which contains in itself structural information about how to build a flying machine. Similarly, the *Giant Crossbow*'s armature is that machine's destructive agent: it is vice, it is a pair of armoured bat wings, and it contains within it the stored energy and structure capable of sending things airborne. This careful consideration of the perception, properties, and functions of the bat's wings is Leonardo's way of visualising a design problem, not a fanciful approach to a decorative design. Most Renaissance artists might consider the forms of the bat's wings ideal for the fanciful depiction of the armature of a giant crossbow, but his main concern appears to have been with the way in which these wings would actually work on a giant terrifying weapon.

For the *Giant Crossbow* and other projects, he looked for various visible and structural benefits to using nature's designs. Valturio's *De re Militari* contains an example of a siege engine designed like an animal. This and other designs in *De re Militari* were well known in many late 15th century Italian libraries. Though the animal happens to be a dragon—a traditional combination of a snake, raptor, cat, and bat—it was likely a strong influence on the *Giant Crossbow* design.[11] He copied similar examples of the *De re Militari* whip-driver (cacciafrussti) and springald drawings on MS B 7v, MS B 8r, and CA 149ar [53va].[12] He likely considered Valturio's dragon cannon an interesting structure that would appeal to the imagination and practical interests of Ludovico Sforza.

Is it possible that Ludovico would have considered Valturio's dragon cannon a practical design? The dragon cannon has buildable components. Also, the earliest cannon (14th century) were rather small and fired the kind of 26-inch bolts visible in the *De re Militari* drawing. Like the whip-driver (cacciafrussti) and springald, the dragon cannon was not a medieval joke or dream, but a serious weapon designed for a wealthy patron expecting reasonable advice on military engineering. Even if the dragon cannon were difficult to build, it is no more "fanciful" than some of the other designs in the military treatise in which it is found. The *Giant Crossbow* should be considered within a similar context and within a series of military drawings.

Georgio Vasari notes Leonardo's interest in bats in a seemingly accurate portion of his biography of 1568. The many details to this story, its reference to Leonardo's early studies of animals and insects, his perfectionist treatement of the shield, the long time that he spent on it, and Vasari's credibility as a very knowledgeable researcher, all indicate that the following statement may be partially based on an actual event. Regarding the following translation of this story by Greorge Bull, it should be noted that a "buckler"

---

[8] MS H 14r, McCurdy (trans), p 1082.

[9] MS *Sul Volo* 15 [14]v, McCurdy (trans), p 416.

[10] MS F 41v, McCurdy (trans), p 472.

[11] Leonardo's copies of Valturio's *De re Militari* appear mainly in his MS B (many of the folios, 4–100) and Codex Atlanticus (especially 149ar [53va] and 181r [64r]).

[12] An enlarged example of the central image whip-driver (cacciafrussti) is on CA 181r [64r] (Fig. I.2), although there are several simpler examples of this design in Leonardo's notebooks. The simplest of these is on MS B 7v, which appears to be a direct copy from the *De re Militari*. He drew a box-framed springald at the top of CA 149ar [53va] (Fig. I.4), which is similar in design to the same siege engine in Valturio's treatise. For visual examples of Leonardo's interest in dragons, see: Windsor folios 12331, 12363, 12369, and 12370r. For an historical approach to the dragon as a merged composite predator image of a snake, raptor, bat, and cat, see: David E. Jones (2000) *An instinct for dragons*, Routledge, New York.

is a small shield normally held by a swordsman's left hand at arm's length, and that such shields were often adorned with the head of Medusa in the 15th century:

> The story goes that once when Piero da Vinci was at his house in the country one of the peasants on his farm, who had made himself a buckler out of a fig tree that had been cut down, asked him as a favour to have it painted for him in Florence. Piero was very happy to do this, since the man was very adept at snaring birds and fishing and Piero himself very often made use of him in these pursuits. He took the buckler to Florence, and without saying a word about whom it belonged to, he asked Leonardo to paint something on it. Some days later Leonardo examined the buckler, and, finding that it was warped, badly made and clumsy, he straightened it in the fire and then gave it to a turner who, from the rough and clumsy thing that it was, made it smooth and even. Then having given it a coat of gesso and prepared it in his own way Leonardo started to think what he could paint on it so as to terrify anyone who saw it and produce the same effect as the head of Medusa. To do what he wantes Leonardo carried into a room of his own, which no one ever entered except himself, a number of green and other kinds of lizards, crickets, serpants, butterflies, locusts, bats, and various strange creatures of this nature; from all these he took and assembled different parts to create a fearsonme and horrible monster which emitted poisonous breath and turned the air to fire. He depicted the creature emerging from the dark cleft of a rock, belching forth venom from its open throat, fire from its eyes and smoke from its nostrils in so macabre a fashon that the effect was altogether monstrous and horrible. Leonardo took so long over the work that the stench of the dead animals in his room became unbearable, although he himself failed to notice because of his great love of painting. By the time he had finished the painting both the peasant and his father had stopped inquiring after it; but all the same he told his father that he could send for the buckler when convenient, since his work on it was completed. So one morning Piero went along to the room in order to get the buckler, knocked at the door, and was told by Leonardo to wait for a moment. Leonardo went back into the room, put the buckler on an easel in the light, and shaded the window; then he asked Piero to come in and see it. When his eyes fell on it Piero was completely taken by surprise and gave a sudden start, not realising that he was looking at the buckler and that the form he waw was, in fact, painted on it. As he backed away, Leonardo checked him and said: "This work certainly serves its purpose. It has produced the right reaction, so now you can take it away." Piero thought the painting was indescribably marvellous and he was loud in praise of Leonardo's ingenuity. And then on the quiet he bought from a peddler another buckler decorated with a heart pierced by a dart, and he gave this to the peasant, who remained grateful to him for tht rest of his days. Later on Piero secretly sold Leonardo's buckler to some merchants in Florence for a hundred ducats; and not long afterwards it came into the hands of the duke of Milan, who paid those merchants three hundred ducats for it.[13]

Apparently, Vasari knew of a story that Leonardo painted a dragon on a shield, using bat and butterfly wings as his models. MS B 100v also contains examples of such comparisons between the wings of bats, butterflies, and "various strange creatures of this nature." At least three examples attributed to Verrocchio provide definite proof of Leonardo's early access to dragon, bat wing, and butterfly wing imagery on military armour. Dragons crouch atop the helmets of Verrocchio studio bas reliefs of Alexander the Great (National Gallery, Washington), Scipio (Louvre, Paris) and a military officer in *The Beheading of St. John the Baptist* (Duomo, Florence). Scipio also has the dragon wing on his shoulder pads, which is the usual combination of bat wing curves and butterfly spots. Leonardo produced a similar drawing of the helmet in his *Bust of a Warrior in Profile*. Dated to around 1480 by A. E. Popham because of its "character", "technique", and "perfection" of draughtsmanship. [14] This silverpoint drawing in the British Museum could have been copied much earlier, while Leonardo was in the

---

[13] Giorgio Vasari (1965–1972) *The lives of the artists*, c. 1568, a selection translated by George Bull, Penguin, Middlesex, pp 259–260. See also: Giorgio Vasari (1927/1963/1970) *The lives of the painters, sculptors and architects*, Everyman's Library, 4 vols., A. B. Hinds (trans), Dent, London, vol. II, pp 158–159. Hinds' translation is perhaps closer to the original Italian format, the English of which Bull revises in an arguably better way.

[14] British Museum no. 1895-9-5-44. A. E. Popham (1946/1994) pp 40 and 129. The exceptional draughtsmanship of the sculpted marble book pedestal in Leonardo's *Annunciation* (1473) shows that he had made careful studies of similar sculptures while in the Verrocchio studio around 1473. D. A. Brown (1998) discusses Leonardo's approaches to dragons, naturalism, and Verrocchio studio sculptures in *Leonardo da Vinci, origins of genius*, 1998, pp 58–73.

Verrocchio studio making other drawings of the studio sculptures. Popham also notes, however, that Leonardo was at work on drawings for an Adoration, Epiphany, and a St. George and the Dragon shortly after March 1481, just before his trip to Milan.[15] Drawings related dragons from this period include the Ashmolean Museum's *A Horseman in Combat with a Griffin* and the Windsor Royal Library (12370r) *Study of a Dragon*. In any event, one can see that he had a particular interest in the animal and insect forms that make a dragon in the years leading up to his trip to Milan. When considering the form of the *Giant Crossbow* armature he used one of these potentially terrifying forms of nature and the dragon, that of the bat wing.

Another unavoidable aspect of this design is its massive scale. The beams and modular laths forming the armature appear to dwarf the crossbowman standing at the trigger mechanism on the upper carriage. Leonardo notes on the second line of text at the right side of CA 149br that the armature is 42 *braccia* (24 m, 80 ft) wide. As will demonstrated below, this might be his proposed width of the spanned armature (when pulled back by the rope), as illustrated. His intended reference to the width of the spanned armature may be the metal stylus line on CA 149br that extends between the two points where the ropes meet the armature. Superimposed over this metal stylus incision is a white line in Fig. 8.20 to show the location of a thin linear indentation in the paper that is visible only when viewing the drawing directly with raking light. As will be discussed in more detail below, this appears to be the second horizontal line added to the drawing with a metal stylus. The length of this metal stylus line is 22.7 cm. If this is to represent 42 *braccia*, the length of each *braccio* on this portion of the drawing is approximately 0.54 cm (22.7 cm ÷ 42 br = 0.536 cm per br). If one applies this module of 0.54 cm per *braccia* to the 20 cm length of the upper carriage, the total illustrated length of the upper carriage would be approximately 37.5 *braccia* (20 ÷ 0.536 cm = 37.5 br). For reasons that will be noted below, the lower carriage length, which Leonardo states is 40 *braccia* in length, extends another 2.66 *braccia* beyond view, under the armature (see Figs. 4.2, 8.2 and Appendix A). Thus, the lower and upper carriage sections, as well as the armature, conform to a 0.54 cm module per *braccio*. These three lengths of the crossbow are relatively proportional to one another, as each of them happen to be drawn at a scale of 108 : 1 (see Fig. 8.2).[16] Leonardo did not necessarily intend to arrange for a scale of 108 : 1, unless he was concerned that the scale be divisible by 3 (108 = 3 × 36). He made use of the rule of three calculation quite often, and would have favoured numbers divisible by three (integers).[17] But it is not known if he intended CA 149br to be a scale drawing. Among the Codex Atlanticus siege engine designs possibly intended for a *De re militari* treatise, the *Giant Crossbow* is the only example with specific overall dimensions on the recto, whereas the few notes on the other drawings are primarily about how the machines operate. Also, the scale is not necessarily easy to calculate for all portions of the design. One could measure the armature on the drawing at positions that are just beyond "where the rope is attached," and would find a measurement of 24.52 cm for the 40 *braccia* armature that is 24.52 m, and thereby estimate a scale of 100 : 1. But this measurement would not agree with other measurements on the drawing, and it is unlikely that Leonardo considered the crossbow's width without considering its proportional relationship to the length.

It is likely, however, that this proportional relationship between the 22.7 cm armature width and 20 cm carriage length—as opposed to the calculated written dimensions of 42 × 40, to be discussed further below—may derive from his visual intuition

---

[15] Popham (1946/1994) pp 27, 112, 113, pl. 60b and 62.

[16] The scale of 108 : 1 is determined, for example, by converting 42 late 15th century milanese braccia to centimetres, which is 2 451.12 cm (58.36 cm × 42), and then by dividing that number by the linear measurement of the illustration (22.7 cm). See: Appendix A.

[17] See: Augusto Marinoni (1982) *La matematica di Leonardo da Vinci*, Philips-Arcadia, Milan. Giorgio T. Bagni and Bruno D'Amore (2006) *Leonardo e la matematica*, Giunti, Florence.

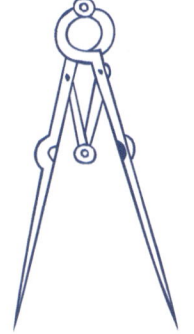

about the placement of his divider and/or straight edge when determining the crossbow's primary dimensions. (The divider is much like a compass, with two sharp ends, such as this copy (◄) of a divider that was originally illustrated by Leonardo.) Metal stylus lines at lengths of 22.7 cm for the width and 20 cm for the length suggest his interest in a spanned crossbow width only a couple *braccia* longer than its length. Thus, these portions of the drawing agree generally with the text at the right of the page, though the measurements are inexact. More on the inconsistencies in the design and written specifications continues below in the section dealing with the carriage dimensions.

Bat Wing Form, Laminated Beams and Blocks

Leonardo considered several structural problems about wings and their proportions at around the same time that he copied Valturio's treatise and wrote about military engineering problems on MS B folios 50v and 62v. He notes with drawings and texts on MS B 89v the proportions of body weight to wing size, the amazing dexterity of bat wings, and the way that small reeds can lift more than 40 pounds of weight. If a coin [*danaro*] weighs about 19.29 modern grains troy—that being 0.00283 lbs—a thousand bound reeds carrying the equivalent of one coin each would only lift about 2.83 modern pounds, or in Leonardo's estimation, 3.5 Milanese pounds.[18] But he knew that any set of bound reeds or twigs could be made to stand on their ends and thereby lift much more than each individual reed or twig of that set could carry if it were standing on one end. The point here is that he is using proportional studies of bird wings, bat wings, and reeds, in order to show that simple proportional calculations are not sufficient to explain the capabilities of certain structures. Also, as a calculated interest—rather than an intuitive estimate—he had a preference on MS B 89v, on the *Giant Crossbow* study and elsewhere in his notebooks, for the number 40 as a dividend of even factors (or divisors). His studies of Euclid and Archimedes with Fazio Cardano in the mid 1480s possibly influenced the numerical estimates: if the geometry of the bow were accurate, it would conceivably work when built, and as a persuasive illustration.

Studies of the armature's form on CA 147av, 147bv, and 57v–69ar (Figs. I.8, 1.2, 2.4) precede the bat wing design on 149br. Thus, the composite form of the armature was initially most important to the design, and would later take the form of bat wings. By 15th century crossbow standards, the proposed structure of this armature is obviously a very ambitious design. At the centre of the armature on CA 149br (Fig. I.1), two beams of wood cross each other and seem to have a round pin fastening them together at the centre. These square posts are thickest at their fastened positions and taper off at their other ends. At rest, these posts would form an X, with the rear portions pinned at the armature centre and the front ends extended 45° straight out to the left and right. According to the statement on CA 149br, each wooden beam would measure well over 21 *braccia*, (<13 m, <41 ft). To form the armature one would presumably soak these beams in water, then secure them to the upper carriage and pull them back to the desired position with rope tied at each end, or with the truss-bending device Leonardo reproduced on CA 888v. Pulled back like this, the arms would bend toward the upper carriage, forming the curved top of a T in relation to the rest of the body. When held at this position, the armature would then have applied to it a front and rear set of toothed laminations for extra support.

---

[18] *Ibid.*, McCurdy, p 438. McCurdy notes that a "*danaro*" is "a small coin, about 20 grains Troy in weight," but he does not mention his source for this. According to Giuseppe Cadolini's Milanese mercantile standards of 1843, the *danaro* is 24 *grani* (grains), an *oncia* (ounce) is 24 *danari*, and a *marco* (mark) is 8 *once*. See: http://www.melegnano.net/pagina004tx.htm. The modern ounce troy is 31 grams or a fifteenth of pound (0.068 lbs). Divide this modern fifteenth of a pound into 24 *danari* to get a *danaro* weighing 0.00283 lbs, or 1.2863 grams troy, or 19.29 grains troy. Therefore, McCurdy's estimate of a *danaro* of 20 modern grains troy is relatively correct.

These modular laminations are the specially cut blocks of wood illustrated at the inside centre and front-end centre of the armature. Details of the design for finely carved toothed laminations appear at the top of CA 147bv (Fig. 1.2). The way in which Leonardo considered applying these toothed grooves is not obvious. It appears, however, that some portions of the wooden laminations would have been smooth, allowing them to better contract and expand with the operation of the bow. One could compare the proposed flexibility of these boards with the hulls of Renaissance Venetian and Turkish galleys. Although nails and pegs attached the strakes (wooden laths) to some medieval Mediterranean and Scandinavian ship hulls, there is evidence that hemp or willow withies (peeled strips) fastened the strakes of some ship hulls at that time.[19] This latter technique would have appealed to Leonardo. It is best described in the account of a replica of the 9th century Scandinavian Gokstad ship, which sailed the Atlantic in 1893. The master of this replica "observed that the ship undulated with the waves; the bottom and keel would rise and fall by as much as 2 cm and the gunwales twist as much as 15 cm out of true. The ship was fast, achieving 10 knots with ease and sometimes more."[20]

Materials: Willow, Iron, Hemp, and a Stone

Leonardo drew studies of ship hulls on W 12650r, a group closely related to the series of early 1480s military engineering studies, as noted further above, and in the set of Windsor drawings from 12647r through 12653r. CA 149br shows the armature laminations pinned and wrapped together, indicating that the boards would not have required nails or glue. Iron bands and linchpins appear to secure the central beams together with their supporting central laminations.

Leonardo's illustration on CA 149br shows that he considered inserting linchpins with looped heads through rectangular holes in central portions of the armature, securing them in place with what look like metal spikes. Though not visible in the crossbow drawings, the rectangular holes are apparently necessary for the insertion of looped eyelets through the body of the armature. As an alternative to this insertion of looped eyelets, Leonardo considered a spike and brace design for the same crossbow project on CA 147bv, where he identifies the spike as a "pole of iron" (*palo di ferro*, Fig. 1.2).[21]

---

[19] The scope of the present study precludes a detailed account of galley construction. Galley construction is an area of expertise for which the Italians excelled almost without rival especially toward the end of the fifteenth century and throughout the sixteenth century. The best source on the construction of Western and Eastern galleys is: Robert Gardiner and John Morrison (eds) (1995) *The age of the galley: Mediterranean oared vessels since pre-classical times*, Conway Maritime Press, London. See also: Frederic Lane (1934) *Venetian ships and shipbuilders of the Renaissance*, John Hopkins University, Baltimore, MD. William Rodgers (1940) *Navel warfare under oars, 4th to the sixteenth centuries: A study of strategy, tactics and ship design*, Naval Institute, Annapolis, MD. L. Tursini (1954) *Navi e scafandri negli studi di Leonardo*, in: *Leonardo, Saggi e Ricerche*, Rom. George Bass (ed) (1974) *A history of seafaring based on underwater archaeology*, Thames & Hudson, London. Angus Konstam (2002) *Renaissance war galley, 1470–1590*, Osprey, Oxford.

[20] The ship was built for a Norwegian exhibit at the Chicago World Fair in 1893. P. G. Foot and D. M. Wilson (1970) *Viking achievement*, Sidgwick & Jackson, London, pp 234, 242, and 148.

[21] As one of his few notes on the *Giant Crossbow* project, this mention of iron was possibly an important one. In one of his earliest notebooks—the MS *Trivulziano*, he made detailed notes on tinned alloys for casting cannon, most likely copying Valturio's 1483 Italian edition of *De re Militari*:

If you wish for the sake of economy to put lead with the metal, and in order to lessen the amount of the tin, which is necessary, first alloy the lead with the tin and then put [this] above the molten copper. … The metal used for bombards must invariably be made with six or even eight parts to a hundred, that is six parts of tin to one hundred of copper, but the less you put in the stronger will be the bombard.

Depending on the materials available, Leonardo could have preferred using an iron alloy for the armature's metal supports. *MS Trivulziano* 49a–50a, McCurdy (1959) pp 1022 and 1024.

Nonetheless, the form of looped pin that he illustrates on CA 149br most resembles the example at the top of CA 90r (Fig. 2.5). Pins in the form of staples with looped ends appear to penetrate boards on this latter drawing. The looped ends extend to the other side of the boards, where spikes inserted through the loops prevent the staples from slipping backwards through the interlocked boards. At the right end of the top example, a toothed lamination in the form of a "dovetail" groove would prohibit any side movements of the boards, but would allow some expansion of the wood from end to end. The types of pins and zigzag forms of toothed lamination in this drawing resemble studies on CA 89r [32va], 90r [33ra], 139r [49vb], 147bv [52vb], 149br [53vb], 888v [324v], 946r [344va], and on MS B folios 5v, 6r, and 27v. CA 90r shows Leonardo's possible early interest in Leon Battista Alberti's *De re aedificatoria*, which deals with the same problems of toothed laminations and pins in Book III, Chap. xii.[22] Calvi and Pedretti date CA 90r to around 1480–1482, as it bears some relationship—in terms of its contents and style—to Leonardo's military architecture, engineering, and canal work for the Medici at this time.[23] For similar reasons, Pedretti dates to the same period 139r and 888v.[24] Leonardo could have found the concepts noted in these drawings helpful for his design of the *Giant Crossbow* armature.

To demonstrate the look of the armature while in use, he shows the giant crossbows of CA 149br and the preparatory drawings (Figs. I.8, 1.2, 2.4) in positions almost ready to launch missiles. Held back by the trigger, the double strands of what would be hemp and/or horsehair rope are near the halfway position of the upper carriage spanning screw (compare Figs. I.1 and 4.2).[25] Only a small portion of this screw is visible behind the trigger in CA 149br. The "span" (draw of the rope) of this crossbow therefore appears almost fully extended. The crossbowman atop the upper carriage stands ready to push down on the trigger's pole. Leonardo has written to the right of the drawing that, "it opens at its arms, that is where the rope is attached, 42 *braccia*." Hence, it would seem that he refers to the illustrated examples as having 42 *braccia* wide armatures in their almost fully spanned positions. These armatures might conceivably extend beyond 42 *braccia* when at rest; but this is not an obvious deduction from the available notes.

The ability to bend the armature's central wooden beams, each of which would be over thirteen metres in length, obviously would depend on a very flexible wood and great technical skill. Anyone requiring this much elasticity on the part of a wooden beam has probably considered the kind of wood necessary for the crossbow's construction. As stated further above, there are numerous references to medieval and early modern composite crossbows in the West and East made of yew or olive wood and supported with horn or whalebone, wrapped with tendon or sinew.[26] Paolo Galluzzi advises that, "in the fifteenth and sixteenth centuries acacia wood was reputed for its flexibility."[27] There is, however, no mention in Leonardo's surviving notebooks of acacia or yew. Still, he was considerably preoccupied with different kinds of materials, mentioning in various places throughout his notebooks woods such as alder, arbutus,

---

[22] L. B. Alberti (1550/1988) L'architettura (De re aedificatoria), tradotta in lingua fioentina da Cosimo Bartoli [...] con l'aggiunta de designi, Florence, 1550. In: J. Rykwert, N. Leach and R. Tavernor (eds and trans from the Latin) (1988) On the art of building in ten books, Cambridge, MA, p 72.

[23] Calvi (1925) *Manoscritti*, pp 33, 41. Pedretti (1978) *Catalogue*, Part I, p 65.

[24] Pedretti (1978) *Catalogue*, pp 84 and 165.

[25] Specialised rope manufacturers in the 15th century made rope usually of hemp, horsehair, or a combination of both materials. Most importantly, these rope shops appropriately tensioned the rope according to the required strength. The *Giant Crossbow* probably required rope of capable of pulling 100 tons of tension.

[26] W. F. Patterson (1990) *A guide to the crossbow*, Society of Archer-Antiquaries, p 35. R. Payne-Gallwey (1903) *The crossbow (with ... an appendix on the catapult, balista and Turkish bow)*, Longmans, Green and Co., London, p 62. Liebel, p 23.

[27] Part of an e-mail message on 11 June 2002. He notes that appropriate documents refer to the popularity of this tree. At present, central and north Italian wine and honey industries use acacia. Its earlier popularity may have led to its extensive deforestation in Italy during the past five centuries.

box, brazil, cherry, laurel, oak, olive, pear, pine, walnut, and willow.[28] On MS B 62v, he quotes Lucan's recommendation of willow for shipbuilding:

> The small boats used by the Assyrians were made of thin laths of willow plaited over rods also of willow, and bent into the form of a boat. They were daubed with fine mud soaked with oil or with turpentine, and reduced to a kind of mud which resisted the water; and [they therefore avoided pine] because pine would split, and [willow] always remained fresh; and they covered this sort of boat with the skins of oxen [when] safely crossing the river Sicuris of Spain, as is reported by Lucan.[29]

Although Leonardo could have found this statement in Latin in Lucan's *Pharsalia* (IV, 130), Caesar's *De Bello Civili* (I, 54) or in Pliny's *Natural History* (IV, 15), it is known that he had direct access to this statement in Paolo Ramusio's Italian translation of Valturio's *De re militari*, published in Verona in February, 1483.[30] From this edition Leonardo could have transcribed the same phrases that appear eight pages of his *Codex Trivulziano* and on fourteen pages of his notebook now known as MS B.[31] Pedretti dates the MS B folio with the Lucan quotation to 1487–1490. There is sufficient evidence to suggest, however, that it was written as early as 1483–1484, when Leonardo had a direct interest in and access to the 1483 Italian and 1472 Latin versions of the widely popular *De re Militari*. Thus the date for this comment would be somethwere between 1483 and 1490/1492. Dates for this project are discussed further above, though it is worth specifying the February 1483 Italian publication of Valturio's *De re Militari*, the comment therein with regard to the flexibility and durability of willow, and the related armature of the *Giant Crossbow* drawing. Leonardo owned a copy of *De re Militari*, which he placed in storage in 1490–1493.[32] He could have found the Italian edition of this book for sale in Ferrara, Milan, or Verona in 1483. At that time, he considered the military advice of this book useful, noting the ancient advice therein about the flexibility of willow for shipbuilding. He could have considered willow ideal for the crossbow's pair of thirteen-metre-long wooden armature beams. Regardless, however, of any consideration of willow as a practical or an impractical choice of wood for the armature of a giant crossbow, he had enough credible written evidence from Valturio's highly regarded military treatise to suggest that this wood was theoretically suitable for large scale wooden structures like ships, as well as giant crossbows.

Preliminary studies of the *Giant Crossbow*, CA 147av and 147bv (Figs. I.8 and 1.2), show Leonardo's early intention to design an armature at least 42 *braccia* in length.[33] At the top of CA 147av, he notes: "42 *braccia* between the end of each arm" (*Braccia 42 nelle istremità di ciascuno braccio*).[34] At the centre of CA 147bv, he writes above the drawing of an armature that, "it is 42 *braccia*" (*É pe[r] braccia 42*). The other note on CA 147bv identifies the drawing of a "pole of iron" (*palo di ferro*), a brace for the arma-

---

[28] Leonardo mentions brazil-wood on MS B 66r. Imported from India and the East Indies since the mid 14th century, artists had used it mainly as a red dye. Cennino Cennini refers to this dye in his *Libro dell'Arte*, written at the end of the 14th century. Cennino d'Andrea Cennini (1933) *The craftsman's handbook* "Il Libro dell'Arte," Daniel V. Thompson, Jr. (trans) Yale, New Haven, pp 39, 103, 117.

[29] This is Richter's translation of MS B 62v, with my notes in brackets. He states that Leonardo's translation is derived from Lucan's *Pharsalia* IV, 130: "Utque habuit ripas Sicoris camposque reliquit, Primum cana salix madefacto vimine parvam Texitur in puppim, calsoque inducto juvenco Vectoris patiens tumidum supernatat amnem. Sic Venetus stagnante Pado, fusoque Britannus Navigat oceano, sic cum tenet omnia Nilus, Conseritur bibula Memphitis cymbo papyro. His ratibus transjecta manus festinat utrimque Succisam cavare nemus." Richter, §1100, p 265.

[30] Valturio copies Lucan's statement on MS B 62v. F. L. Finger (1971) *A catalogue of the Incunabula in the Elmer Belt Library of Vinciana*, Friends of the UCLA Library, Los Angeles, pp 71–75, 77–78. E. Solmi (1908) *Le fonti dei manoscritti di Leonardo da Vinci*, Giornale Storico della Letteratura Italiana, p 289. Pedrettin (1977) *Literary works … commentary*, vol. II, §1100, p 211.

[31] Finger, pp 77–78. Finger credits Augusto Marinoni with finding the eight pages on which Leonardo copied the lists of Italian words from Ramusio's 1483 translation of the Latin Valturio of 1472, p 74.

[32] This is noted in his *Codex Atlanticus* list of forty books on f. 559r [210ra].

[33] CA 147av [52va] and 147bv [52vb].

[34] Transcriptions by A. Marinoni (2000) *Codice Atlantico*, Tomo I, pp 198–199.

Fig. 4.3. Detail of the central portion of the upper carriage, near the 100 pound stone ball, on Codex Atlanticus f. 149br

ture. Since these are the only statements on CA 147av and 147bv, and the designs on these sheets compare most closely to the "42 *braccia*" wide design of a *Giant Crossbow* on CA 149br (Fig. I.1), an armature of 42 *braccia* appears to be one of the first requirements for this latter design.

One other study of the giant crossbow's armature, CA 57v–69ar (Fig. 2.4) shows the use of a sling or basket, loaded with a large ball.[35] Since this sling and stone appear on this drawing as well as on 147av, 147bv, and on 149br (Figs. I.8, 1.2, 4.3), it would appear that the bow was consistently envisaged as a stone ball shooter. CA 888v may also contain this sling (top left). Thus, a primary requirement for this project seems to be that the 42 *braccia armature* shoot the "100 pounds of stone" mentioned on CA 149br.

Before discussing this requirement for the bow to shoot a 100-pound stone, it should be noted that there is a visible trajectory of development in the design of the armature from CA 57v-69ar to 147av, and then from 147bv to 149br. The toothed lamination at the centre of the armature on CA 57v-69ar is drawn in more detail on CA 147av (Fig. I.8), the latter lamination taking the form of an arrowhead and supported with various crossing straps. The angular cuts of wood show a type of lamination method for holding the armature's wooden strips in place at its centre. As noted further above, presentation drawings of this lamination appear on CA 90r [33ra] (Fig. 2.5) and 139r [49vb], and there are useful examples on CA 888v [324v], 946r [344va], and on MS B 27v. The toothed lamination on CA 57v-69ar resembles the two zig-zag lamination studies at the top right of CA 147bv (Fig. 1.2). These laminations might have been small details for each laminated portion of the armature, though they do not appear in the design of central crossing beams on CA 147av, 147bv, and 149br. From the top right of CA 147bv, it is possible to see an evolving design for attaching the armature to the upper carriage

---

[35] CA 57v [17ra].

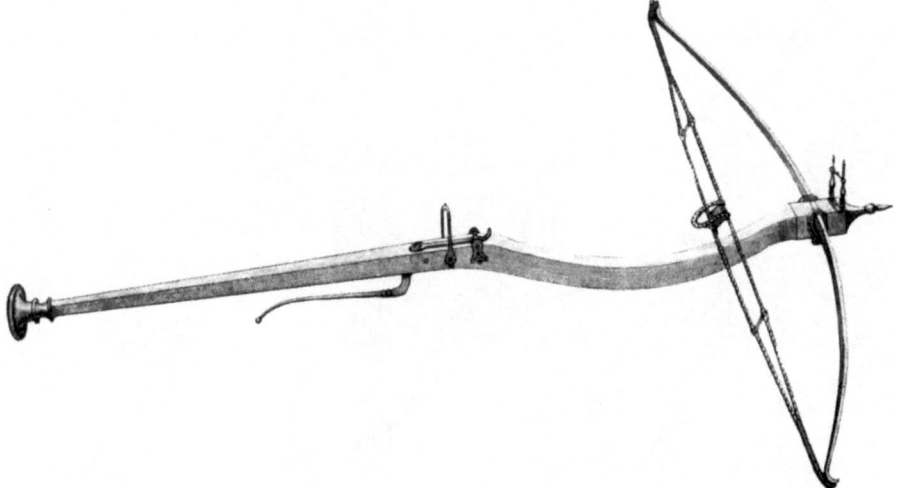

Fig. 4.4.
Illustration of a typical stonebow
(a stone throwing crossbow),
from R. Payne-Gallwey, *The Book
of the Crossbow*, London: Longmans, Green and Co., 1903, p. 157

(the central stock of wood perpendicular to the armature). Leonardo starts with a
central horizontal beam and triangular grooves at the top of CA 147bv. After this, he
appears to have drawn a central beam on a square groove at the centre of 147bv. In the
next sketch at the bottom right corner of 147bv, he then considers using the two crossing beams that are visible in the *Giant Crossbow* presentation drawing. He started with
a simple compound armature with basic laminated boards on CA 57v-69ar and eventually designed a complex compound armature that forms an X in the untreated and
unstrung position, with attached laminations cut in the forms of bat wings on CA 149br.

Something Leonardo removes from the final design, though he added them to
CA 57v-69ar, 147av and 147bv, are the supporting posts on the ropes between the
armature and missile basket. This vertical post, at the centre of the pair of ropes
attaching the armature to the central basket, keeps the ropes and the central basket
upright. These posts were common on "stonebows" (stone throwing crossbows) used
for hunting birds in the 16th century (Fig. 4.4).[36] Though the *Giant Crossbow* drawing
is a large version of a stonebow, the drawing lacks this support on its ropes. Given the
possible length of rope—almost twenty *braccia* on each side of the basket—a post
separating the ropes would have been beneficial to the final design. It is doubtful that
Leonardo forgot to add them to the final design (if he had them in all previous designs). So he might have excluded the posts as unsightly detractions from the
presentation's overall design. Nonetheless, available information cannot explain this
exclusion of the rope's posts, unless the basket or sling were attached to a frame at the
top of the upper carriage.

Regardless of what may be missing from the presentation on CA 149br, the armature appears to be the most carefully considered portion of the bow's design. One
reason for this could be that it had to look like it could have enough force to launch a
100-pound stone a reasonable distance. Leonardo notes in the sixth and seventh lines
of the text on CA 149br that the crossbow "carries 100 pounds of stone." In this case, he
appears to refer to a 100-pound stone ball barely visible at the centre of the ropes
above the upper carriage. The ball appears with more emphasis at the centres of preliminary sketches 147av, 147bv, and 57v-69ar (Figs. I.8, 1.2, 2.4). Also, the apparent size
of the stone ball remains relatively consistent in each crossbow drawing. The ability to
shoot something this size more than a few *braccia* from a low trajectory siege engine
must have seemed absurd to Leonardo's contemporaries. That would normally have
been the task of a trebuchet, mangonel, or other large catapult. His incendiary missile

---

[36] Payne-Gallway, pp 157–160. Patterson, pp 92–95.

Fig. 4.5.
Illustration of Leonardo's
drawing of an explosive dart on
Windsor folio 12651r

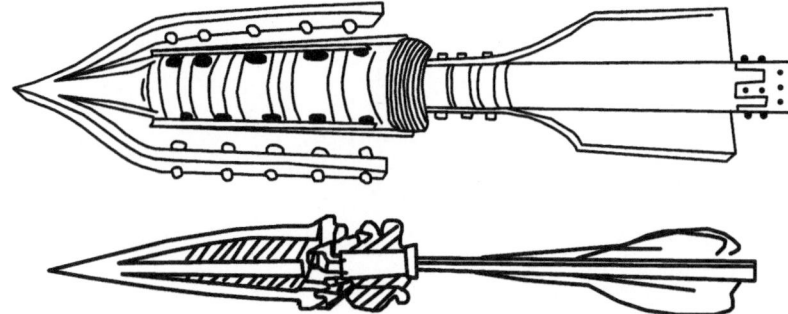

Fig. 4.6.
Illustration of Leonardo's
drawing of an explosive dart on
Manuscript B folio 50v

designs were better candidates as *Giant Crossbow* ammunition, the most detailed studies of which are on Windsor folio W 12651r (Fig. 4.5) and MS B 50v (Fig. 4.6). Regarding the latter, he notes:

> This is a dart to be shot by a
> great crossbow laid flat, and the 2 corners
> have the things which produce fire bound in linen cloth; and when
> the point buries itself the corners are pressed closer together and
> set fire to the powder and the tow that is soaked in
> pitch. And this is good as a weapon for use against
> ships and wooden bastions and
> other similar constructions; and no one
> will make good work in this business of burning
> unless the fire is kindled only after the dart has struck,
> because, if you should wish to light the fire
> before, the violence of the wind will extinguish it on its way.[37]

In any event, two of the few consistent features in three giant crossbow studies, 147av, 147bv and 149br, happen to be the very large stone and the stated preference for a 42 *braccia* armature. Whatever Leonardo's reasons for the proportional assessment of a 100-pound stone ball, launched with a 42 *braccia* armature, this ability to launch such a heavy object with wooden beams would not have been easily accepted by professional engineers in his time, any more than it would be believed today.

---

[37] Questo • e uno dardo daessere tracto da uno / gran balestra • a panca e i 2 • corni / ano iloro focho • in panolino ecquando / lapunta • sificha ichorni sistringano e / danno focho apoluere estopa ipe[-] / golata • ecquesta • e bona perna[-] / vili • bastioni di legniame e / altri simili strumenti enessuno / fara bona • operatione inbrusare sel focho non saciende dopo ilcolpo / perche seuorai aciendere ilfocho inna[-] / anzi lafuria deluento lospegnie travia. MS B 50v.

5

Proportional Design and Operation of the Armature

As a way of addressing this problem of the theoretical functionality of the design, the following discussion offers Leonardo's initial considerations for the armature's proportion in relation to the rest of the crossbow's design and operation. For example, what were his likely reasons for designing a crossbow with a 42 *braccia* wide armature when ready to shoot, which could be elevated 14 *braccia*, and could launch a 100-pound stone ball a reasonable distance? For an artist with some civil engineering experience (as at Verrocchio's studio), but with little or no training in military engineering, this was indeed a bold proposal to prepare for a powerful duke known for his team of military engineers. The following examples will use the information already given about the *Giant Crossbow* to briefly propose that Leonardo's earliest approaches to practical problems of military engineering were based on theories of proportion. Since the *Giant Crossbow* was never built according to his design, however, the following examples will concentrate on theoretical proportion studies possibly considered for the design of the *Giant Crossbow*.

First, one finds a multitude of studies in Leonardo's notebooks on the theoretical proportions of the movement of force in crossbows and other wooden spring engines. A summary of the main ideas, along with a few examples, is as follows.

Among his notes on Valturio's *De re Militari* in MS B, perhaps as early as 1484, Leonardo states at the bottom of folio 4v:

> On Motion and proportion
>
> If you want to know where or at what part of its path the thing drawn with violent cause—make a mass of fresh earth, and with a crossbow, shoot an arrow into it from various distances, and where the same arrow drives in the most, that distance will be strongest. Let us consider an example: The arrow shoots 400 *braccia* in an arch; make the first shot from 25, then from 50. And so, move backwards, 25 *braccia* at a time, until 400, shooting at each 25; and thus you will see at what part of the course the arrow is the most vigorous, whether at one-third, or one-fourth, or one-fifth of the way.
>
> If you want to see an experience in proportion: If a stone is cast twenty *braccia* from the small catapult using a counterweight of ten, see whether or not a counterweight of twenty will cast it twice as far. (MS B 4v)[1]

This statement, dated to around 1485–1489, appears to have been Leonardo's earliest surviving reference to the subject of catapult "motion and proportion." He requested an observable solution to this problem illustrating it conceptually as a proportional "rule of three" estimate: 10 20 20, or rather, 10 lbs = 20 br : [is proportional to] 20 lbs = c, such that $(\frac{10\,lbs}{1} \times \frac{20\,br}{1} = \frac{20\,lbs}{1}) = \frac{400}{10} = 40$; 10 lbs = 20 br : 20 lbs = 40 br.[2] A common

---

[1] Translation by Venerella (2002) *MS B*, pp 7–8.

[2] See: the example given at MIT for the merchant, Michael of Rhodes, c. 1401–1445: http://dibinst.mit.edu/DIBNER/Rhodes/math_toolkit_three.html. The first notation for Leonardo might have been simply, 10 20 20, or the consequence of this, $10/1 \times 20/1 = 20/1$.

medieval method, this "rule of three" applied basic proportional principles to arithmetical problems. In this case, Leonardo outlines the problem rather than the solution, since he has yet to apply his theory to an observable model. One reason that he appears to take a lateral (or indirect) approach to problems in his notebooks, instead of a more direct approach, is that these notes often record for himself the general principles in preparation for an eventual direct approach to a problem when confronted with it in practise. Luck favoured the prepared. Most of his comments on the rules of nature at the time of the *Giant Crossbow* design and afterward refer to problems of proportion rather than solutions to specific problems. With regard to his early plans for treatises, for example, many of his records refer less to these solutions than to problems deemed worthy of the "demonstration" of practical "experience". There are therefore few definite personal claims on engineering in his notebooks before 1495. Additionally, few entries before 1489 appear to have been intended for an audience outside the studios in which he worked, or outside the small group of Ludovico Sforza's military engineers. Thus most of his surviving written opinions on motion and crossbow proportions—much of them written for treatises—seem to post-date 1487, especially in Manuscripts A, I, M, Arundel 263, Madrid I, the Codex Atlanticus, and the Forster Codices I and II.[3]

A good example of this later knowledge of crossbow proportions is noted with respect to Leonardo's careful study of Albert of Saxony's *De proportionibus* (c. 1368).[4] For example, he states in MS I around 1497, the following interests:

> I ask the crossbowman what difference there is in holding the hand at *a* or at *b* or at *c*, for making the arrow shoot higher or lower, and where this comes from.

On Motion

> Albert of Saxony says, in his *On Proportion*, that if a power moves a mobile object at a certain velocity, that it will move half of this mobile object at twice that speed, which does not seem to me to be so, since he does not consider whether this power operates at the extreme of its effectiveness. For, if it did operate in such a manner, something that weighed less would not be in proportion to the force of the motor, or of the medium through which it passed, so it would be a fluttering thing, and not one of straight motion, and it would not go as far.[5]

In this case he refers to three positions—a, b, c—along the upper carriage of the crossbow at which one would withdraw the armature's rope to different stages of resistance and then release it. He had already considered part of this problem and that of an object's weight in "proportion to the force of the motor" in MS A around 1492, if not earlier.

On MS A 29v, he had tried to explain the proportional relationship between the force on an object and the distance that it might travel, referring to this relationship between "force" and "distance" as the "quality proportional to its [the object's] power," (*proportionevole qualita assua potentia*):

---

[3] Some of the notes most relevant to the present study are in the following folios: MS A 29v, 30r, 35r–v, 36r; MS I 101r–v, 120r, 135r, 136v; MS M 90r–v, 93v, 94r; Arundel 54r; Madrid I, 50v–59v; CA 78r [27rb], 754r [278rb]; Forster I 44r; Forster II 33v, 57v.

[4] Leonardo refers to the "di proportione" of "Albertuccius" in MSS I 120r and F, where he had copied Italian translations of the Latin original. Pierre Duhem and Edmondo Solmi first made note of this. P. Duhem (1906–1913) *Études sur Léonard de Vinci, ceux qu'il a lus et ceux qui l'ont lu*, Paris, 3 vols., vol. 1ff. E. Solmi (1908) *Le fonti dei manoscriti di Leonardo da Vinci*, Giornale storico della letteratura italiana, Turin: Loescher, Suppl. 10–11, pp 44, 59–57. Various Latin editions of *De proportionibus* were printed in the 1480s in Padua (1482, 1484, and 1487) and Venice (1487). Finger, *Elmer Belt Library*, pp 6–9. Venerella, *MS I*, p 168, fn. 1.

[5] Translated by J. Venerella, in: Leonardo da Vinci (2000) *Manuscript I, The Manuscripts of Leonardo da Vinci in the Institut de France*, Ente Raccolta Vinciana, Castello Sforzesco, Milan, pp 167–168.

On the nature of movement

[Drawing of a double crossbow, loaded with an arrow.]

Make two nuts, that is two releases, one on the crossbow and the other on the bow, and both will release at the same time. [...]

For what reason does the bow shoot its projectile farther, by a great deal, than the crossbow?

[...]

The weight hurled by the fury of a force will not make the course that it ought, if it is note of a quality proportional to its power.

This is clearly confirmed by experience, since if you thrust away from yourself with rapid arm movement an object whose lightness is no match for your force, that thing will be of little movement. Again, if you thrust something that, by its heaviness, exceeds your force, this will have a brief course.

A force moving a weight proportional to its quality will move the weight away from itself as much more, one time than another, as the time required for one operation of the motor enters into that of the other.[6]

While examining differences between the forces applied by the crossbow and the bow, Leonardo argues on this folio that an object hurled by a force that is proportional to the quality of its power will travel the course that it is capable of travelling.[7] He refers to the "quality" of the object's "power" because of his belief in the impetus theory that the force applied to the object would be embedded in that object until it had exhausted itself. His reference to "a quality proportional to its power" appears to be an original explanation, since he has marked with thick horizontal lines through a very similar statement immediately above this one. This is presumably because the former statement is poorly phrased with respect to the "proportional quality" of the object's "power", instead noting the condition that, "if the object that is thrust is not proportional to this force."[8] In this case, he appears to be distinguishing between the "quality" of the "power" embedded in the object and the "force" that had been applied to it. The "proportional quality" of the object's associated power is at issue, as opposed to a simple assessment of the "proportion" of the associated force or power. He had been trying to explain the "proportional quality" of an object with respect to the experience demonstrated by said object, especially when the application of a "rule of three" proportion theory would offer an incorrect answer with regard to launching disproportionate objects from disproportionate weapons, such as a pebble from longbow, or a cannonball from a handheld crossbow. "Proportional quality" in this case refers to the optimal weight and shape of an object, which is determined with the help of the mechanical experience of the force applied by the weapon to the object. He uses this proportional quality instead of Aristotelian arithmetic for this calculation.

---

[6] The Italian transcription is from Ravaisson-Mollien (1881) *MS A*, 29v. The English translation is by Venerella, *MS A*, pp 85–6. Leonardo's use of "*potentia*" appears to refer to "power" in its broadest sense as "potency" and as a "potential." *Potentia* in his 15th century context seems to be used as a combination of the modern terms *potenza* and *potenziale*. For additional discussions on Leonardo's interests in the mechanics of movement, see: M. Kemp (1981) *Leonardo*, pp 142–148. Ivor Hart (1963) *The mechanical investigations of Leonardo da Vinci*, University of California Press, Berkeley, pp 109–114. F. Schuster (1910) *Zur Mechanik Leonardo da Vinci, Erlangen*, p 34. P. Duhem (1906) vol. 1, p 145. E. Mach (1988) *The science of mechanics*, Open Court, Illinois, p 100.

[7] This is my general paraphrase of Leonardo's statement, "La forza che movera proportionevole peso asua qualita tanto la lontanera dasse più una volt ache unaltra quant oil tempo delloperare del motore entra luno nell altro." The transcription is from Ravaisson-Mollien (1881) *MS A*, 29v. Venerella offers a relatively awkward translation: "the weight hurled by the fury of a force will not make the course that it ought, if it is not of a quality proportional to its power." Venerella, *MS A*, p 85.

[8] See: Ravaisson-Mollien (1881) *MS A*, 29v.

A few lines after Leonardo's reference to "a force proportional to its quality," he concludes at the bottom of the page that:

> This is demonstrated by experience, since if you shoot a projectile from a bow using a force of fifty pounds and it flies 300 *braccia* away from you, and then with a force of three hundred pounds you shoot a projectile of the same weight from a crossbow, this will not exceed a distance of 250 *braccia*, because of the disproportionate force. The reason that the bow shoots farther is that the motor accompanies the moved thing…."[9]

Kemp notes that this is not one of Leonardo's standard objections to Aristotle's rules, that a disproportionately large force applied to a modest object will result in something that lies outside the normal rule of force, outside the range of "moderateness" (like extreme light and dark).[10] This statement and those noted above are in general agreement with Thomas Bradwardine (b. 1295/d. 1349) and Nicholas Oresme (b. 1320/d. 1382), who criticised Aristotle's proportional rules for force, resistance, and velocity.[11] Leonardo was well aware of some of the arguments of these philosophers, but there is no evidence to suggest that he understood their proportional solutions to the Aristotelian problems before 1495.[12] Hence there are the numerous remaining questions in his notebooks about the observable quality of an object that is proportional to the power applied to it.

Proportional Form

In the following investigations of the proportions and dimensions of the crossbow, and in Appendix A, the detailed measurements do not necessarily imply that Leonardo proceeded with such measurements. These measurements represent data for analysis with a view to disclosing the extent to which he was using dimensional and proportional strategies in his designs.

The division of the bow into third portions gave him great flexibility when calculating the sizes of its sections and the proportions of its draw strengths. Consider, for example, that an armature span of 42 units is ⅔ of a 63-unit armature perimeter and that a carriage length of 40 units is ⅔ of 60, a number divisible by three. Also, at the Verrocchio studio, Leonardo could have known the method for calculating the circumference or perimeter of a circle by multiplying its diameter by the Archimedian fraction of ²²/₇. He estimated that a 40-*braccia* length of carriage, multiplied by ²²/₇, equals a maximum armature perimeter of 63 *braccia* (40 · ²²/₇ are equal to the number 125¹⁷/₂₅, half of which is 62²¹/₂₅, and this half circle is theoretically the maximum extent to which the armature would bend). These are calculations undoubtedly similar to those used to determine the circumference of the *palla* (gilt copper ball) atop the dome of the Florence Cathedral, measuring approximately four *braccia* in width, with a half circumference of nearly 6⅞/₅ *braccia*.[13] Around 1515, he wrote about the *palla* on MS G 84v and 82r. It is possible that his estimate of a 42-*braccia* armature conveniently represents ⅔ of the length of its 63-*braccia* outer edge, which also happens to be half the circumference of a circle with a diameter or carriage length of 40 *braccia*.

The proportional relationships between bow measurements of 40, 42 and 63 are not coincidental, as they appear to have been part of the initial layout of the drawing. If the armature's ropes and restraining pins were detached, its arms—if still in an untreated state—would hypothetically spring forward to a V-shape. The curve of the armature could have been initially designed with the rotation of the upper end of a divider around its outer edge, the other end of the divider resting at the bottom centre of the sheet (Fig. 5.1).

---

[9] Translated by Venerella, *MS A*, p 86.
[10] Kemp discussed this with me in June 2004.
[11] This late medieval approach to Aristotle is discussed in: Kemp (1981) *Leonardo*, pp 142–143.
[12] *Ibid.*, p 143.
[13] For a thorough discussion of Verrocchio's treatment of the *palla* commission, see: Dario A. Covi (2005) *Andrea del Verrocchio, life and work*, Olschki, Florence, pp 63–69.

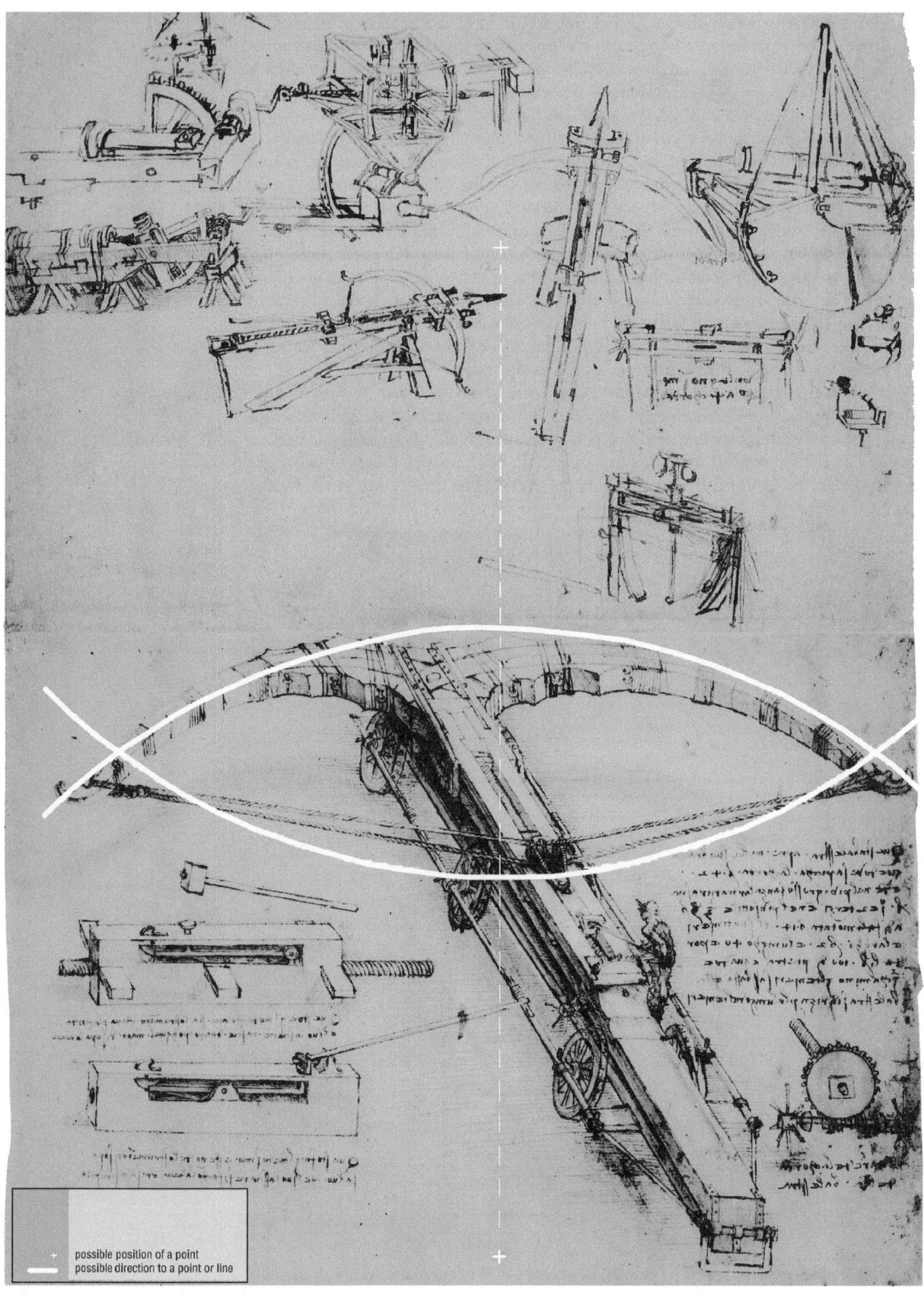

possible position of a point
possible direction to a point or line

A circle created with the divider in this position would have had a radius of 40 *braccia*, noted with a horizontal white line across the bottom of the sheet. Twice this radius (80), multiplied by ²²⁄₇, is a circumference of 251 and ⁹⁄₂₅ *braccia*. Inscribe in this circle a hexagon and the resulting six sections of the circumference will measure approximately 42 *braccia* each (41 and ¹⁷⁄₁₉). Thus, one can see in Fig. 5.2 that the partial straightening of a 42 *braccia* front section of the armature would straighten that section to a width extending to both ends of the armature, where the rope is attached and where parallel metal stylus incisions form invisible horizontal lines between these ends. These numbers represent another coincidence between specified proportions, 42 and 40, and the Euclidian and Archimedean geometry Leonardo possibly studied with Fazio Cardano in Milan and/or Pavia in the mid 1480s. A simple lesson in geometry would have revealed that a circle with the radius of 4 has a circumference with six equal sections measuring 4 and ⅓ each. A great crossbow stock 4 *braccia* in length, a sixth of the circumference of a circle, formed around the edge of a spanned armature at the front of this stock, would measure 4 and ⅓ *braccia*. At a scale of ten times this example, the *Giant Crossbow* might have been measured in this way.

Alternatively, Cardano could have introduced Leonardo to the fourth book of Euclid's *Elements*, proposition 15, wherein it is shown that any angle of the equilateral and equiangular triangles that compose the equilateral and equiangular hexagon inscribed within a circle is one third of two right angles.[14] In other words, this ⅓ of two right angles—which is 60°—is also the angle of the carriage to the picture plane on the

Fig. 5.2.
Superimposition of black and white lines over CA 149br, showing the proportions of the span with respect to a hexagon inscribed in a circle

---

[14] I would like to thank Dr. J. V. Field for drawing my attention to this proposition in Euclid's *Elements*.

drawing, two thirds of which is the carriage length (40), as well as the radius of the circle in which Euclid inscribes the hexagon. The geometry of Euclid and possibly Archimedes apparently helped with the formatting of CA 149br and its modular components of third-part proportions.

The extensive use of three as a factor in his proportion theories was as fundamental to his work as it was to the first section of one of the first printed arithmetic textbooks, *l'arte de l'abbacho*, published in Treviso in 1478, which uses this factor to explain the differences between addition, subtraction, multiplication and division.[15] This book is most likely the "*abacho*" Leonardo recorded among a booklist of five on Codex Trivulzianus 2r, around 1485–1487. Consistent with this approach to factors of three is the plausibility that the outer edge of the armature would measure 63 *braccia*—⅓ greater than 42. Also, each half of the armature is 21 *braccia* in length, or three sections of seven *braccia*. To determine measurements of armature proportions one would simply multiply modules of seven by three (21), or seven by six (42), or seven by nine (63). By repeatedly noting a spanned armature width of 42 on CA 147av, 147bv, and 149br, a number that has factors of three and seven, he indicates that this is not a random number, and that it may be one of the most important calculated measurements of the crossbow.

Leonardo's insistence on a number with factors of 3 and 7 to explain the proportions of a curved object is probably his earliest reference to Archimedes' "Measurement of a Circle," propositions 2 and 3. Proposition 3 states that "The circumference of any circle is greater than three times the diameter and exceeds it by a quantity less than the seventh part of the diameter but greater than ten seventy-first parts" $(c = d(3\frac{1}{7}) < n < d(3\frac{10}{71}))$.[16] Proposition 2 (which logically follows proposition 3) states that "The circle bears to the square on the diameter the ratio $11:14$."[17] In other words, the ratio of the area of a circle to that of a square with a side equal to the circle's diameter is close to $11:14$. Thus, the area of a circle is $(\frac{22}{7})r^2$, as noted in proposition 3, and this is proportional to the area of a circle that is $(\frac{11}{14})d^2$, where "d" is diameter. To repeat the result of the estimate further above, twice the radius of 40 *braccia* (80), multiplied by $\frac{22}{7}$, is a circumference of 251 and $\frac{9}{25}$ *braccia*, a sixth of which is approximately 42 *braccia* (41 and $\frac{17}{19}$).

Note that these numbers and fractions refer to inexact, continuous, geometrical proportions, rather than to the problems of exact, discontinuous arithmetic. Heron of Alexandria, Ptolemy, Eudoxus and Euclid addressed with various large fractions the axioms of Archimedes, making proportional proofs of this geometry that were necessarily inexact mathematically.[18] Improving on inexact proofs such as these was an academic game for mathematicians of ancient Greece as well as 15th century Italy. Ptolemy, for example offered a close approximation with sexagesimal (base-sixty) fractions, such that the perimeter of a circle is more than three times the diameter, exceeding by a quantity of $\frac{8}{60} + \frac{30}{60^2}$ (in modern notation: $p = 3 + \frac{8}{60} + \frac{30}{60^2}$, or 3.1416). To attempt to square the circle with exact numbers was a respectable science, a riddle that Leonardo believed he could solve around 1508. His direct references to this specific problem tend to post-date his 1504 record of the "quadratura del circulo" in his Codex Madrid II book list on folios 2v–3r.[19] For the purposes of the present study, however, his repeated references to the theories and inventions of Archimedes and Euclid between the mid 1480s and the mid 1510s suggest that he was well aware of their general

[15] See: the translation and discussion by David Eugene Smith (1924) *The first printed arithmetic*, Isis VI(3):311–331.

[16] Ivor Thomas (trans) (1939/1998) *Selections illustrating the history of Greek mathematics*, vol. I, Loeb Classical Library, Harvard, Cambridge, MA, p 321.

[17] *Ibid.*, vol. I, p 333.

[18] *Ibid.*, vol. I, pp 317–334. Solmi, *Scritti*, pp 63–70.

[19] Leonardo da Vinci (1974) *I Codici di Madrid (The Madrid Codices)*, 5 vols., Ladislao Reti (ed), McGraw-Hill, New York, vol. 3 (Commentary), p 92.

principles possibly from the time of his association with the Pavian mathematics professor, Fazio Cardano in the early to mid 1480s.[20] In fact, Leonardo's early geometry studies coincide in this case with the time of his MS B 33r study of Archimedes' steam-powered canon, the "*Architronito*".

The Motion of Missiles Cast … Will Be in Proportion to the Angle of the Propelling String

Now that Leonardo's possible interests in the crossbow armature's proportional dimensions have been briefly addressed, his studies of the proportional angles of a crossbow's "propelling string" will continue in the following discussion, especially with regard to the perceived effect of the "span" of armature on the "motion of missiles". In modern terms, a "span" is the draw of a bowstring. Leonardo referred to this action with phrases about the "draw" of the rope. In the case of the *Giant Crossbow*, the "span" is the extent to which the crossbow "opens its arms, that is where the rope is attached," as he states on CA 149br. For reasons that will be discussed immediately below and again toward the end of this study, it appears as though Leonardo intended to set the illustrated and written span of the armature at an approximate maximum width of 42 *braccia*.

Although very little evidence exists from the time of the *Giant Crossbow* drawing about his reasons for a 42 *braccia* span of its armature, there is evidence that at an early stage in Leonardo's study of crossbows that he used a pyramidal proportion theory to determine the optimal span widths of crossbow armatures. There are, for example, armature span proportion studies from the early 1490s on Codex Atlanticus 78r [27rb] Fig. 5.3 and Codex Madrid I 50v-51r Fig. 5.4, noting his expectation that a pyramidal law governed the extent to which a projectile's embedded impetus would last, or would keep the projectile airborne.[21] Referring to the image of Codex Madrid 51r, (Fig. 5.4), he states:

> The power of the mover drawing the string of the crossbow increases as much as the angle created at the centre of the string increases.
>
> …
>
> The motion of the missiles cast by the crossbow will be in proportion to the angle of the propelling string and to the motion that generated this angle.[22]

These comments claim that the force of the string is proportional to the internal angle between the centre of the pulled bowstring and the centre of the straight line between the points of attachment of the string on the bow arms. Noting that Leonardo was convinced of this solution, Kemp finds in these notes "that there was an incrementally proportional correspondence (inversely) between the angle of the taut bowstring and the force required to withdraw it, based upon a 'pyramidal' law."[23] Leonardo says that these pyramidal "degrees of the descent" of the string "will almost appear to look like the foreshortenings of those who practise perspective."[24] He refers to this perspective pyramidal law immediately after stating that an increased pull on the rope by "weights in units of 10" will show that "the degrees of descent will not be equal."[25] Thus he appears to recognise that the proportion of draw weight to that of the string's

---

[20] Solmi, *Scritti*, pp 111–115.

[21] MS Madrid I is generally regarded as a product of the early 1490s and Pedretti dates CA 78r [27rb] to the mid 1490s. There is sufficient reason to suggest, however, that the wrench and clasp studies on CA 78r are so similar to the same examples on Madrid I, 53r that both folios must date to around the early 1490s. This would be in agreement with the crossbow pyramidal law studies that are similar on Madrid I, 50v–51r and CA 78r. Kemp (1981) p 22. Reti, *Madrid Codices*, vol. IV, pp 100–102. Pedretti (1978) *Catalogue*, vol. I, pp 57–58.

[22] MS Madrid I, 50v. Reti (trans) *Madrid Codices*, vol. IV, p 100.

[23] Kemp (1981) *Marvellous works*, p 144.

[24] MS Madrid I, 51r. Translation by Reti, *Madrid Codices*, vol. IV, p 102.

[25] *Ibid.*

Fig. 5.3.
Illustration of Leonardo's diagram of the proportional draw strengths of a crossbow on Codex Atlanticus f. 78r [27rb]

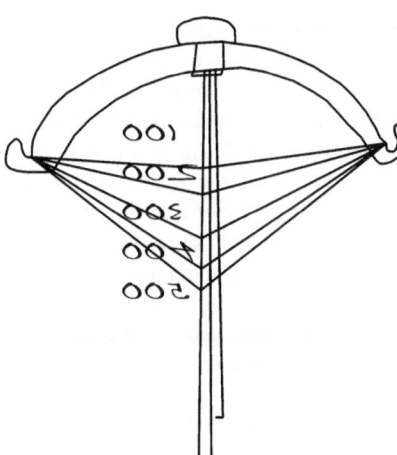

Fig. 5.3.
Illustration of Leonardo's diagram of the proportional draw strengths of a crossbow on Codex Atlanticus f. 78r [27rb]

Fig. 5.4.
Illustration of Leonardo's diagram of the proportional draw strengths of a crossbow on Codex Madrid I, f. 51r

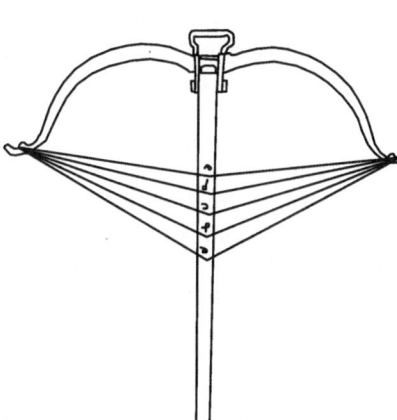

visible withdrawal happens to be as unequal as the proportional diminution of distances between horizontal lines progressing toward the top of a perspective pyramid. In other words, evenly measured positions of a string's withdrawal along a crossbow's upper carriage will bear no relationship to the actual amount of draw weight necessary for the pull of that string. He notes instead that the ratio of draw weight to that of the string's visible withdrawal is almost as uneven as the "foreshortenings of those who practise perspective," and he therefore concludes that the "motion of the missiles" will increase to the extent that the internal angle of the "propelling string" decreases. Thus, several years before he wrote these comments in what is now called Codex Madrid I, he could have designed the *Giant Crossbow* with this theory in mind: that internal angles produced by the draw of the "propelling string"—and therefore the proportion of the 42 br armature with respect to the 40 br upper carriage—had to be sufficient for the launch of a heavy stone ball. His assessment of this problem is a kind of diagrammatic and proportional reasoning, as his reasons for this finding may have been limited to written diagrams, or even to conditions observed at crossbow target practise.

Knowing that the illustrated angle of string withdrawal and the draw weight cannot represent in equal portions the equal distances of a projectile's journey, his statements on Madrid I 51r also refer to an inverse proportional rule that is based on the impetus theory of dynamics.[26] According to his impetus theory on MS A 30r, "the more the moving force accompanies the thing moved, the longer the movement will be."[27] He understood the rule for the diminishment of this force on MS A 35r, stating that, "the more weight falls, the more it increases; and the more force falls, the more it diminishes."[28] In addition, therefore, to what he might have known about the internal angle of the propelling string, he nonetheless understood in the early 1490s the basic principle that the greater the distance that the string and projectile travel on the upper carriage—before the projectile launches from the string—the greater the impetus stored in the projectile, allowing it to travel further. Still, he also knew that by pulling the string back twice as far, one was not guaranteed that the projectile would travel twice as far. As stated on MS Arundel 54r (at least twenty years after MS A), "Excessive force against a like resistance profits the projectile nothing. But if the force of the mover

---

[26] Franciscus de Marchia (fl. 1320) and John Buridan (c. 1295–c. 1358) formulated impetus theories as a way of explaining how an object keeps moving after it leaves the thing that throws it. This was first explained by the 6th century Alexandrian, John Philoponus, who corrected Aristotle's poor definition of impetus. One of Buridan's definitions of impetus theory that Leonardo appeared to know is that a falling body was believed to generate additional impetus embedded in that body: as impetus increases with the fall of the body, the body's velocity increases. A standard source for Marchia and Buridan is: David Lindberg (1992) *The beginnings of western science*, Chicago U, Chicago, pp 302–303.

[27] Translation by Venerella, *MS A*, p 87.

[28] Translated by Venerella, *MS A*, p 105.

should find itself in proportion to its projectile the movement made by the projectile will be in the first stage of its strength."[29] The problem here appears to be that Leonardo apparently had no way of illustrating this "force of the mover… in proportion to its projectile," unless of course he had a consistent dimension with respect to the proportions of the crossbow, such as the angle of the "propelling string". The span of the bow (by withdrawing the bowstring) alters the key dimensions of the width of the bow opening and the distance from the centre of the bow to the centre of the line across the points of attachment. These points of attachment move in an arc as the string is withdrawn, so there is no stable point of reference in the base of the triangle made by the line across the points of attachment and the centre of the bow string.[30] Lacking a measurable quantity with which to illustrate the bow's capabilities, Leonardo considered the angle at the centre of the string, where one can barely see the rear projectile placed on Madrid I 51r. This solves his problem of finding stable positions of reference with respect to the crossbow pyramidal law illustration on that folio. He labels the withdrawn strings with letters a through e and states that, "if angle a is the double of angle e, angle e will shoot twice as far as angle a."[31] To double the draw weight of the string, he cut in half the interior angle of that string. He says on MS A 35r that, as "force falls," "it diminishes," and as "weight falls," "it increases."[32] He therefore finds the inverse properties of these two impetus principles on Madrid I 51r: the bowstring's force is greatest where the nut releases it and the projectile's weight is greater at the end of its journey with the bowstring than it is at the beginning of that journey.

In short, it would appear that his various comments in the early 1490s on impetus theory and crossbow proportions recall his earlier interests in crossbows, and thus his possible knowledge of these principles at the time of his *Giant Crossbow* project (throughout the period, 1483–1492) should not be excluded from a study of the proportion theories likely associated with its design. In general, proportion theories that he discussed and illustrated for the purposes of crossbow design and operation seem to indicate that he intended to use a structure ten or twenty times the size of the largest great crossbow (10×4 br, or 20×2 br) in order to increase the distance and size of the projectile in use by nearly ten or twenty times the normal distance or weight. What he had understood by the early 1490s, however, had been that all of these factors were not easily explained with simple proportional comparisons, and that one had to observe the proportional changes in the crossbow's structure in order to see how it might behave in practise.

Evidence that Leonardo might have estimated the force of the *Giant Crossbow*'s armature seems to be apparent in its design. Figure 5.5 illustrates two hypothetical positions of the armature's rope along the upper carriage. At the upper carriage, I have marked with white lines two positions that the bowstrings could reach if the crossbowmen were to pull them back with the help of the trigger mechanism and central winding screw. Crossing the centre of the armature is a white horizontal line that marks the straight edge of the armature when the bowstrings are slack. (Before the ropes were attached and the armature was formed, its central beams would have hypothetically sprung forward into a V-shape, as illustrated in Fig. 5.6.) The bottom horizontal line crosses a position on the upper carriage that represents maximum distance that the trigger can pull the bowstrings. This horizontal line is also at the centre of the 37–40 *braccia* diagonal line that marks the length of the carriage. Thus, we can see that Leonardo estimated the maximum span of the armature to extend back, down the centre of the carriage by approximately 20 *braccia*. At the right end of the metal stylus line that is drawn between armature ends, I have drawn a vertical

Fig. 5.5. ▶

Superimposition of white lines over CA 149br, illustrating three possible spans of the armature with respect to the likely positions of the ropes at those positions

---

[29] Translated by McCurdy, p 586.
[30] I am grateful to Martin Kemp for his advice on this.
[31] MS Madrid I, 51r. Translated by Reti, *Madrid Codices*, vol. IV, p 102.
[32] Translated by Venerella, *MS A*, p 105.

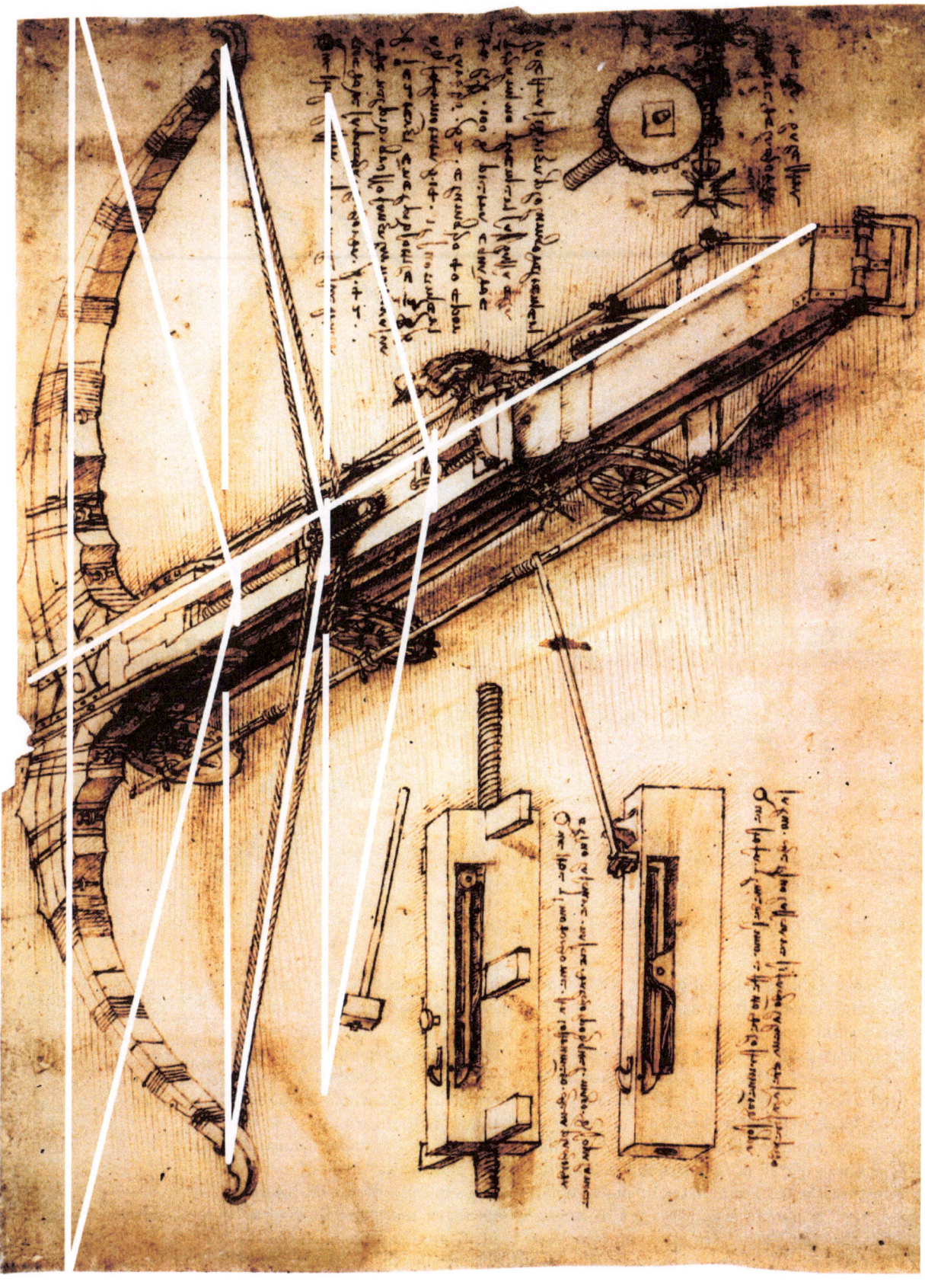

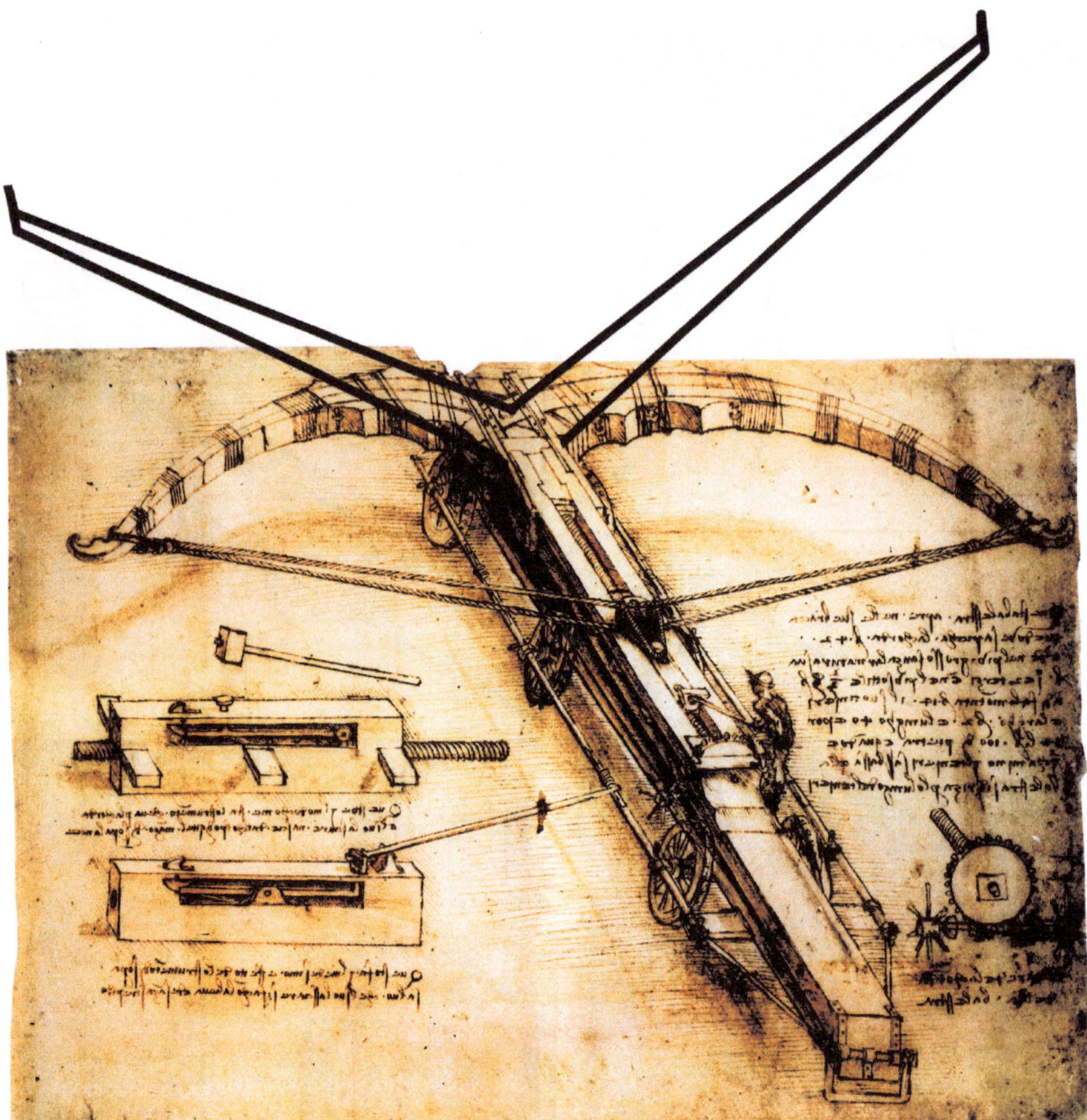

Fig. 5.6. Superimposition of black lines over CA149br, showing the extent to which the armature would have extended forward before it had been bent, treated, and attached to ropes

white line, marking the right edge of the 42 *braccia* illustrated armature width at the bowstring points of attachment. This vertical line meets both the upper horizontal armature line and the lower horizontal bowstring limit line at a 90° angles. It is possible to draw a line at a 45° angle from the centre of the carriage diagonal line up to the right corner of the upper horizontal armature line. If one associates this angle with the draw strength of half the armature, it would follow that the whole armature has an "angle of force" proportional to 90° (or twice that of half the armature). Furthermore, Leonardo's illustrated position of the bowstrings on the upper carriage, where I have placed white horizontal and vertical lines, denotes an "angle of force" proportional to 60°. From that point on the upper carriage, one can draw a line to the upper right horizontal line that is at a 30° angle. Leonardo has illustrated the bow at ⅔ of its shooting potential, if one compares the 90° and 60° angles of the positions of the bowstrings with regard to the straight horizontal position of the armature when the strings are slack. The "angle of force" is zero when the bowstrings are slack, and at 100% when they are at a 90° angle, and at 66.67% when they are pulled to their illustrated position. The amount of impetus presumably stored in the projectile that is launched from the illustrated position on the crossbow might have been considered to be as much as ⅔ that of the impetus stored in a projectile released from the position at the centre of the carriage.

We cannot be guaranteed that Leonardo considered these reliable proportional estimates for the string's "angle of force" on the *Giant Crossbow*, any more than he was sure of his crossbow proportion theories on Madrid I 51r—where he stated at the top of the sheet, immediately above the crossbow diagram—"Test it first and state the rule afterward."[33] Immediately to the right of the crossbow diagram, he wonders if his inverse pyramid theory, "would be incorrect, because—should angle *a* shoot as far as one span—*e* would shoot to the distance of 2 spans."[34] Even if he considers the halving of the bowstring's angle incorrect for a determination of a projectile discharge "to the distance of 2 spans," he confirms that the proportional pyramid law is useful at the bottom left of the page. He states that: "the arrow's weights must be proportional either to the angles of the string that thrust them or to the weights that generated such angles."[35]

The problem with this pyramidal law of Leonardo's, as Martin Kemp has told me:

> ...is that relating this crossbow angle to the other dimensions in the geometry of the withdrawn bow (given the known length of the bow string and the right angle where the line from the centre of the string meets the centre of the line between the points of attachment) requires trigonometry, which Leonardo did not have. So he just used the angle, since he needed some measurable quantity which could conform to his desire for a proportional law governing the behaviour of all dynamic and static systems. The consequence is that he did not have any precise way of relating the width of the bow to its performance.[36]

In any event, Leonardo's stated inability to solve this problem with numerical accuracy—knowing that angle *e* would not provide the force of two crossbow spans—did not deter him from his numerous discussions of the problem. He was looking for a way to visually demonstrate, and thereby numerically calculate, the way in which the doubling of a crossbow's power could send a missile twice as far.

At the bottom right of Madrid 51r, he states:

> Suppose you draw a crossbow with a weight whose power balances the resistance of the string exactly; that is, when the string is placed precisely at the nut of the crossbow. If the weight is 400 pounds, the arrow weighs 3 ounces and the motion of the arrow is 400 *braccia*. I affirm that if you double the weight, the resistance of the string, and the weight of the arrow, and the primary motion is the same, such an arrow will travel twice the distance regardless of whether the arrow's length is the same or doubled.[37]

Leonardo estimates that, if a crossbow with a 400 pound draw could throw the 3 ounce arrow a distance of 400 *braccia*, a crossbow with a draw tension of 800 pounds would throw a 3 ounce arrow a distance of 800 *braccia*. Since a simple theory like this would not have accounted for the effects of the resistance of air—as Leonardo would have found when observing a demonstration—he might have copied the theory from another source, knowing that this could have been inaccurate in practise. He was aware of the resistance of air to flying arrows, noting with respect to this on Madrid I 50v that one should "arrange trials in different kinds of air: that is, foggy, rainy, and clear..."[38]

---

[33] MS Madrid I, 51r. Translated by Reti, *Madrid Codices*, vol. IV, p 102.
[34] *Ibid.* I added the dashes for clarification.
[35] *Ibid.*
[36] As stated in an e-mail of 13 June, 2003.
[37] MS Madrid I, 51r. Translated by Reti, *Madrid Codices*, vol. IV, pp 103–104.
[38] MS Madrid I, 50v. Translated by Reti, *Madrid Codices*, vol. IV, p 100.

In Manuscripts A, B, I, Madrid I, the *Codex Atlanticus,* and the Windsor folios, he made numerous proportional calculations with an interest in multiplying the strengths of machines. Add a giant crossbow's 100 pound stone missile to the estimate above and one would theoretically need at least an eighty ton draw tension to throw that stone a distance of 400 *braccia*.[39] Additionally, considering that the estimate above refers to the crossbow drawing on Madrid I 51r, one could make a calculation based on that folio's approximately one *braccio* wide span of the armature. This span (when pulled) could be multiplied by 42 and would then supposedly have 8.4 tons of draw tension, theoretically capable of throwing a 100 pound stone almost a distance of 42 *braccia*, or a 10 pound stone a almost distance of nearly 420 *braccia*.[40] These proportional calculations are nonetheless purely theoretical, and Leonardo would have wanted such approaches to be demonstrated and measured in different types of weather.

In theory, it would appear that he designed the *Giant Crossbow* to perform as well as bronze cannon or mortars. Various 15th century illustrations and literary sources refer to the placement of cannon within thirty *braccia* of a castle moat, the position most suitable for causing the most damage to castle walls or doors.[41] A *Giant Crossbow* at this distance from a castle could have been quite damaging if Leonardo had calculated that it would launch a 100-pound stone almost 42 *braccia*. By designing a crossbow rather than a cannon, he thereby avoided the expense and weight of the bronze, avoided the dangers of gunpowder, and proposed that his crossbow could launch missiles "without noise" (*sanza strepido*), as specified in the fifteenth line of text on CA 149br.[42] This capability of "quietly" shooting very heavy stones accurately along a low trajectory (higher trajectories being less accurate) would have surprised any enemy accustomed to the more common occurrence in the late 15th century of loud cannon fire. Also, those cannon were heavy and expensive to operate. From the late fifteenth through the eighteenth centuries, the Turks produced cannon weighing nearly four tons each, referred to as "18 pounders" because they fired 18-pound balls.[43] The famous Dardanelles gun, produced in Turkey in 1464, weighed 16 tons and fired a stone weighing 1 000 pounds using 80 pounds of gunpowder.[44] The expense of producing a cannon that shot 20 pound, or even 100 pound stones, would have been enough motivation for Ludovico Sforza to have Leonardo design alternative siege engines, or develop a treatise on them. The availability of bronze could not have been taken for granted, especially in 1494, when the French had invaded Italy and Ludovico had had to send most of his bronze to Ferrara so that Ercole d'Este could produce anti-French cannon.[45] ITN Factual Television had 100-pound stone balls produced for their reproduction of the *Giant Crossbow*, as seen in Fig. II.3. Though their nearly

---

[39] A Milanese pound was traditionally twelve ounces; so a 100 pound stone would weigh the same as 400 three ounce arrows. This weight of 400 arrows multiplied by the 400 pound draw strength is 160 000 pounds (80 tons).

[40] 400 lb draw strength × 42 (*Giant Crossbow* armature width) = 8.4 tons, capable of throwing 100 lbs a distance of 42 *braccia*, or proportionally, a 10 lb stone a distance of 420 *braccia*.

[41] Gravett, pp 53–55.

[42] Beginning in the fourth quarter of the 15th century, the practise of casting cannon in bronze replaced the tradition, since c. 1370, of using welded and "hooped" iron bars to form these "bombards" (*bombos*) as they were called. "Hooped" bombards were more dangerous than their bronze successors. One of these exploded at the siege of Roxburgh (Scotland) in 1460, killing James II of Scotland. Philip the Good of Burgundy recorded the use of 900-pound stones in the iron bombards in 1451. The stones were cut with precision. Sometimes they were covered in lead to reduce gun barrel friction. C. Gravett (1990) pp 52–53. Philip Warner (1968) *Sieges of the Middle Ages*, Barnes and Noble, New York. Ian Heath (1982) *Armies of the Middle Ages*, 2 vols., WRG, Worthing, Sussex, UK.

[43] *The Royal Armouries: Fort Nelson*, (guide book), London, 2002, p 13.

[44] *Ibid.*, p 12.

[45] Kemp (1981) *Marvellous Works*, pp 211–212.

full-scale model differed considerably from Leonardo's specifications, in the ITN engineer's attempt to make it more reliable, it could not even launch a 20 pound stone more than a few feet.[46]

Following the reference to a 100-pound stone ball, Leonardo states in lines eight and nine of the CA 149br text that the carriage "lowers itself and the crossbow extends along the length of the carriage." This might appear to indicate that the upper carriage could be lowered to a level beneath the illustrated level of 3⅔ *braccia* above the ground, just above the 3⅓ *braccia* tall front wheels. But only by removing the armature would have it be possible to lower the upper carriage any further. The benefit of detaching the armature would be to prepare the bow for a long journey. By 1500, it was common practise for French and German armies to transport batteries of three or four dozen cannon long distances in order to set them up within thirty paces of castles and city walls, firing thousands of rounds within a few days.[47] There is a record from 1477 of two Italian bombards requiring 48 horse drawn wagons to transport the upper carriages, shot, gunpowder, and other equipment.[48] As noted further above, Ludovico Sforza had an interest in transporting siege weapons when, as Machiavelli states, the armies directed by Ludovico "plundered" the Venetian territories of Bergamo, Brescia, and Verona in 1483.[49] Hence, Leonardo illustrates visually and verbally that his crossbow's armature, as it appears attached to the upper carriage with rectangular iron supports, could detach from the upper carriage, making it easier to transport both the armature and the upper carriage separately.

This note on the theoretical detachability and transportation of the armature concludes the present discussion on the armature's design, construction, and theoretical function. Possible reasons for Leonardo's approaches to these problems were at issue here. It would not be possible to draw definite conclusions about his proportion theories from most of the available evidence discussed herein. Instead, it should be stressed that his records refer mainly to problems in the progress of being investigated, rather than to definite solutions. To restate the general argument of the present study, one way of understanding his numerous investigations and unresolved problems is to compare the proportional methods common to these studies. First, human and animal proportions were important to nearly every study, such as military engineering, perspective construction, drawing, painting, sculpture, architecture, treatise arrangement, anatomy, natural philosophy, cartography, etc. With regard to the armature, the bat wing structure and proportion studies of weight and motion in MS B bear some relationship to the design on CA 149br and proportion studies in Manuscripts B, A, I, Madrid I, the *Codex Atlanticus* and among the Windsor folios. Second, referring to the "degrees of the descent" of the bowstrings on Madrid I 50v-51r, Leonardo states: "I propose that the nature of this power shall be pyramidal…" and that these pyramidal forms "will almost appear to look like the foreshortenings of those who practise perspective."[50] This is one of the earliest references to a pyramidal law of proportion in his notebooks. It is also proof of his comparison of an aspect of perspective construction with the look and action of a crossbow, to see if the proportional adjustment of the internal angle of the "propelling string" of a crossbow would bear some relationship to the

---

[46] David Hepworth, a lecturer in 2002 in Biomimetics at the University of Bath Mechanical Engineering Department, advised ITN that the composite portions at each tip of the armature could snap under the expected 100 tons of tension if built according to Leonardo's specifications. I had discussed this with him at the ITN filming on 22 June 2002. Though I am indebted to Dr. Hepworth for his generous advice on this matter, I would still argue that Leonardo's design for the armature on CA 149br should work better than the ITN reconstruction if built according to Leonardo's illustrated and written specifications.

[47] Gravette (1990) p 54.

[48] *Ibid.*, pp 53–54.

[49] N. Machiavelli, *Istorie fiorentine*, book 8, chapter V, paragraph 8.

[50] MS Madrid I, 50r–51v, Translation by Reti, *Madrid Codices*, vol. IV, p 102.

force with which it could launch a missile. Third, numerical divisors (factors) in the *Giant Crossbow* design and various other studies happen to be in third and fourth portions of the "divident" of twelve. These divisors appear in different forms, including the three offices of force (pulling, pushing, and stopping), the three violent passions (force, motion, and blow), and in the third and fourth portions of the designs of each crossbow component. Fourth, the geometry of Archimedes' "Measurement of a Circle" and Euclid's Elements assisted with the crossbow's third-part modular proportions, as well as the its measurements of 40, 42, and 63 *braccia*. Additionally, the demonstrated armature width of 42 *braccia* is three times the written upper carriage elevation of 14 *braccia*. If the stated arm-length on Windsor folio 19134r is 14 units and Leonardo knows of Alberti's statement that an arm is one third of a man, then the armature elevation or draw one third of its own width could correspond to the proportions of a person's arm-length, compared to that of his or her body length.

Now, after looking generally at the design and theoretical operation of the armature, discussions continue on the lower carriage and wheels, and then the upper carriage. A detailed examination of the proportionate layout of the CA 149br folio and its crossbow design follow afterward. These discussions will show some of the basic aspects of Leonardo's design skills. One of the main reasons to point out the proportional relationships between the parts of his design is to demonstrate some of his methods for producing the *Giant Crossbow* presentation drawing. This drawing also demonstrates his ability to make a design that could have theoretically worked if built to scale. If any part of the crossbow design were to have been out of proportion to any other part, the crossbow would not have been worthy of the attention of an engineer, or a patron knowledgeable about siege weapons, such as Ludovico Sforza. Engineers, carpenters, and patrons of this design, or the treatise for which it was intended, would have checked this crossbow design for its possible life-size measurements. If critical portions of the bow were to have been visibly out of proportion to the mechanical needs of the bow, the result would be a rejection of the faulty design. As designers, engineers, and architects like Leonardo were aware, the *Giant Crossbow* drawing would have possibly met with initial approval if one could have seen how each part of it would have been measurable in actual *braccia*, and how each of these parts would have worked proportionally with one another.

6

The Lower Carriage, Side Poles, and Wheels

Leonardo's reference to the crossbow's "carriage" (*tenieri*) on CA 149br, refers generally to the crossbow's main body or "stock" on wheels. The carriage is in two parts: a wheeled cart and an upper portion that holds the trigger mechanism and armature. Upper and lower portions of the carriage connect at the back end of the crossbow with a large hinge, visible at the bottom right of the drawing. Leonardo illustrates the upper carriage propped up at the front end with a wooden block visible just behind the front left wheel (Fig. 6.1). This prop elevates the armature above the front wheels and represents in the drawing the lowest position at which the upper carriage could rest. As stated at the right of the drawing—on the fifth line—this upper carriage "has an elevation or draw of 14 *braccia*." The cant of wheels on the lower carriage would have provided stability for the elevated upper carriage especially when the crossbow had launched a missile. Poles along the outside of the lower carriage appear to be designed to help hold the canted wheels in place. Leonardo also designed a cannon carriage with canted wheels, on CA folio 154br [55vb] (Fig. 6.2), including vertical side posts (top left of sheet) with which part of the carriage rests on the posts and wheel hubs. Trigger mechanisms on the left and

Fig. 6.1.
Detail of the front end of the crossbow on CA 149br

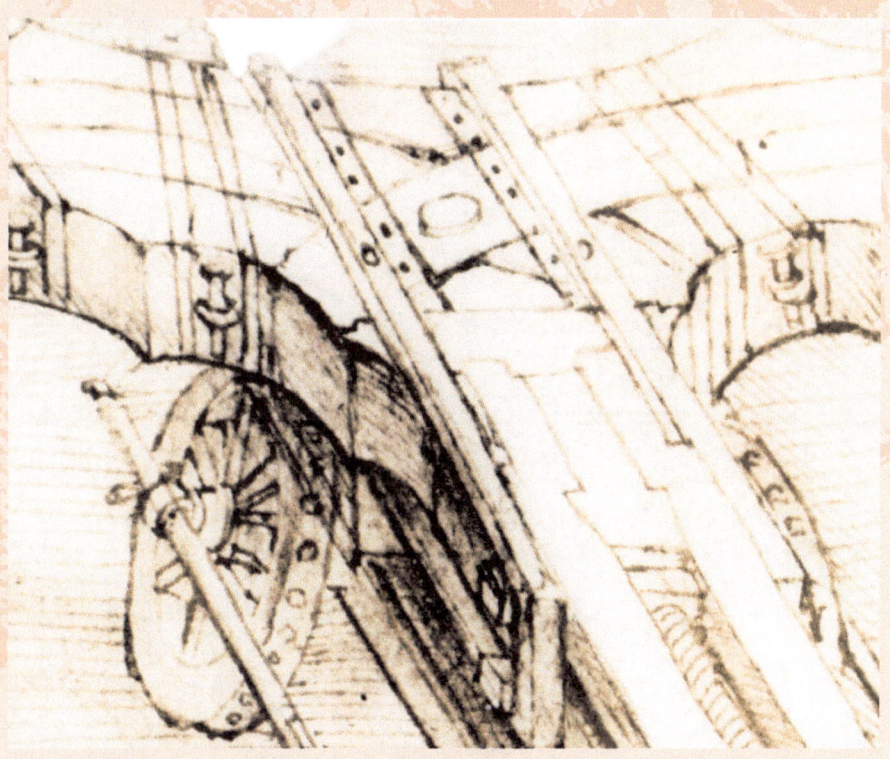

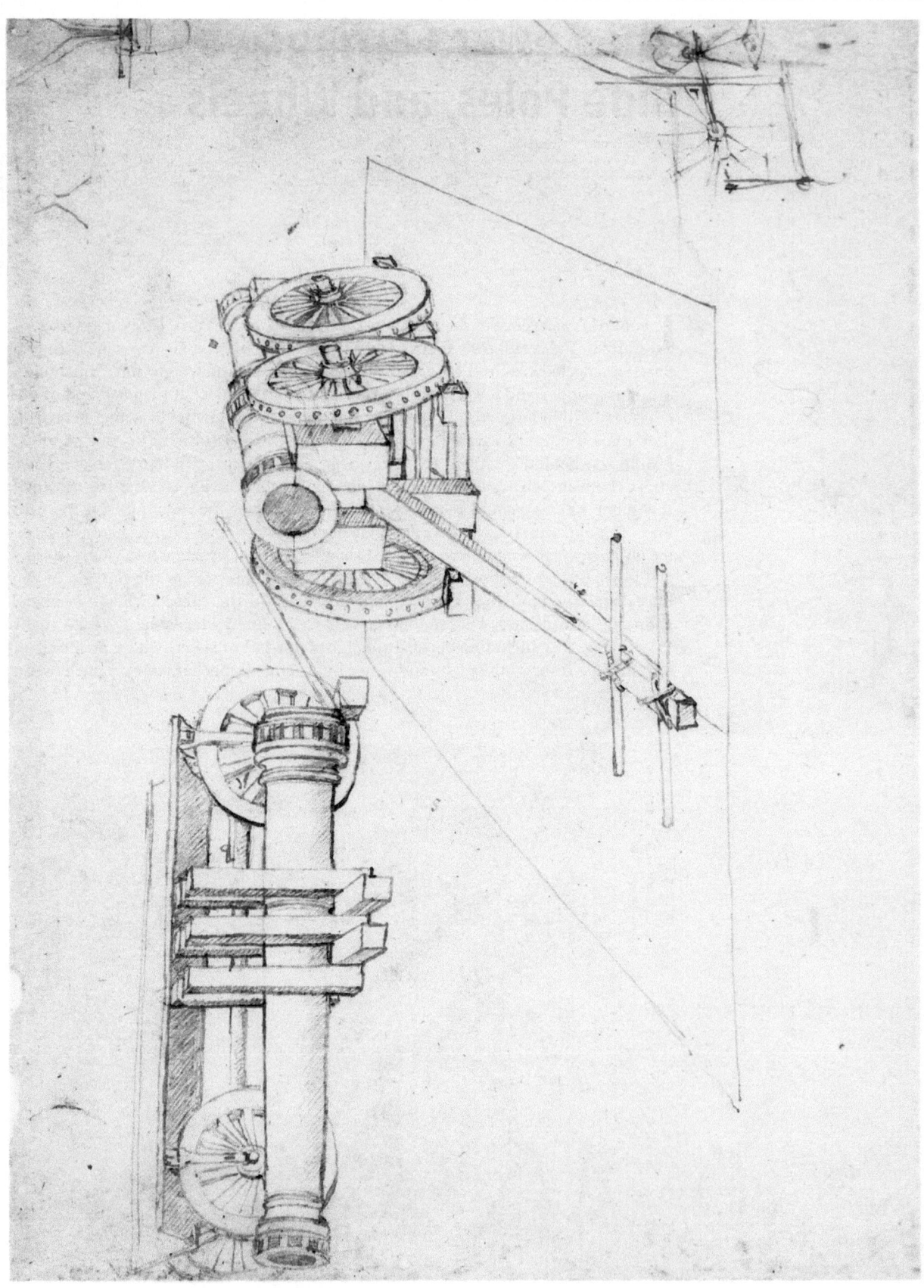

◀ **Fig. 6.2.**
Leonardo da Vinci, Cannon and carriages, Codex Atlanticus f. 154br [prev. 55vb], Biblioteca Ambrosiana, Milan

the worm-screw mechanism on the right illustrate isolated, exploded views of mechanisms embedded at the centre of the upper carriage. Based on the exploded view of worm screw and capstan on the right, capstan bars to the left of the lower carriage should instead extend from the upper carriage, if the worm screw is to meet the giant gear appropriately. This mistaken location of the capstan bars on the lower carriage suggests that CA149br was possibly recognised by Leonardo as a flawed design. More on the features of the lower carriage continues below, along with a discussion of side poles and wheels. At issue is the legibility of the *Giant Crossbow* as a diagram, especially in terms of the proportional relationships between its illustrated sections. With these proportional comparisons, one sees the extent to which the diagram functions as a viable theory or proposal.

This study will also demonstrate the hypothesis that Leonardo produced the earliest known dimetric orthographic projections of scaled designs of military engineering projects. For the closest historical parallel to this kind of design, one would have to examine the architectural studies of Giuliano da Sangallo (c. 1445–1516) and possibly Antonio da Sangallo the Elder (c. 1455–1534). Leonardo met Giuliano in Milan or Vigevano around 1492, and the two corresponded considerably thereafter.[1] It is unlikely that one could overestimate the extent of engineering advice mutually shared by Leonardo, Giuliano, Bramante, and Francesco di Giorgio Martini in Milan and elsewhere. In any event, Leonardo's invention of the orthogonal, scaled design seems to pre-date his association with Giuliano and Francesco di Giorgio by approximately five years. Moreover, Leonardo's detailed attention to scale on CA 149br distinguishes his drawing from those of Francesco di Giorgio (especially the latter's *Trattato I*, MS Ashb. 361, f. 37v-44v). The present study will deal primarily with Leonardo's drawings, though there is still much to reconsider with regard to his work in the context of the extensive 15th century engineering treatise and notebook tradition. The best examples include the codices of Taccola, Giovanni Fontana, Guido da Vigevano, Conrad Kyeser, Paolo Santini (copies of Taccola), Antonio Averlino detto Filarete, Lorenzo and Buonaccorso Ghiberti, Filippo Brunelleschi, Donatello, anonymous Sienese engineers, and manuscript illustrations of Vegetius, Vitruvius, and other ancient Roman sources. Indeed, as part of this tradition, Leonardo, Bramante, and Giuliano da Sangallo continued a relatively intellectual trend of research to measure ancient buildings in their spare time.[2]

---

[1] See: Solmi, *Le Fonti*, pp 260–261. For a recent, detailed summary of the available resources that place Leonardo and Giuliano in the same locations between 1492 and 1515, see: D. M. Budd (2002) *Leonardo da Vinci through Milan: Studies on the documentary evidence*, PhD Thesis, Columbia University, pp 432–449. Ludovico Sforza, Leonardo, and Bramante were supposedly in Vigevano in October 1492, at the time of a letter from Piero di Lorenzo de'Medici announcing the agreement to send "Giuliano ingegnero" with a wooden model of the Medici family villa at Poggio a Cajano. See: L. H. Heydenreich (transcr) (1977) *Giuliano da Sangallo in Vigevano: ein neues Dokument*, Cologny, Sammlung Giannalisa Feltrinelli, in: In: Ciardi Dupré Dal Poggetto MG, Dal Poggetto P (eds) *Scritti di storia dell'arte in onore di Ugo Procacci*, Electa, Milan, pp 321–323.

[2] The practise of recording the measurements and proportions of ancient buildings, especially those in Rome, was not necessarily a utilitarian effort in the 15th century. This was the intellectual research and cataloguing of ancient evidence by artist/engineers such as Brunelleschi, Donatello, Sangallo, Bramante, Leonardo and others. Leonardo's measurements of San Paolo fuori le mure in Rome are noted on CA 471r (172ra–vb). Sangallo studied San Paolo's column bases, as noted in Codice Barberiniano, f. 15. Vasari notes Bramante's interests in taking time off work in favour of measuring ancient buildings when he and Leonardo left Milan in 1499. See: Creighton Gilbert (1983) *Bramante on the road to Rome (with some Leonardo sketches in his pocket*, Arte Lombarda 66:5. Tuccio Manetti's *Life of Brunelleschi* notes the interests of Brunelleschi and Donatello in making excavations and measurements of Roman buildings. See: Isabelle Hyman (1974) *Brunelleschi in Perspective*, Prentice-Hall, Englewood Cliffs, NJ, pp 68–69; and see: Budd, *Leonardo*, pp 448–449.

Lower Carriage

Because the lower carriage's oblique projection is at a 60° angle to the picture plane, it appears to be slightly longer than the armature's width. In fact, however, the measurement on the drawing of the orthogonal illustration of the lower carriage is shorter than the armature width. Leonardo could have opened a divider so that the ends would have reached each side of the armature's ropes and then he could have rotated the divider so that one end touched the rear end of the lower carriage and the other end extended in the direction of the planned front end of the lower carriage. Figure 6.3 illustrates the diagonal lines that start at the upper carriage's rear hinge and extend toward the upper left of the sheet. The possible reasons for this line's placement and its measurements will be part of the discussion on proportional page formatting further below. It may be relevant that the estimation of the lower carriage's length and location on the drawing bears some relation to the 42 *braccia* measure between the armature's arms and the previous notes on CA 147av and 147bv specifying armature's 42 *braccia* in width. The estimation of the lower carriage's length appears to have followed the estimation of the armature's width.

Leonardo notes that the lower carriage is "2 *braccia* wide and 40 long," which would be just over a metre wide (1.17 m) and nearly 24 metres long if it were built to full scale. A problem with these written specifications, however, is that they do not agree with the illustrated dimensions in the drawing. First, measurements of the illustration show that the lower carriage and upper carriage are the same width at the rear of the upper carriage as at the front. There is no consistent diminution of scale between the foreground and the background. The carriage is therefore an oblique projection, rather than perspective construction. Possibly added for a builder's convenience, the comparable dimensions at both ends of the upper carriage could enable someone to easily measure the portions of this design, which is a 15th century version of a machine schematic, plan, or blueprint. Second, if the illustrated width of the rear of the lower carriage represents a unit of two *braccia*, 21 of those units, placed side to side, would exceed the illustrated width of the armature. If the illustrated armature is to measure 42 *braccia* in width as stated on three crossbow studies (147av, 147bv, 149br), then the illustration at the end of the lower carriage is much wider than a ½₁ (2 *braccia*) unit of the armature's width. Third, if one were to place two ends of a divider at the two top corners of the lower carriage's rear end, and then rotate the right side of the divider up to the left, along the 60° angle line of the lower carriage's upper left edge, then 21 equal portions of that diagonal two *braccia* measurement along the 60° angle line would extend well beyond the front end of the illustrated crossbow. As noted in Appendix A, the carriage's width would only be consistent with the dimensions of its length and the armature width if it were 66% the size of the illustrated example. Most primary sections of the crossbow measure approximately 0.54 cm per *braccio*. The carriage width at the back end is 1.62 cm, therefore measuring 0.81 cm per *braccio*, according to Leonardo's written specifications that this portion equal 2 *braccia*. A *braccio* of 0.54 cm is approximately 66% of the illustrated 0.81 cm *braccio*. If the carriage width were illustrated at the appropriate scale—consistent with the scale of its length and the armature's width—then it would represent a three *braccia* width, since it measures 1.62 cm, three times the 0.54 cm module of the carriage length and armature width. Nonetheless a proposed two *braccia* wide carriage measuring 1.62 cm in a drawing is at a scale of 72 : 1 to the drawing. Although the back end of the carriage is slightly out of proportion with the rest of the crossbow drawing, the front and mid sections of the carriage width measure 1.23 cm, which is consistent with the 80 : 1 scale of the carriage "thickness," or side dimensions.

The horizontal metal stylus incision illustrated on Fig. 6.4 that marks the (22.7 cm) distance between each armature ends is just 2.7 cm longer than metal stylus incision marking the (20 cm) length of the carriage. I note the locations and measurements of these metal stylus incisions on Fig. 6.4. There are no pinholes or indented spots made

Fig. 6.3. ▶
Superimposition of black lines over CA 149br, illustrating the carriage proportions

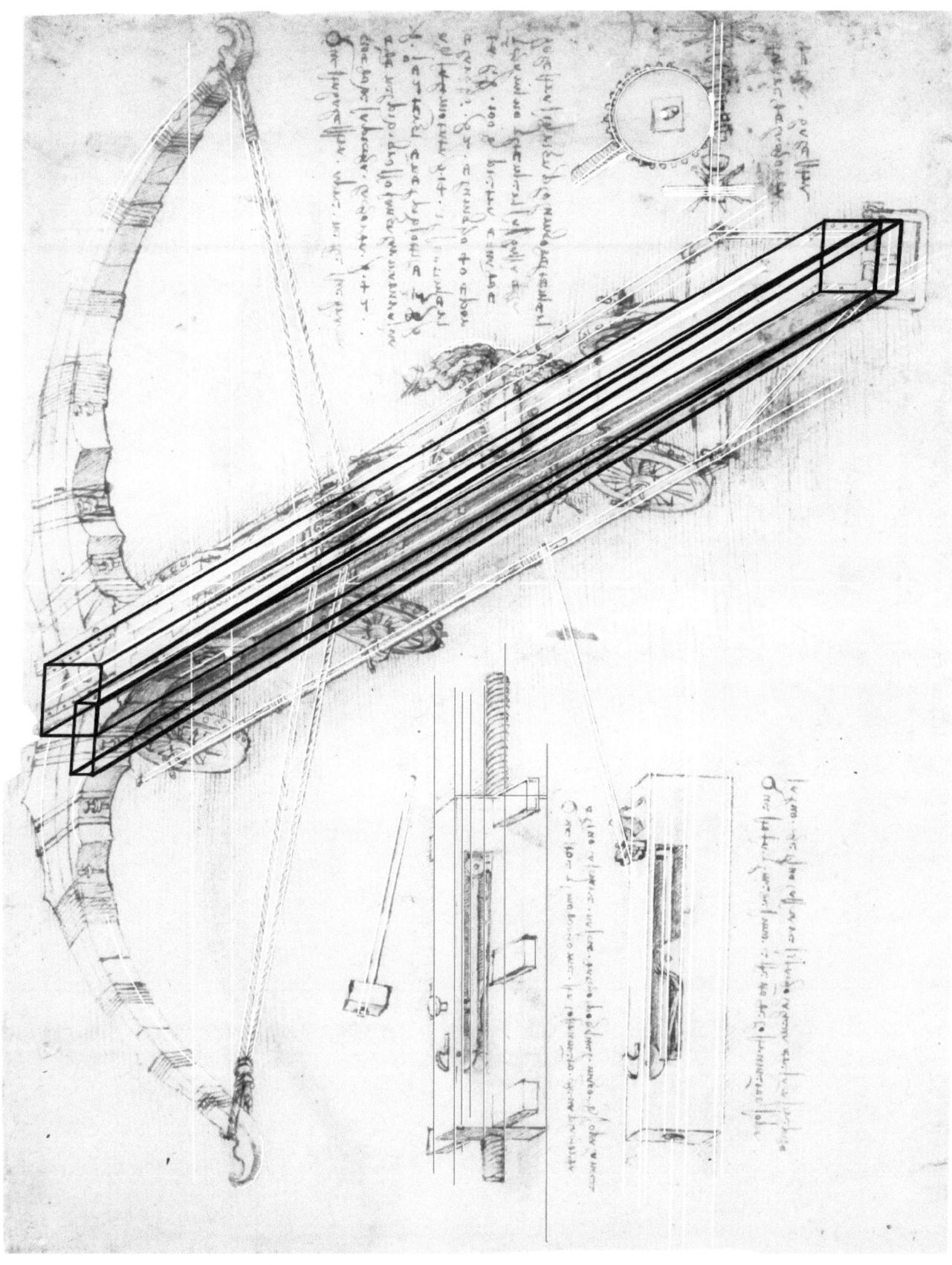

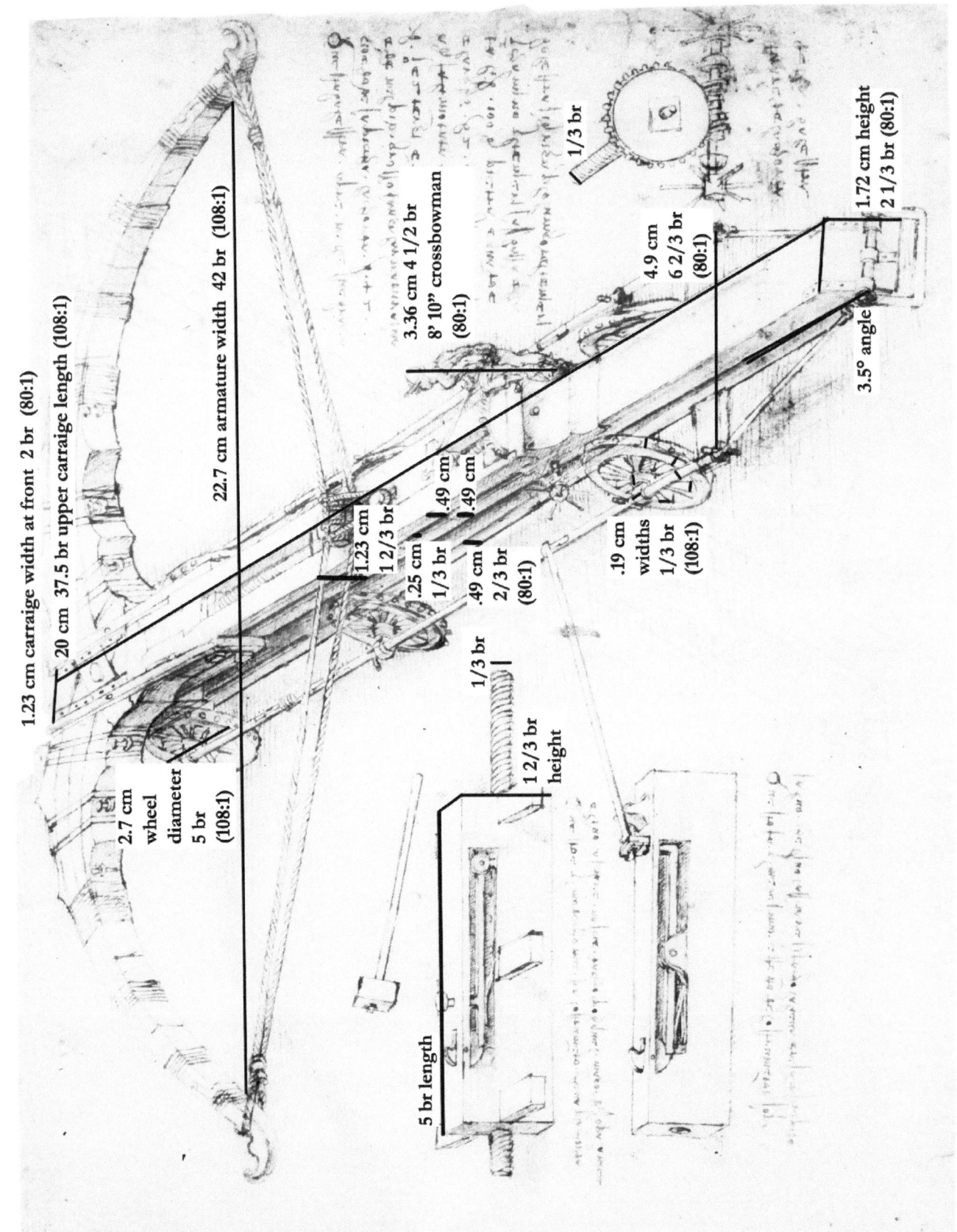

1.23 cm carraige width at front 2 br (80:1)

20 cm 37.5 br upper carraige length (108:1)

22.7 cm armature width 42 br (108:1)

1/3 br

3.36 cm 4 1/2 br
8' 10" crossbowman
(80:1)

4.9 cm
6 2/3 br
(80:1)

1.72 cm height
2 1/3 br (80:1)

3.5° angle

1.23 cm
12/3 br

.25 cm
1/3 br

.49 cm
2/3 br
(80:1)

.49 cm

.49 cm

.19 cm
widths
1/3 br
(108:1)

2.7 cm
wheel
diameter
5 br
(108:1)

1/3 br

12/3 br
height

5 br length

with a metal stylus tip near the ends of these two lines on the folio. Thus it would appear that a straight edge determined the positions of both metal stylus lines, possibly with the help of a protractor or a similar tool for determining the 60° angle of the second line along top edges of the carriage. It is possible that the 20 cm line along the carriage was intentionally shorter than the 22.7 cm line between armature ends, in order to agree generally with the 42 br width and 40 br length specifications at the right of the sheet. Although the crossbow illustration preceded the written specifications on the right (since the statement is wedged into its space, almost framing one side of the crossbowman), prior consideration of the general crossbow dimensions possibly determined the initial estimates of 42 br and 40 br metal stylus line-lengths. This is a necessary distinction if one is to consider the few areas in the illustration that might agree with the written specifications. Furthermore, some of these agreements between text and image prove that Leonardo produced a design that one could consider building to scale. Setting aside the usual scepticism that the life-size version of this design might not work, at issue in the present study is whether Ludovico Sforza's engineers could understand any agreement between the illustrated and written specifications of the crossbow design.

For example, if one of Ludovico's engineers had compared the measurement of the armature width with a measurement of the carriage length he would have found the upper carriage length to be shorter than the estimated 40 *braccia*. How would he or Leonardo explain this slight discrepancy? One could argue that the front end of the lower carriage extends below the armature to a length of 40 br. Beneath the armature extends a diagonal portion of the front end of the left-side wheel support pole. As illustrated at the back of the carriage, the diagonal pole extends from the side pole over to the end of the lower carriage. The armature covers the lower carriage extension at the front end. Assuming that the front-end diagonal side pole attaches itself to the front end of the carriage in the same way that it attaches at the rear end, the lower carriage would extend approximately 2⅔ *braccia* beyond the estimated length of metal stylus line visible along the upper carriage, as described above. That metal stylus line, compared with the width of the spanned armature, would measure approximately 37⅓ *braccia*. It is therefore possible that the lower carriage would appropriately measure 40 *braccia*, if one adds the upper carriage length of 37⅓ *braccia* to the lower carriage extension of 2⅔ *braccia*. The lower carriage length and spanned armature width thus appear to be of a similar scale in the drawing and according to the text. Measurements indicate that this width and length are at a scale of ¹⁄₁₀₈, as noted in Appendix A (see Figs. 6.4 and 8.2). If the crossbow design were proportional throughout, then one could build each of its sections 108 times larger than the illustration in order to produce a life-size example of the crossbow. A comparison of the written specifications and the illustrated dimensions, however, reveals various inconsistencies in the design's proportions. Nonetheless, a scale of ¹⁄₁₀₈ is a convenient because 108 is divisible by 3 and 36, which means that—since components of the crossbow tend to be measured in ⅓ proportions—it can be easier to calculate any measured adjustments to one component, and then apply that adjustment equally to the other proportional modules. The scale and proportions of the crossbow can therefore be adjusted with a unit of measure that is a proportional module of one third, rather than several specific numbers.

Some proportional inconsistencies are not unexpected for this kind of mechanical illustration, and there is no evidence among Leonardo's other drawings of a rigorous approach to mathematical accuracy in drawings with basic geometrical elements. His proportional use of third-portions, as will be demonstrated, did not require exact mathematical or geometrical accuracy. Although a study of the methods of other painters will not be possible in the present study, it is worth noting that his proportional uses of third portions was a common technique among 15th century painters. Examples of this use of third portions, instead of rigorous mathematical or geometrical approaches, have been identified in the following works: Piero della Francesca's *Baptism of Christ* and *Flagellation*, Masaccio's *Trinity*, Masolino's *Disputation of St. Catherine*, Donatello's *Feast of Harod*, and Domenico Veneziano's *Carnesecchi Tabernacle* and his *Annunciation*.[3]

Figure 4.2 displays the lower carriage's extension. An illustration of this kind of extension appears at the top left of CA 147bv (Fig. 1.2). On the latter drawing, however, the front wheels extend well beyond the front end of the upper carriage, presumably to keep the heavy armature from tipping the crossbow forward when the upper carriage is elevated to higher positions.

On CA 149br, at least two possible reasons for its front-end extension are obvious. First, to keep the bow from tipping forward onto its heavy armature, a prop or mound of soil formed beneath the 2⅔ *braccia* extension would hold the front end up, especially when one elevates the upper carriage. Additionally, if a hoist were to lift the upper carriage's back end in order to transport or reposition the bow, the front-end extension would help brace or even pivot the crossbow during the manoeuvre. If the armature "has an elevation [or draw] of 14 *braccia*" (8.17 m, 26.80 ft), as stated by Leonardo on CA 149br, the U-bolt on the carriage's rear hinge would be the only portion of the bow attaching it to something heavy enough to keep the bow from tipping forward when it is elevated.

As discussed further below with regard to the upper carriage, the angle of the upper carriage, with respect to the lower carriage, at this 14 *braccia* elevation would be around 16° at the rear hinge.[4] Figure 6.5 illustrates this angle and Appendix A notes the calculation for this extension. The point here is that the proposed height, along with a look at the upper carriage angle (with the help of modern trigonometry), shows that the crossbow would tip forward with a heavy armature at that height. This is not to suggest that Leonardo understood trigonometry, or that he had a way to mathematically calculate the upper carriage angle with respect to the lower carriage. The determination of the angle of elevation is a way of checking the likelihood that the crossbow would tip if elevated 14 *braccia*. No evidence exists of the form of guesswork necessary for his estimation of this relationship between the upper carriage elevation and the tipping of the front end.

The exact purpose of the rear U-bolt is not obvious in the design, though it would be useful as a hitch or a tow bar attaching a hoist or a munitions cart. The U-bolt is,

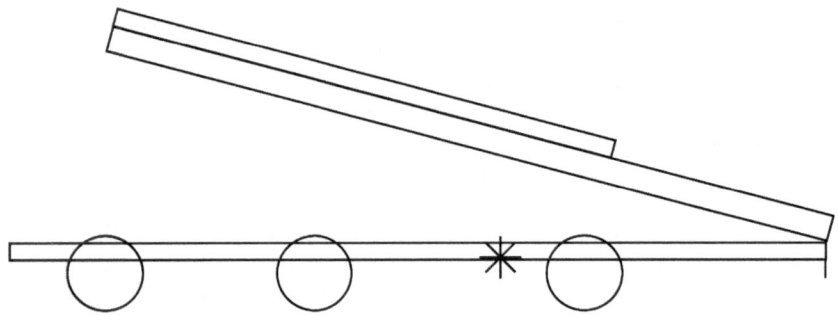

Fig. 6.5.
Diagram of the appearance of the upper carriage at the elevation of fourteen *braccia*, at an angle of approximately 16°

---

[3] James Elkins sums up some of the research on this problem in: J. Elkins (1991) *The case against surface geometry*, Art History 14(2) June:143–174. He discusses the inconsistent uses of proportions, modules and third portions, using these examples, as well as Leonardo's *Last Supper*.

[4] The $\sin \theta$ equals height—between upper and lower carriage, divided by the upper carriage length (hypotenuse). To determine the distance between the upper carriage bottom and lower carriage top when the upper carriage is elevated 14 *braccia*, the following formula applies:

14 br – (2⅓ br distance between the ground and the cart +⅓ br axle thickness +⅔ br cart thickness)
= 10⅔ br

This 10⅔ br distance, divided by the 37.5 *braccia* upper carriage length (the hypotenuse) equals
= 0.284 $\sin \theta$ of 16°.

Fig. 6.6. Albrecht Dürer, *Landscape with Cannon*, Metropolitan Museum of Art, New York

however, more of a hitch than a tow bar because of its rectangular shape. Note, for example, the triangular form of tow bar on Dürer's engraving, *Landscape with Cannon* (Fig. 6.6). According to CA 147av [52va] (Fig. I.8), Leonardo designed the earlier giant crossbows so that oxen or horses could pull them forward at the front end. In this way, the crossbow would be in a forward position as it approached its target. This machine does not appear to be a long-range weapon—because it shoots a massive 100-pound stone, so it would have to get into position quickly. To lift the crossbow at the back end and deliver it to the battlefield in the backwards position would require a quick 180° turn of the entire crossbow before it would be useful. Dürer's *Landscape with Cannon* illustrates the traditional form of cannon cart, with a tow bar for transportation at the back end. Since the rear end of the crossbow extends approximately 12⅔ *braccia* from the rear axle and the front end extends 2⅔ *braccia* from the front axle, the rear end is more likely to drag the ground if the front end were lifted. Hence, the crossbow steering, and the ability to haul it long distances might depend on the use of the rear hitch, which oxen could pull forward during a siege.

The second reason for the extension at the front end of the lower carriage is that the side poles supporting the wheels would necessarily extend, with short linking poles, to the front and rear portions of the lower carriage. Thus, there has to be ample room between the poles' attachments at the sides of wheels and the ends of the lower carriage. On CA 149br, at the front of the lower carriage one can see the left wheel's side pole extend to a joint, where another portion of the pole extends under the armature at nearly a 45° angle to the picture plane. This joint and extension nearly equal in size the same joints and extensions visible at the rear of the lower carriage, at the bottom of CA 149br. Examples on CA 147av (Fig. I.8) and 147bv (Fig. 1.2) show Leonardo's early interest in adding a set of wheels to the front end of the lower carriage in order to support the heavy armature. Vertical and lateral support for the armature include—on CA 147av and 149br, side poles securing the wheel axles to the lower carriage.

Extending the lower carriage 2⅔ *braccia* beyond the front wheel would provide essential support for the side poles. If the short poles that link the side poles to the ends of the carriage were removed, lateral support would depend on the rectangular crossing beams located before the front wheels and behind the back wheels. If built to scale, those crossing beams and the side poles, as depicted in the drawing, would not hold such weight themselves, at least not without bending considerably. The short poles that link the ends of the long side poles to the ends of the lower carriage would redirect the outward forces imposed by the wheels in toward the lower carriage and its ends. This redirection of the outward weight on the side poles in toward the lower carriage would take some of the strain off the rectangular crossing beams. The triangular format of crossing beam, the short connecting poles at the end, and the lower carriage would conceivably provide strong, stable, efficient, and lightweight support against thousands of pounds of potential outward pressure on the side poles.

The Wheels

Three axles attached to the lower carriage hold three pairs of large wheels. Leonardo has managed to place the central and rear wheels at relatively equal distances from each end of the lower carriage. Because the carriage is an oblique projection, rather than perspective construction, the scale of the front end does not diminish in size, and the sets of wheels are not further along the length of the carriage than they appear to be in the drawing. If the crossbow were built to scale, the distances between the lower carriage ends and the central and rear wheel axles would be approximately 12⅔ *braccia* each, as noted on Appendix A. Moreover, the distance between the central and rear wheel axles would be around 12⅔ *braccia*. Thus, central and rear wheels are equidistant from one another and from each end of the cart. This distance happens to be nearly ⅓ the total length of the carriage, a third-portion division consistent with similar divisions in the rest of the crossbow design.

The approximate diameter of each wheel is 5 *braccia* (2.92 m, 9.57 ft), for reasons discussed further below in the section dealing with the upper carriage. A comparison of the illustrated ⅔ *braccio* thicknesses of the carriage sections, and the illustrated ⅓ *braccio* portions of each wheel section, suggests that the wheel sections above and below the hub are roughly 2⅓ *braccia* each, and that the axle diameter is around ⅓ *braccio*. The sum of these portions is 5 *braccia* (2⅓ + 2⅓ + ⅓), as noted on Fig. 6.4 and in Appendix A.

The wheels tilt in toward the body of the crossbow, at an angle of around 72°, according to their illustrated angles as well as their calculated dimensions noted on Appendix A. Using Adobe Photoshop® software with a scan of CA149br, one can find with the "measure tool" the angle of each wheel in the drawing. The illustrated angles range from 72° to 74°. These angles also compare with the angles of calculated dimensions of the wheels, which range between 69° and 74°. The illustrated distance of each wheel from the carriage is approximately one *braccio*, although the amount of clearance between the wheel rim and the carriage is difficult to detect in the drawing. For a five-*braccia* diameter wheel, the height of which diminishes to 4⅔ br when tilted, a safe distance of the rim from the carriage would be around ⅓ br, or 7½ inch. If the half wheel height is 2½ br, and half the tilted wheel height is 2⅓ br, and the distance along the axle between the inner spokes and the rim is ⅔ br, then the wheel has a tangent angle of 74°, a cosine angle of 74°, and a sine angle of 69°.[5] The average of these angles is 72°. Because the hub for each wheel is approximately one *braccio* wide (⅓ br outer portion, ⅓ br portion holding the spokes, ⅓ br inner portion), the hole cut at its centre

---

[5] The hypotenuse (*h*) is 2½ br, the opposite side (*o*) is 2⅓ br, and the adjacent side (*a*) ⅔ br. The tangent of a 73.96° angle is 3.48 (*o*/*a*). The cosine of a 74.45° angle is 0.268 (*a*/*h*). The sine of a 68.75° angle is 0.932 (*o*/*h*). See: Appendix A.

would be at a 72° angle to the axle. Additionally, the wheel axles would probably require more lubrication than the axles for normal perpendicular wheels.[6] Virgil writes that heavy carriages have "screeching axles" and he may thereby refer to ancient tendencies to let the axles wear down without lubrication.[7] Leonardo illustrates similar canted wheels on CA 154br [55vb] (c. 1485), though they tilt at less of an angle. Such canted wheels might have been common on late 15th century cannon carriages.

There are two possible reasons for the illustration of tilted wheels. First, Leonardo would have known that an old carriage wheel tilts when the continued friction between the axle and wheel rim wear down the wooden or iron materials between them. It is unlikely, however, that he was interested in showing the effects of age on a new machine design.

Second, canted wheels would support the crossbow in the event of lateral movement. The bow's recoil, when shot, could pull the lower carriage to one side or to both sides. The draw strength of each side of a giant, complex compound armature would be somewhat disproportionate at all times; and the canted wheel and side bar support would prevent the likely structural warp produced when several tons of draw tension differ slightly at each side of the massive armature. Periodically, the armature would require a structural adjustment in order to recalibrate the direction of the draw tension. Due to the size and complexity of the armature, however, this recalibration would not be an exact science. Presumably, to help solve this problem, canted wheels and taught side poles would help stabilise the massive structure. The side poles are also convenient handles with which soldiers could move the carriage. These poles would also provide structural support for the carriage if shafts were attached at the front end and the crossbow were pulled by horses, as illustrated on CA 147av [52va] (Fig. I.8). According to the order of the giant crossbow drawings, the idea of the side poles on CA147av preceded the idea of the canted wheels on CA 149br, since one can see straight wheels on the former drawing, as well as on 147bv [52vb] (Fig. 1.2) and 57v–69ar (Fig. 2.4). This ingenious combination of side pole supports and canted wheels appears be the first invention of its kind.[8]

Held to an axle with a stub in the form of a cylinder, the canted wheel hubs appear to be relatively thick in order to hold the downward and outward forces imposed on them. This might also be the reason that Leonardo has drawn such large wheels. At an estimated full-scale measurement of nine feet, six inches in height, these wheels would be impressive portions of the design. Each section of a wheel and axle appear to represent ⅓ of a *braccio* (19 cm, 7½ inch), including the thickness of the wheel, width of the felloe (rim), outer width of the hub, spoke width and thickness near the felloe, stub ring width, and the axle thickness. In this case, the total width of the hub would be three times the ⅓ br module, if the outer hub width is ⅓ br, the spoke width is ⅓ br and the inner hub width is ⅓ br. The total hub width would therefore measure a *braccio* (58.36 cm, 1.91 ft) at full scale. An engineer of the Sforza Court would probably recommend that an iron bar be used for the axle, because of the expected pinch of the canted wheel hub.

The most likely reason for the canted wheels would be that they could sit in that position especially at the time of the crossbow's operation. A logical option would be for the central and rear sets of wheels to provide a form of quadruped stand in conjunction with the front wheels acting as a pivot. The sidebar reinforcements would be

---

[6] See: H. A. Harris (1974) *Lubrication in antiquity*, Greece and Rome XXI/1 (April):32–36. Landels, pp 180–182.

[7] Landels, p 181.

[8] This invention does not appear to be in any publication to date (2004). For an early list of inventions in the fields of applied mechanics, hydraulics, and military engineering, see: Abbott Payson Usher (1929) *A history of mechanical inventions*, Harvard, Cambridge, MA, pp 222–224. William Parsons (c. 1939/1968) *Engineers and engineering in the Renaissance*, M.I.T. Press, Cambridge, Mass. See also: Cooper, *Inventions of Leonardo*; Gibbs-Smith, *Inventions of Leonardo*; Cianchi, *Machines*; Galluzzi, *Art of Invention*; Pedretti, *Leonardo, The Machines*.

able to hold these wheels in place. The rear carriage could be lifted with a hoist attached to the rear U-bolt.

This U-bolt, extending from the rear hinge, could not reach the ground if built to scale. Charles Gibbs-Smith states that the lower carriage's design has "a hinged tailpiece that could be driven into the ground to hold the bow against recoil."[9] This is not possible if the hinge only appears to drop a distance of ⅓ *braccio*, below a carriage that is 2⅔ above the ground. As noted in the comparison of image measurements and written specification of Appendix A, the rear end of the carriage appears to be much larger than its two *braccia* width, measuring 1.62 cm wide by 1.73 tall. According to those measurements, the illustration of the back end is at least ⅓ of its total size too large, compared to the other measurements on the sheet. For example, the representation of a *braccio* at the back end is 0.81 cm, whereas the representation of the average *braccio* is 0.54 cm for a recommended 42-*braccia* armature width measuring 22.7 cm on paper. The average illustrated *braccio* width of 0.54 cm at the armature is exactly ⅔ of the average *braccio* width of 0.81 cm illustrated at the back end of the carriage. Since the U-bolt extends 0.29 cm from the bottom of the carriage, it is reasonable to suggest that its measurement is ⅓ of its total size too large in the drawing, given the over-size representation of the carriage back end. Thus, the ⅓ *braccio* length between the bottom of the U-bolt and the carriage bottom would measure 0.19 cm, if its measurements were consistent with measurements for the width of the armature, or the lower carriage length, or the wheel dimensions. All of these measurements represent an average of 0.54 cm per illustrated *braccio*. The 0.19 cm illustrated representation of a ⅓ *braccio* is ⅔ the length of the 29 mm measurement between the carriage and the lower inside edge of the U-bolt. At full scale, position of this U-bolt would be 2⅓ *braccia* above the ground. Its purpose is likely to provide a wide grip for a hoist, crane, and/or a munitions cart.

According to the design, it might be possible to re-aim the crossbow relatively quickly using a simple hoist to lift and move its rear portion, along with rear and central wheels, pivoting the carriage on the two front wheels. To straighten the front wheels in preparation for the cart's transportation, it might also be possible to lift the front end with a hoist, fitting a U-bolt or clamp between the stub rings and the wheel hubs. In this case, the weight of the armature would cantilever the large proportion of the weight of the bow toward the front end, making the rear end relatively easy to lift. This is the same cantilever principle used with the wheelbarrow. In Milan, Leonardo had access to manuscripts that might have contained drawings of great crossbows mounted on large wheelbarrow style lower carriages, used and illustrated in 15th century Germany (as illustrated in Fig. 3.5b).[10] As discussed above in the section on "Leonardo's sources", he made use of his unusually privileged access to libraries—especially in Pavia and Milan, containing manuscripts on arms and military engineering in Germany and France.[11] Having studied early manuscript sources, he could have modelled his *Giant Crossbow* on the German wheelbarrow crossbow design. Dürer's 1518 print of a *Landscape with a Cannon* (Fig. 6.6) offers perhaps the a good example of the German model of a well-crafted cannon carriage that could have influenced Leonardo's designs. It has rear wheels on a swivel, which would allow the quick manoeuvrability of the cannon lower carriage. At around the same time as the *Giant Crossbow* study, Leonardo had drawn this kind of swivelling transport carriage at the bottom left of his giant cannon foundry drawing, W 12647r.

---

[9] Gibbs-Smith, *Inventions*, p 35.

[10] The illustration of Fig. 3.5b is of drawing from the 15th century, MS lat. 599 f. 34v, at the Bayerische Staatsbibliothek, Munich. See also: Liebel, pp 34–36.

[11] Regarding the large Ducal library at Pavia, see: M. G. Albertini Ottenghi (1991) *La biblioteca dei Visconti e degli Sforza: gli inventari del 1488 e del 1490*, Studi Petrarcheschi n.s. 8:1–238. For further information, see: S. Cerrini (1991) *Libri dei Visconti-Sforza; Schede per una nuova edizione degli inventari*, Studi Petrarcheschi n.s. 8:239–282; and E. Pellegrin (1955) *La Bibliothèque des Visconti des Sforza Ducs de Milan au XV' Siecle*, Paris; and G. d'Adda (1875) *Indagini storiche sulla Libreria Visconteo-Sforzesca*, Milano.

As the crossbow's description states on CA 149br, "when it is moving, the carriage lowers itself and the crossbow extends itself along the length of the carriage." This suggests that it would be possible to "lower" the upper carriage into the position illustrated on CA 149br, just above the front wheels. When ready for siege warfare, a hoist would raise the upper carriage to as much as "14 braccia" (8.17 m, or 26 ft 10 inch) and place it on blocks. A heavy lower carriage used to haul the crossbow and to anchor it could hold extra ammunition as ballast. The set-up procedure would require a hoist, crane, and/or wedges in order to lift the upper carriage to the desired height. At the start of this set-up, the back end U-bolt could be attached to a munitions cart or a similar heavy weight. The canted wheels could hold the carriage in place in the event of movement from side to side. The lynch pins visible on the side poles, just next to the stub rings and crossing beams would be there to prevent the forward of backward shift of side poles and wheel axles. If the crossbow moved during set-up, or if the armature happened to jerk the machine to one side, the entire structure would not warp, thanks to those taught side poles.

Another feature contributing to the stability of the crossbow is the series of bolt heads along the wheel rims. These bolts or studs are clearly visible on the wheels in the drawing. Ideal for traction and for holding metal rims on carriage wheels, the bolts were often necessary on 15th century cannon carriage wheels. The Edinburgh Castle's "Mons Meg" cannon has these bolts on its wheel rims. Built in Belgium in 1449, it could hurl a stone ball of 150 kilos nearly 260 metres.[12] The bolts help the wheels grip the ground, keeping them from accidentally rolling forward or backward on hard surfaces, and from slipping from side to side while in transit. If one had to push the wheel spokes forward, in order to assist draught animals pulling the carriage through the mud, the bolts on the rim would help with traction. The bolts are visible in Durer's *Landscape with a Cannon* (Fig. 6.6). In the case of Leonardo's *Giant Crossbow*, the stability of its wheels would be essential to the lift and operation of the armature on the upper carriage.

Wheel rim bolts were especially helpful on hard surfaces, such as on the stone floors of fortifications. Great crossbows and cannon were more effective when stationed atop fortifications, hills, or earthworks, from which they could launch missiles along low, and therefore more accurate, trajectories.[13] Whereas trebuchets could lob stones in general directions into and out of fortresses, the large fortress crossbows and springalds could take out the gunners of attacking siege engines with such accuracy as to significantly demoralise approaching troops. Medieval manuscripts and documents contain various illustrated and written accounts on this effectiveness of fortress crossbows against attacking siege engines. A good illustration of this use of a great crossbow atop a fortress—against besiegers with large siege engines, is on the City Charter of Carlisle, dated 1316, in which the city thwarts the Scottish siege of 1315.[14]

Most features illustrated on CA 149br—from the bolt heads on the wheels to the elevation and draw of the armature—appear to be essential components of the crossbow's design. Leonardo carefully considered the combination of features necessary for the machine, as well as the proportions of those features to the whole design. He thoughtfully designed each piece of the armature, each portion of the lower carriage, and every feature of the wheels. His design of the upper carriage was no less meticulous, a discussion of which continues in the next section.

---

[12] Robert D. Smith and Ruth Rhynas Brown (1981) *Bombards: Mons Meg and her sisters*, Royal Armouries Monograph 1, London. Gravett (1990) p 30.

[13] See for example Leonardo's studies for fortifications, including especially Codex Atlanticus 767r [282r] (c. 1500), which illustrates three alternative plans for a fortified wall with lines tracing of the extended trajectories of cannon and possibly crossbow shots.

[14] Nicolle (2002) *Medieval siege weapons*, p 7. Paul Chevedden has found some of the best accounts in Spanish medieval manuscripts of the devastating effects of great crossbows used in defence during sieges. Chevedden PE, Kagay DJ, Padilla PG (eds) (1996) The artillery of King James I the conqueror. Iberia and the Mediterranean world of the Middle Ages: Essays in honor of Robert I. Burns. E. J. Brill, Leiden, pp 47–94.

7

The Upper Carriage

The present section continues the discussion about the upper carriage and offers concluding remarks about the general structure of the crossbow. Of particular interest is the extent to which Leonardo has divided the *Giant Crossbow* and its upper carriage into proportional units of ⅓ and ⅔ portions.

Lines of text on the right side of CA 149br, from the third line through the eleventh line, refer to the upper portion of the "carriage", its "100 pounds of stone," and its winding screw and trigger mechanisms. The third and fourth lines refer to the crossbow "at its thickest, without its armature, 1 and 2 thirds *braccia*, and at its thinnest, ⅔ of a *braccio*." This can only refer to the thickness of the carriage without the armature attached. We know from the sixth line of text that the width of this carriage should measure "2 *braccia*" at full scale. The carriage thickness is therefore a vertical measurement rather than an estimate of the width.

In order to explain the relationship between the carriage thickness and the dimensions noted in the text, it is first necessary to mention that the carriage illustration is an oblique projection, rather than a perspective projection. As illustrated in Fig. 6.3, dimensions are similar at each end of the upper and lower carriage. Metal stylus incisions along the length of the upper carriage, noted in Fig. 8.20, reveal the extent to which Leonardo estimated its position. Tracing these lines, one can make several boxes within the oblique projection, all of which have slightly different dimensions at both ends. There is no indication, however, that the front ends of these boxes should be any smaller than the back ends. There is no significant diminution of scale from the foreground into the background, no perspective construction. Thus the viewer, and 15th century engineer, can see that the measurements at the back end of the design are consistent with the measurements at its front end. Due to the lack of a significant diminution in scale between the supposed foreground and background, a builder would not have to guess at a calculation of measurements at the front end. This is the benefit of using an oblique projection rather than a perspective projection. Leonardo hereby offers a design that is useful for an engineer or a builder, rather than a promotional display drawn in perspective of the crossbow's actual appearance.

Proportional Consistency

If the crossbow is "at its thickest… 1 and 2 thirds *braccia*, and at its thinnest, ⅔ of a *braccio*," then it would have an upper carriage with these measurements, or a combination of the upper and lower carriage with such dimensions. A comparison of measurements at the sides of the upper and lower carriage, however, suggests that Leonardo could be referring to upper carriage measurements as well as the combination of upper and lower carriage measurements. Compare for example the dimensions of the carriage side with the carriage wheels. As discussed further above, illustrations of the carriage wheels are in ⅓ *braccio* modules. The rear wheels appear to be slightly larger than the central and front wheels, although one can expect that the

⅓ *braccio* modules are consistently applicable to all sets of wheels, since the dimensions do not diminish between the base of the sheet and the top. The reference to the "thinnest" portion of the carriage possibly refers to the three carriage sections between the rear and central left wheels. These sections equal one another in thickness, representing the lower carriage, lower portion of the upper carriage, and upper portion of the upper carriage. If the felloe (rim) width of the centre wheel represents ⅓ *braccio*, twice that thickness equals approximately the thickness of the thickest carriage portions along the left side (Fig. 6.4). In other words, each of those carriage sections—the lower carriage, the lower portion of the upper carriage, and the upper portion of the upper carriage, represent "⅔ of a *braccio*." Any of these carriage sections could be the "thinnest" portion of the crossbow to which Leonardo refers.

Another way to check the carriage thickness would be to compare the felloe width of the rear wheel with the lower carriage thickness at its back end. Twice the rear felloe width equals approximately the lower carriage thickness at that end. The central and rear carriage measurements on the drawing differ by several millimetres, which is relatively easy to see without observing the exact drawing measurements (Fig. 6.4 and Appendix A). It is possible that Leonardo determined the upper edges of the carriage and its rear thickness—as one can see among the metal stylus incisions of Fig. 7.1—before determining the angle by which the upper carriage extends along the lower carriage (see Fig. 8.20, with the next phase of metal stylus incisions), or before determining the area needed for the trigger mechanism. Thus, he necessarily decreased the thickness of the carriage at its centre, making room for a two-part upper carriage and the proportional continuation of 3.5° angle of upper carriage incline (see Appendix A for the drawing measurements). Most probably, he initially misjudged the extra space needed for the top portion of the upper carriage, since the two-part upper carriage is missing in the preparatory designs on CA 147av and 147bv (Figs. I.8 and I.2). In any

Fig. 7.1. Superimposition of white lines over CA 149br, illustrating adjustments of the first phase of diagonal carriage metal stylus lines, with an outline of 30°/60°/90° triangle template

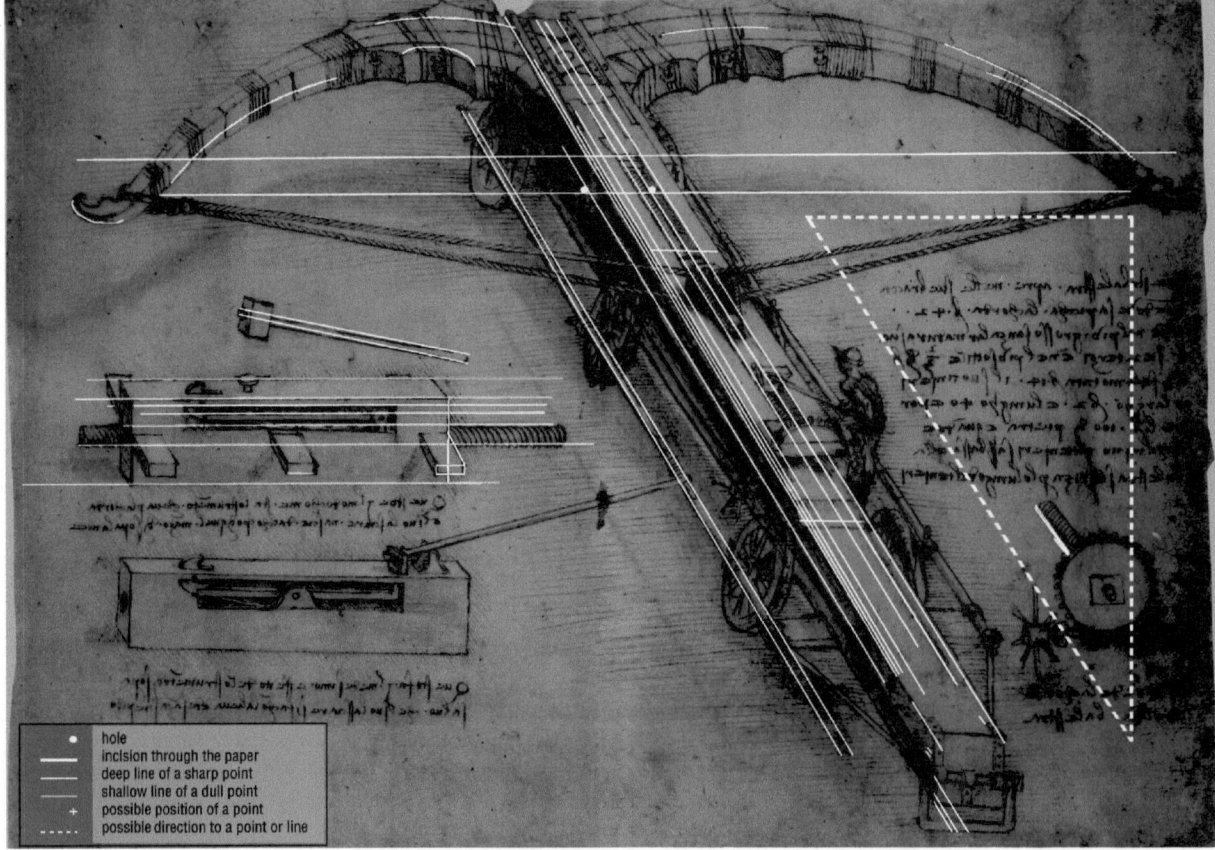

hole
incision through the paper
deep line of a sharp point
shallow line of a dull point
possible position of a point
possible direction to a point or line

event, proportional comparisons of the wheel portions and carriage side portions indicate that the latter sections are generally twice that of the former sections.

The measurements of these items on the drawing, however, differ by as much as 33% because of Leonardo's pictorial adjustments. Even if he had planned to make a drawing with a proportionally consistent set of dimensions throughout, there is no evidence that his adjustments to certain portions of the drawing conform to a particularly measured approach. His design is composed of proportional and pictorial modules, rather than specifically measured sections. Most of these modules are proportionally divisible by, or divisions of, one third, such as the third of a *braccio* that is a consistent module throughout the design. Leonardo also maintains a remarkable consistency to scale throughout the design, with only two minor scale variants from the norm. These minor variations include two measurements of the scale at the back end of 1/72, and seven measurements along the side that are at a scale of 1/80, whereas most of the crossbow is at a scale of 1/108 (see Appendix A). And as it happens, 108 is divisible by three (and 36). As noted elsewhere in the present study and in Appendix A, the illustrated *braccio* of the back end carriage width measures approximately 0.81 cm, whereas the illustrated *braccio* at the lower carriage near the worm screw measures 0.73 cm, and a *braccio* of the carriage width near the worm screw measures 0.54 cm. Therefore, the estimated scale of the drawing at the back end is roughly 1/72, and the scale of carriage thickness near the worm screw is around 1/80, and the scale of the armature, the wheels, and the carriage width near the worm screw is approximately 1/108 scale. The back end is 33% larger in scale than the armature, wheels and carriage width near the worm screw, though this is likely the result of alterations to the carriage: splitting it into two portions (upper and lower) and lifting the front end to its present angle.

With these estimates, one can see on Fig. 6.4 that upper and lower carriage sections at the centre of the carriage and at the back end are proportionally consistent, though they differ in scale by 33%. The lower carriage represents 2/3 *braccio* in thickness. Top and bottom sections of the upper carriage would each be 2/3 of a *braccio* thick, and the space between those sections would be 1/3 of a *braccio*. The total thickness of the upper carriage between rear and front wheels would be 1 2/3 *braccia*. At the back end, the thickness of the upper carriage, without the top portion—located between the rear and front wheels, would be one *braccio*. The total carriage thickness (with the addition of the 2/3 *braccio* lower carriage and without the 2/3 *braccio* top portion of the upper carriage) would be 1 2/3 *braccia*. These dimensions are proportionally consistent with Leonardo's statement that the crossbow "at its thickest, without its armature, 1 and 2 thirds *braccia*, and at its thinnest, 2/3 of a *braccio*." As one of the most sophisticated features of this drawing, this proportional consistency required some form of mechanical (divider and straight edge) or measuring tool with which to gauge the appropriate positions of the carriage lines.

The illustration of the crossbowman (*balestriere*) (Fig. 7.2) happens to be remarkably inconsistent with the scale of the other portions of the bow, as he would measure more than twice the normal height of a man. He stands almost directly above the worm screw (hidden under the upper carriage), ready to press down a bar that can release the nut of the trigger mechanism illustrated at the lower left of the sheet. He measures approximately 3.36 cm in height, which would represent 4 1/2 *braccia* at full scale, or about eight feet, ten inches.[1] If one takes into account that he bends forward

---

[1] Ivan Williams, Designer and Project Manager for ITN's *Giant Crossbow* reconstruction, told me that he used the crossbowman in the drawing as the standard for determining the bow's dimensions; at the filming on 22 June 2002. He made these preparations before I had a chance to tell him of Leonardo's written directions. ITN's reconstruction of the crossbow is therefore somewhat smaller than Leonardo's recommended design (since the bow is smaller than its recommended dimensions by comparison to the tall man). Leonardo likely knew that the man in his CA 149br drawing was not to scale and therefore would have required an engineer to compare the verbal size specifications for the dimensions with the drawing's visual portions of these dimensions. Thus, the visible crossbow upper carriage, with written specifications that it is two *braccia* in width, is only about half the illustrated height of the crossbowman in the drawing.

Fig. 7.2.
Detail of the central portion
of the upper carriage, near
the crossbowman, on Codex
Atlanticus f. 149br

slightly, then his illustrated height may be even taller. His location in the drawing makes his size comparable to that of the carriage "thickness," where an illustrated *braccio* is around 0.74 cm, or at 80th of full scale. If Leonardo expected the crossbowman to be six feet in height, then the man measures a 54th of full scale in the drawing. This is proof that he added the crossbowman to the drawing without careful attention to, or interest in, the body's scale or illustrated measurements.

The crossbowman's proportions, though unrelated to the crossbow's scale, are accurate only with respect to his demonstrated activity. This is apparent when one examines the way in which his figure is drawn on the upper carriage. There are only three metal stylus marks visible on the crossbowman, as noted in Fig. 8.20. Other metal stylus incisions may be covered with ink, and are therefore difficult to distinguish on the page. First, at the top of the helmet, toward the back of the head, a pinhole is at the centre of a black spot. This does not appear to be the locus of a significant position for a divider, as it appears to be unrelated to the rest of the design. It is relatively equidistant from the upper left corner of the trigger mechanism lever and the crossbowman's left boot heel, but it was not necessary for the placement of the crossbowman with respect to the trigger lever. The pinhole is also the centre of a circle that extends to two inside curves of tips of the inner armature, just to the right of the upper carriage. This may simply be a coincidence. As one can see in the diagram of white lines on Fig. 8.20 and black circles and dots on Fig. 7.3, however, the centres of circles that compose the armature do not correspond to any holes on the sheet. The pinhole on the crossbowman may be the result of a divider placed in the area to determine the initial proportional placement of the man.

Two metal stylus incisions along the outside of the crossbowman offer more information about the proportions of his figure. These lines show that this outer edge was an important early tracing for his position on the drawing. Before drawing the crossbowman in ink, it appears as though Leonardo sketched with the tip of a metal stylus deep lines of the curves of the man's back, buttocks, and hamstring. Other metal stylus lines on the figure may be masked by ink, but the two un-inked lines along the outside show that the man's position above the carriage was carefully studied in ad-

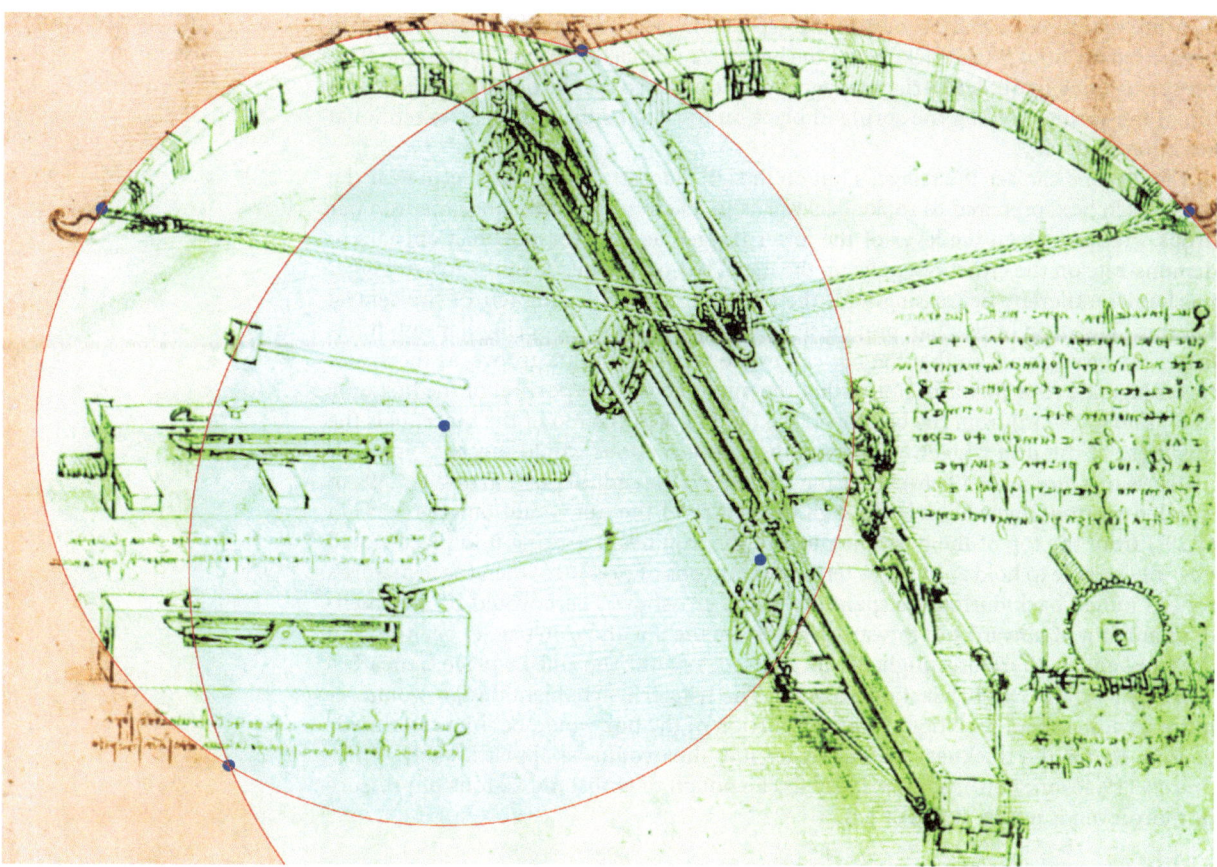

Fig. 7.3. Superimposition of blue dots and red and green circles CA 149br, illustrating directions of a divider when plotting the armature curves

vance of the freehand ink drawing of his figure. The outer line refers to the general position of the body and right leg, whereas the inner line shapes the right shoulder blade, lower back, and buttocks. Of greatest importance in this case is the curve of the man's form when he is about to press the trigger lever. If his size or proportions with respect to the carriage were of any importance, there would be a better indication on the figure of the divisions between certain portions of his body. Instead, there is only a general sense of four equal portions of the figure, as one sees in Leonardo's *Vitruvian Man* drawing. This is merely the artist's instinctive approach to the motion of a body when pressing a trigger lever, not a measured approach to human proportions in relation to the carriage. The figure is therefore an illustration of the mechanical functions of the design rather than indicative of portions of the design's size or form.

Trigger Mechanism

Proportions of the trigger mechanisms to the left of CA 149br appear to agree with corresponding proportions of the trigger and screw in the upper carriage. The illustrated carriage width between the rear and front wheels is approximately three *braccia*, which agrees with the illustrated representation of the *braccio* measuring 0.54 cm at the centre of the carriage. Thus, the trigger mechanism occupies the central third portion of upper carriage front end, and would measure a *braccio* in width, approximately five *braccia* in length, and 1⅔ *braccia* in height. The screw at its centre would measure ⅓ *braccio* in diameter, according to the illustrations on the left of the sheet. Wooden beams that hold the upper trigger between the top and bottom portions of the front-end upper carriage would measure slightly less than ⅓ *braccio*. A proportional comparison with the thicknesses of these beams and the inner trigger arm that releases the trigger nut indicates that this iron beam—within both trigger mechanisms, would measure ⅓ of a ⅓ of a *braccio*, or ⅑ of a *braccio* (6.49 cm). The spring beneath the iron trigger bar is about ⅓ the thickness of the bar, or ¹⁄₂₇ *braccio* (2.16 cm). Within the trigger spring housings of both trigger diagrams, one can also see the bolt or nail heads on top of the spring, holding the spring in place on the right side. This is an exceptional attention to detail.

Clearly, one can see the trigger's nut on the left side, attached at its centre near the top of each box, prepared to rotate backwards in its circular groove once the iron bar drops. Pressing down the lever of the lower trigger, as the crossbowman appears to demonstrate on the crossbow, pulls up the metal bar on the inside of the lower trigger mechanism, thereby depressing with the bar a thin metal spring left of the central fulcrum. The drop of this bar, previously held in place with the spring beneath it, releases the circular nut so that the nut can rotate forward into its groove, as the armature slings the rope forward. Notice that the nut attaches to a portion of the box only ¹⁄₁₈ *braccio* (3.24 cm) from the box top. If the thickness of the top of the box equals the thickness of the iron trigger bar within, then both sections would measure approximately ⅑ of a *braccio* (6.49 cm). If the nut attaches to the central portion of this 6.49 cm wooden section, then the pole through the centre of the nut would only be 3.24 cm (1.28") from the top of the box. The nut and the iron beam locking it in place would possibly be able to hold as much as ten or twenty tons of pressure when the armature's rope is pulled back during the spanning of the crossbow. There would be a tremendous amount of upward and forward pressure on the nut and iron trigger beam within the box, and it therefore might be ill advised to hold the nut in place beneath a 3.24×58.37 cm (¹⁄₁₈×1 br) section of wood. The trigger mechanism design would require an adjustment such that the upper portion of the box would be thicker, or made of iron. As for the thickness of the trigger nut, this would be approximately ⅓ of a *braccio* (19.46 cm, 7.66"), appropriately thick enough to withstand 20 tons of pressure or more, even if it were made of oak.

> Beneath the upper trigger mechanisms, Leonardo states:
> This is the way in which the instrument behaves. It works with the rope,
> its release commencing with the blow of this mallet above the nut.

> This makes the same effect of the instrument above,
> except that its release is done with the lever and it is without clamour.

These are thoughtful approaches to the design, offering the patron extra options and allowing him/her the ability to design a portion of the crossbow. There is no indication that Leonardo changed his mind and made another trigger mechanism for the sheet. The sheet is formatted in such a way that a trigger mechanism only at the bottom left, or at the centre left of the sheet would have left negative (empty) space in the design. Alternatively, one could expect that the explanatory text on the right could have gone in the negative space above the lower left trigger. The metal stylus lines on that lower trigger are cut into the paper at the same depth as metal stylus lines elsewhere in the design, whereas the metal stylus lines on the upper trigger are relatively shallow incisions, compared to metal stylus lines on the rest of the sheet (Fig. 8.20). Thus, the upper trigger was drawn shortly after the lower trigger. Otherwise, metal stylus lines estimating the placement of the upper trigger were in place before any other metal stylus lines were added to the sheet. For reasons explored further below, this latter option is improbable, due to the initial formatting of the sheet according to the width and length of the crossbow design. Leonardo was interested in displaying both trigger designs, since they appear in the preparatory design on CA 1048br (Fig. 1.6). Both ideas for the trigger are incorporated into the crossbow design. The side beams of the top trigger appear between the top and bottom portions of the front upper carriage. The lever that operates the lower trigger mechanism is visible at the top of the crossbow. Both trigger mechanisms were to have the side beams, which are also part of the lever-activated trigger on CA 1048br.

One bit of evidence suggests that the lower trigger mechanism was an addition to CA 149br before the addition of the carriage's side poles, the lower trigger mechanism's lever extends over the left side pole. It is therefore possible that one or both of the trigger mechanisms were on the sheet before Leonardo added side poles to the upper carriage. The reason for this later addition of the side poles is uncertain, however, since a preparatory drawing, CA 147av (Fig. I.8), clearly illustrates his earlier intention for the side poles.[2] Sometime after he designed the armature and the upper carriage, it might be reasonable to assume that he continued with designs of the upper carriage's mechanisms, such as the trigger and gearing noted on CA 1048br (Fig. 1.6). This drawing might have been the next step in the design after CA 147av, and the step before CA 149br. In any case, the carriage's design appears to have been considered at a different phase of the design process than that of the consideration of the carriage's side poles.

Another design on CA 1048br, that of the gear, ratchet and pawl mechanism appears at the right of CA 149br. A pawl is a spring loaded triangular piece that snaps down onto a wheel of angular teeth as those teeth pass under the tip of the triangle, thereby keeping an extended angular tooth from falling back in the opposite direction of its movement under the triangular piece. The portion of this mechanism that is visible on the left of CA 1048br would have been on the front side of the gear and worm screw mechanism on the right side of CA149br. Another example of the ratchet and screw mechanism on CA 1048br appears at the bottom of CA 148ar (Fig. 1.3). But as one can see at the right of CA 149br, Leonardo appears to abandon the complicated inner ratchet mechanism for a relatively simple worm screw mechanism. If there is to be a way on CA 149br of locking the large gear in place with the help of a ratchet and angular teeth inside the large gear, then it is hidden at the front side of this gear on the drawing. Something would have to keep the large gear from rotating when the screw has pulled back the trigger mechanism, spanning the bow. The *Giant Crossbow* drawing contains no visual evidence about a way to lock the large gear in place once one has spanned the bow.

The reason is also unknown for the location of capstan bars and their corresponding worm screw pole at the lower carriage (rather then their plausible location in the upper carriage—at the bottom of the large gear). This distance of the worm screw from the large gear does not agree with the illustration at the right of the sheet. The text below the mechanism on the right states:

> To pull the rope
> of the crossbow.

Crossbowmen could rotate the capstan bars on both sides of the central worm screw, thereby moving the cogs of that gear and turning the screw at its centre. Nonetheless, the text only addresses the purpose of the worm screw mechanism. It may be that the inclusion of the right corner mechanism was to correct the mistaken location of the capstan hub at the lower carriage. The corner drawing also offers a view of the way a worm screw rotates the large gear and central screw. Nonetheless, its position in such a narrow space, along with the text above, suggests that Leonardo added it and the text only after he inserted the metal stylus lines for the crossbow design and trigger mechanisms. One could also argue that he was not worried about the depiction or function of this portion of the crossbow, since he had studied several years before in Florence similar forms of gearing, large screws for tensioning devices, and windlasses. The Uffizi

---

[2] Proof that CA 147av precedes 149br appears in the form of the large capstan wheel at the end of its crossbow and the same feature on the crossbow at the top left of 147bv (Fig. 1.2). The gear and worm screw mechanism on CA 149br replaced as a winding mechanism the large rear end capstan wheels on CA 147. The mechanical reason for this is that the worm screw with two sets of capstan bars on each end can pull much more weight than the rear capstan. Also, the armature design at the bottom of CA 147bv and at the top of 149br would pull more tension than the armature designs preceding it.

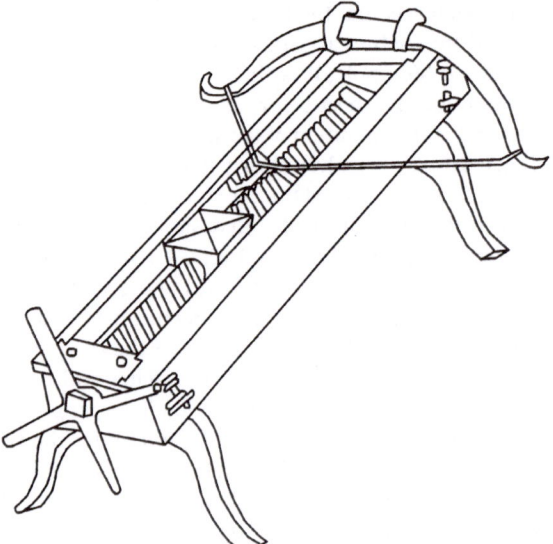

Fig. 7.4.
Illustration of Leonardo's
drawing of spanning bench,
from Uffizi, no. 446r; Gabinetto
dei Disegni e delle Stampe degli
Uffizi, Florence

drawing, inventory 446, dated to 1478, offers evidence of these early mechanical studies. A copy of the spanning bench on that sheet is in Fig. 7.4. Illustrated with this bench on Uffizi inv. 446 is the block that travels along the central screw (▶).

The fifth and sixth lines of text at the right of CA 149br state that the crossbow "carriage is 2 *braccia* wide and 40 long." Thus, one can assume that the lower carriage has these dimensions. The upper carriage appears to have the same width, though it would lack the 2⅔ *braccia* front-end extension, as noted further above, with regard to the lower carriage. Thus, the upper carriage would be 37⅓ *braccia* long (see Appendix A).

Referring to the crossbow, Leonardo indicates on the fifth line of text at the right that, "it has an elevation [or draw] of 14 *braccia*." As noted and illustrated above, regarding the lower carriage, this would lift the armature as much as 8.17 metres, or 26.8 feet. This relatively low level of 16.5° angle of elevation would be important for two reasons: for reaching distant targets and/or relatively high-level distant targets. Medieval and 15th century drawings of siege engines show that they were often used on hills and fortifications from which they would shoot objects along relatively low trajectories. In fact, Leonardo copied on CA 149ar [53va] several of the best visual examples of what are now known of the Italian 15th century tilting chassis for great crossbows, cannon and springalds. Given its design for shooting stones along a low trajectory, the *Giant Crossbow* might have fit best above the walls of the Sforza Castle or Milan's city walls. It would not have been too big to fit on the roof of the Sforza Castle's largest corner crenulations, or especially on the large corner fortifications of the city's walls.[3] The top floor of one of the front towers measures an estimated 30 metres (98 ft) in diameter, within which the Giant Crossbow would fit. This is not to propose a specific use of the *Giant Crossbow*, which could have served several purposes: from that of a launcher of a variety of projectiles, to that of a show piece in pageants. To date, one of the largest crossbows ever built—with a ten-foot armature—was used in 1894–1897 by the Flying Zedoras circus acrobats to shoot Pansy Chinery (Alar the Human Arrow) through a paper target to her sister, Adele, who would catch her on the other side of the circus tent. Medieval trebuchets launched human and horse cadavers into and out of castles, and the *Giant Crossbow* could have been designed to have this capability. Its optional elevation to 14 *braccia* made it useful in more places, above or below fortifications.

---

[3]   A 17th century French engraving of the plan of Medieval Milan is reprinted in: Hart, pl. 24. This gives an initial impression of the proportions of the city wall and castle fortifications. The Italian print of Fig. 3.3 has the date, 1523, at the bottom corner of the page.

The Crossbowman

The crossbowman's proportions are comparable to those in Leonardo's *Vitruvian Man* (◄) drawing of c. 1490. In that drawing and on CA 149br, there are relatively equal lengths between the feet and knees, the knees and division of the legs, the division of the legs and middle of the breast, and between the middle of the breast and top of the head.[4] Both men consist of four equal lengths, and this happens to agree with the Vitruvian principle that the height of a "finely shaped human body" measures four cubits, or four *braccia*.[5] This symmetrical division of a man is useful in architectural illustrations. The modular sectioning of people and buildings could refer to convenient proportional measurements common in early modern and medieval architectural illustrations. This division into four large equal portions is helpful to an architect who must use the largest possible proportions—rather than average human proportions—in order to build spaces into and onto which anyone would fit comfortably. Leonardo's reason for using this four-part proportional division for the present study is not obvious, though it would appear that he instinctively preferred this four-part scheme in order to exhibit proportionally the bends of the left leg, left arm, and the torso, in relation to the straight portions of the body. Moreover, he instinctively illustrates the proper angles and proportions by which the crossbowman would demonstrate the operation of a trigger lever from above the *Giant Crossbow* upper carriage. Some of his mechanical illustrations in the Codex Atlanticus appear to be no less schematic or diagrammatic in their apparent references to proportional measurements with the help of modular forms. A long Roman cubit was normally a *braccio* in most Italian cities during the 15th century and it might have appeared to Leonardo that a crossbowman four *braccia* tall, standing on an upper carriage 1⅔ *braccia* thick and two *braccia* wide, might appear to be the size of the man illustrated on CA 149br. This is impossible to prove, however, since this is relatively rapid sketch of a figure whose task appears to be the illustration of the bow's operation, rather than its scale.

If, however, Leonardo had considered the use of an Italo-Byzantine division of the figure into three *braccia*, then he would have clearly overestimated its illustrated size.[6] The illustrated and specified thickness and width of the carriage—between two and three braccia (see estimates in Appendix A), is nearly half that of the man, suggesting that his height would be between eight and twelve feet. In any event, it is not likely that Leonardo considered an Italo-Byzantine cannon when illustrating the crossbowmen. First, the scale of the man is unrelated to the sizes of other features in this mechanical schematic. Second, in early modern mechanical and architectural diagrams the exact measurements of their illustrations were less important than the display of the modular proportions; the poses of human figures were more important for demonstrations of the illustrated equipment than their illustrated dimensions. Third, as Panofsky notes about Leonardo:

> When dealing with human—as opposed to equine—proportions, he mostly resorted, after the model of Vitruvius and in sharp contrast to all other Italian theorists, to the method of common fractions without, however, entirely rejecting the 'Italo-Byzantine' division of the body into nine or ten face-lengths.

> In Leonardo's studies both types—one corresponding to the Vitruvian proportions, the other to the Cennini-Gauricus canon—coexist without differentiation so that it is often difficult or impossible to connect a particular statement with either the one or the other.[7]

---

[4] The words of Vitruvius for the "middle of the breast" are "a medio pectore." Vitruvius (1931/1998) *On architecture*, book III, chapter 1, 2; Frank Granger (trans), Harvard, Cambridge, Mass, pp 158–159.

[5] *Ibid.,* "Pes vero altitudinis corporis sextae, cubitum quartae, pectus item quartae." (The foot is a sixth of the height of the body; the cubit a quarter, the breast also a quarter.)

[6] Cennino Cennini, Ghiberti, Alberti, Filarete, and other Italians used the Italian/Byzantine canon in the 14th and 15th centuries. See: Erwin Panofsky (1955) *The history of the theory of human proportions as a reflection of the history of styles*, in: Meaning in the Visual Arts, Doubleday, New York, pp 104–105.

[7] *Ibid.,* p 127, no. 84.

Fourth, the crossbowman is not a study of human proportions; he illustrates the way one would release the crossbow trigger's nut. One reason for this giant man might have something to do with the way in which he was drawn. Leonardo use of the four-part architectural human proportions of Vitruvius rather than the three-part Italo-Byzantine human proportions of Alberti shows his preference for the use of architectural proportions in a mechanical design.[8] Still, the crossbowman is not in proportion to what one might consider "architectural" features of the crossbow design. His physical size is incorrect with respect to the rest of the machine. At issue instead are the ways in which the man operates the machine. Hence, this figure does not conform strictly to Italo-Byzantine or Vitruvian human proportions. He conforms to a kind of "mechanical proportion": a scheme of proportions in agreement with the operation of a machine, rather than with the sizes of any object in the composition. Leonardo's likely interest in crossbowman's illustrated ability to "demonstrate" the *use* of the crossbow, rather than its *size*, might explain his preference for drawing this man out of proportion to the crossbow's dimensions as specified in the text on the right.

His mechanical and proportional draughting techniques show his interest in producing useful designs that an engineer could rely on for basic measurements. Proportions of the illustrated crossbow reveal an approach to the use of third-portions throughout the design, so that each portion of the design is relatively proportionate and the scale of each portion easily determined by the builder. On the other hand, the design is mostly the product of the mechanical techniques of draughting instruments, with no obvious—or at least no rigorous, application of geometric proportions. The text on CA 149br appears to have been its last addition and therefore offers only a rough estimate of dimensions. Any practical link between the dimensions in the text and image is merely a general estimate. Nonetheless, Leonardo's approaches to the *Adoration* perspective drawing (as discussed above) show that he was accustomed to producing oblique projections and perspective constructions with proportions divisible by three.[9] Some portions of the *Giant Crossbow* appear to have exact measurements and thereby fit the third-portion proportions of the rest of the bow. This is evident in the chart on Appendix A. Other portions of the crossbow are the product of mechanical skill with draughting tools. The folio's text appears to be a combination of original intentions for the crossbow's dimensions and proportional estimates of the final drawing. More proof of this will continue below, as part of the discussion on mechanical draughting techniques used in the drawing.

Leonardo's intended measurements for the parts of the crossbow are unknown. Measurements of the following procedure serve as proportional guides for extrapolating his methods and possible reasons for detailed approaches to the sizes and functions of the great crossbow's portions. If Piero della Francesca had produced the drawing on CA 149br, the exact measurements for this design might be more consistent, as in Piero's painting, *The Flagellation*. A primary finding in the present study, however, is that some of Leonardo's surviving presentation drawings exhibit a combination of proportional solutions and problems that offer more information about the fundamental elements of his designs than one can find with an assessment of any evidence of possibly relevant rigorous measurements of these designs. In any event, the present study uses measurements—without evidence in each case of whether they may or may not be relevant to the original design—to help verify the most consistent proportional aspects of the designs. This tendency to find consistent proportional solutions and problems amid relatively inconsistent measurements continues in the next section on mechanical draughting techniques.

---

[8] Vitruvius III: 1.1–3. Alberti, *De Pictura*, I: 19. Vitruvius, *On Architecture*, F. Granger (ed, and trans) (1998), books I–V, Harvard, Cambridge, MA, (1931 1st ed.), pp 158–161. Leon Battista Alberti (1991) *On painting*, Cecil Grayson (trans), Martin Kemp (notes), Penguin, London, p 54.

[9] Other studies in the Codex Atlanticus that contain oblique projections—initially marked with metal stylus incisions, include: 149br [53vb], 143r [51rc], 145r [51vb], 147av [52va], 148ar [53ra], 148br [53rb], 149ar [53va], 152r [54va], 158r [56vb], 159br [57rb], and 160br [57vb]. See also: Elkins, *The case against surface geometry*.

8

Mechanical Draughting Techniques

This study will discuss some of the basic aspects of Leonardo's design skills. At issue is the possible sequence of marks on the surface of the *Giant Crossbow* drawing and the perceived proportional estimates instrumental in the determination of those marks. This study of the sequence of marks on a mechanical study offers a prime example of an advanced form of mechanical draughting in 15th century Italy.

As designers, engineers, and architects like Leonardo were aware, the *Giant Crossbow* drawing would have probably met with approval if one could have seen how each part of the bow corresponds to measurements in actual *braccia*. This is not to suggest, as discussed further above, that these artist/engineers had to reproduce each portion of the drawing to scale. Instead, there appears to be more attention given to the proportional relationships between each portion of CA 149br. Thus, regardless of the final preferences of a builder—who could change portions of a design significantly, in order to comply with the limitations of the materials, other practical considerations, or the patron's wishes—the artist/engineer's design could retain much of its original character in final production if proportions of the design were relatively consistent. Slight differences in portions of a drawing's scale would not alter the final product as much as differences in that drawing's proportions. It is unknown if Leonardo was aware of this specific benefit of proportional design components, as opposed to the benefits of exact scale replicas. His preference for the former, however, indicates that he had some knowledge of this benefit of relative—rather than exact—proportional consistency in design. If significant portions of the crossbow design were disproportionate, it would not have been worthy of the attention of another engineer, or a patron knowledgeable about siege weapons, such as Ludovico Sforza. An engineer, carpenter, and patron would have checked this crossbow design for its possible life-size measurements. A design that was not consistently to scale throughout would not have been as problematic for a builder as a design without proportionate components. An engineer was in a position to build a design's components at different scales, though he would still need to see convincing display of the proportions or functions of an overall design before considering changes in scale.

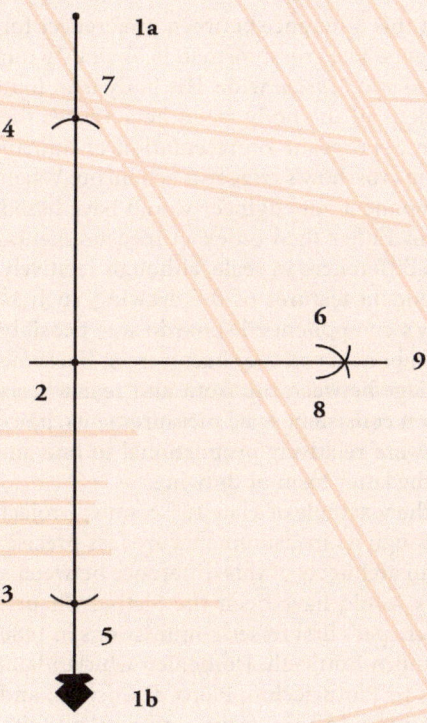

Fig. 8.1.
Diagram of the crossed arc method

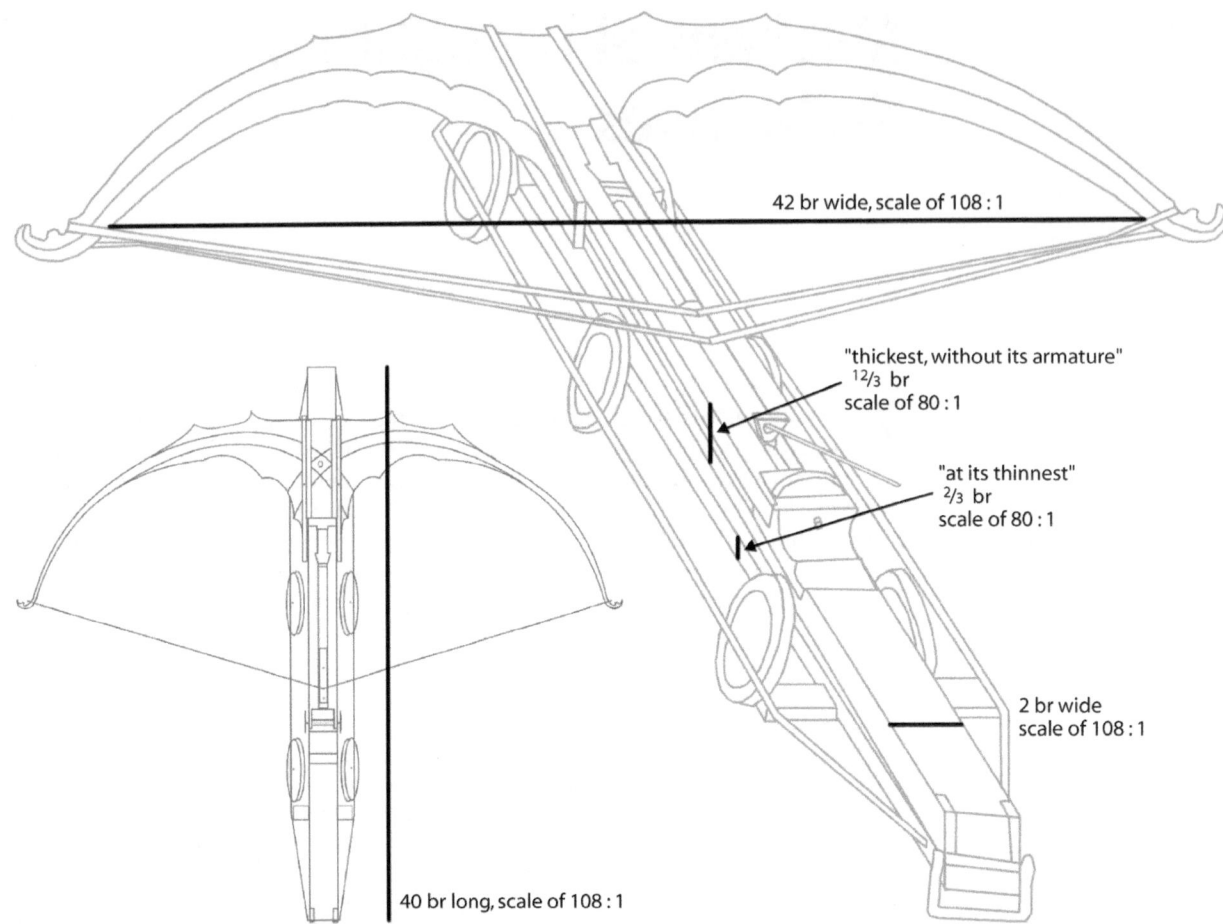

42 br wide, scale of 108 : 1

"thickest, without its armature"
$1^2/_3$ br
scale of 80 : 1

"at its thinnest"
$^2/_3$ br
scale of 80 : 1

2 br wide
scale of 108 : 1

40 br long, scale of 108 : 1

Fig. 8.2. Illustration of the Giant Crossbow, with notes about its scale and proposed dimensions

CA 149br's crossbowman is an example of this difference between preferences for the bow's scale and its mechanical requirements. The crossbowman is obviously too tall for a crossbow forty *braccia* in length and two *braccia* wide. His body is in four equal proportions, demonstrating the positions of the body when he activates the trigger mechanism. The crossbowman is an appropriate representation of human movement with respect to the function of the crossbow's trigger mechanism. When determining the sizes of each crossbow component, an engineer would have based those estimates on proportional comparisons rather than on exact measurements. He would have seen in the drawing obvious differences in scale, although relatively consistent proportional estimates of independent features of the drawing, such as the crossbowman's figure and the crossbow's components. Leonardo was possibly capable of redrawing CA 149br so that all components on the sheet were at the same scale. His meticulous treatment of the carriage between the front and rear wheels refers to this capability. Instead of working on consistent scale measurements, however, he produced design components that were relatively proportional to one another. An engineer would have understood the latter form of drawing.

It might be useful to note at this point another example of a late 15th century project that has relatively accurate proportions, although its measurements are less precise. At that time, a popular example among Italian architects of this difference between a design's proportions and its measurements would have been the Sistine Chapel. Leonardo was well aware of the award of the chapel's first fresco commissions in 1480 and 1481 to the workshops of Perugino, Sandro Botticelli, Domenico Ghirlandaio, Cosimo Rosselli, Luca Signorelli, as well as to Pinturicchio, Piero di Cosimo, and Bartolomeo della Gatta. The first three painters in this group had worked with

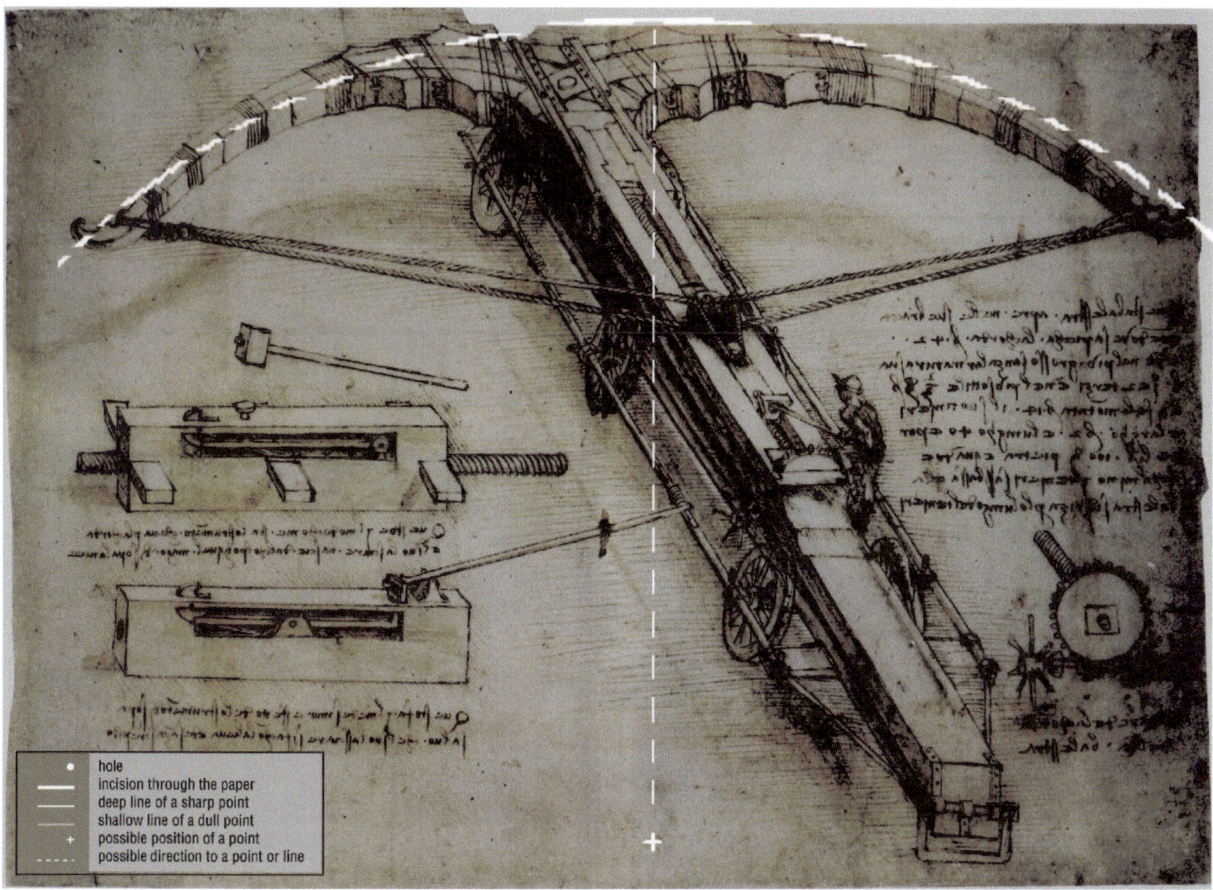

hole
incision through the paper
deep line of a sharp point
shallow line of a dull point
possible position of a point
possible direction to a point or line

Fig. 8.3. Superimposition of white lines over CA 149br, illustrating the possible directions of divider when estimating the position of the armature

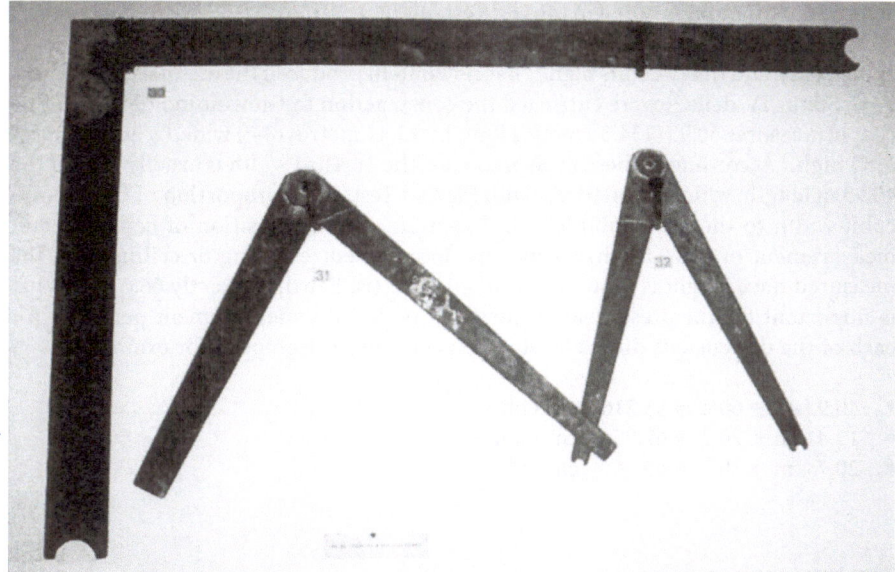

Fig. 8.4. ▶
Photograph of the sectors supposedly used by Brunelleschi to build the Florence Cathedral, Cathedral of Sta. Maria del Fiore, Florence

Leonardo in Verrocchio's studio. When they left Florence for the Pope's commission in 1480–1481, and when Verrocchio went to Venice in 1480 to start work on the equestrian monument of Bartolomeo Colleoni, Leonardo was at work on the *Adoration of the Magi* and had possibly considered applying for work in Milan (as discussed above). Built between 1477 and 1480, the chapel was to have the exact dimensions as the

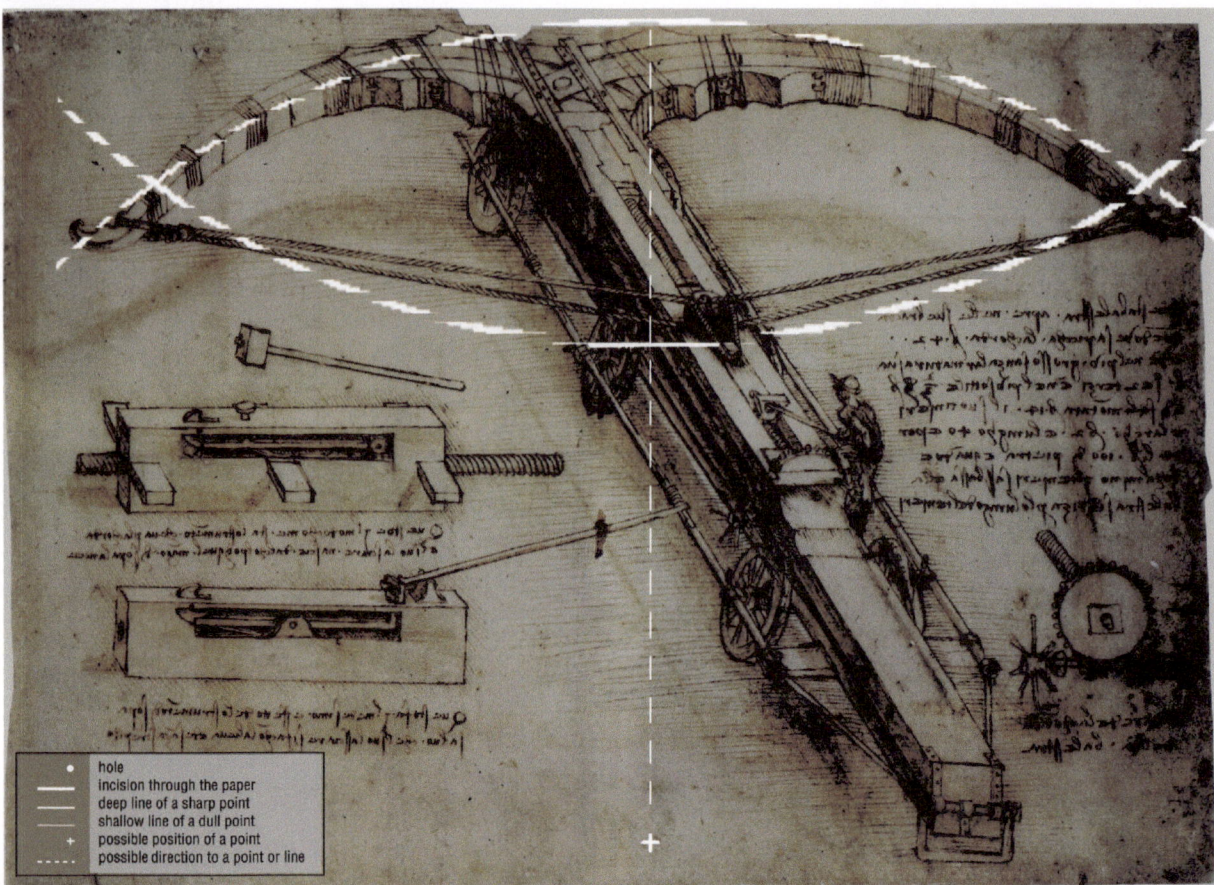

hole
incision through the paper
deep line of a sharp point
shallow line of a dull point
possible position of a point
possible direction to a point or line

Fig. 8.5. Superimposition of white lines over CA 149br, illustrating the possible directions of divider when estimating the position of the armature

Temple of Solomon, as noted in the Old Testament, 1st Kings, 6:2. This sentence states that, "The house that King Solomon built for the Lord was sixty cubits long, twenty cubits wide, and thirty cubits high."[1] Baccio Pontelli produced the architectural plans.[2] Pope Sixtus IV della Rovere entrusted the construction to Giovannino de' Dolci. The chapel measures 40.93 (134.3') metres long, by 13.41 metres (44') wide, by 20.70 metres (68') high.[3] According to these measurements, the 13.41 m width is exactly 33% of the 40.93 m length, which is consistent with the Old Testament proportion of the twenty-cubit width to the sixty-cubit length. Depending on the position of ceiling height measurement of 20.70 m, that being the floor to centre ceiling *or* ceiling side, the measured nave height is around 65% of 40.93 m (rather than exactly 66.67%). What is important for the present study, however, is that the measurement per cubit for each of the dimensions differs by at least a centimetre. Compare, for example:

- 40.93 m ÷ 60 c = 68.2167 cm/cubit
- 13.41 m ÷ 20 c = 67.05   cm/cubit
- 20.7 m ÷ 30 c = 69   cm/cubit

---

[1] Bruce Metzer and Roland Murphy (eds) (1962/1991) *The New Oxford Annotated Bible*, Oxford University Press, New York, p 432.

[2] See: R. Salvini and E. Camesesca (1965) *La Cappela Sistina in Vaticano*, Milan. E. Steinmann (1901–1905) *Die sixtinische Kapelle*, 2 vols., Munich. Musei Vaticani Online: <http://mv.vatican.va/2_IT/pages/CSN/CSN_Main.html> (online, January 2008).

[3] *Ibid*. A guide posted on the wall of the Sistine Chapel (as of July 2000) notes the chapel measurements.

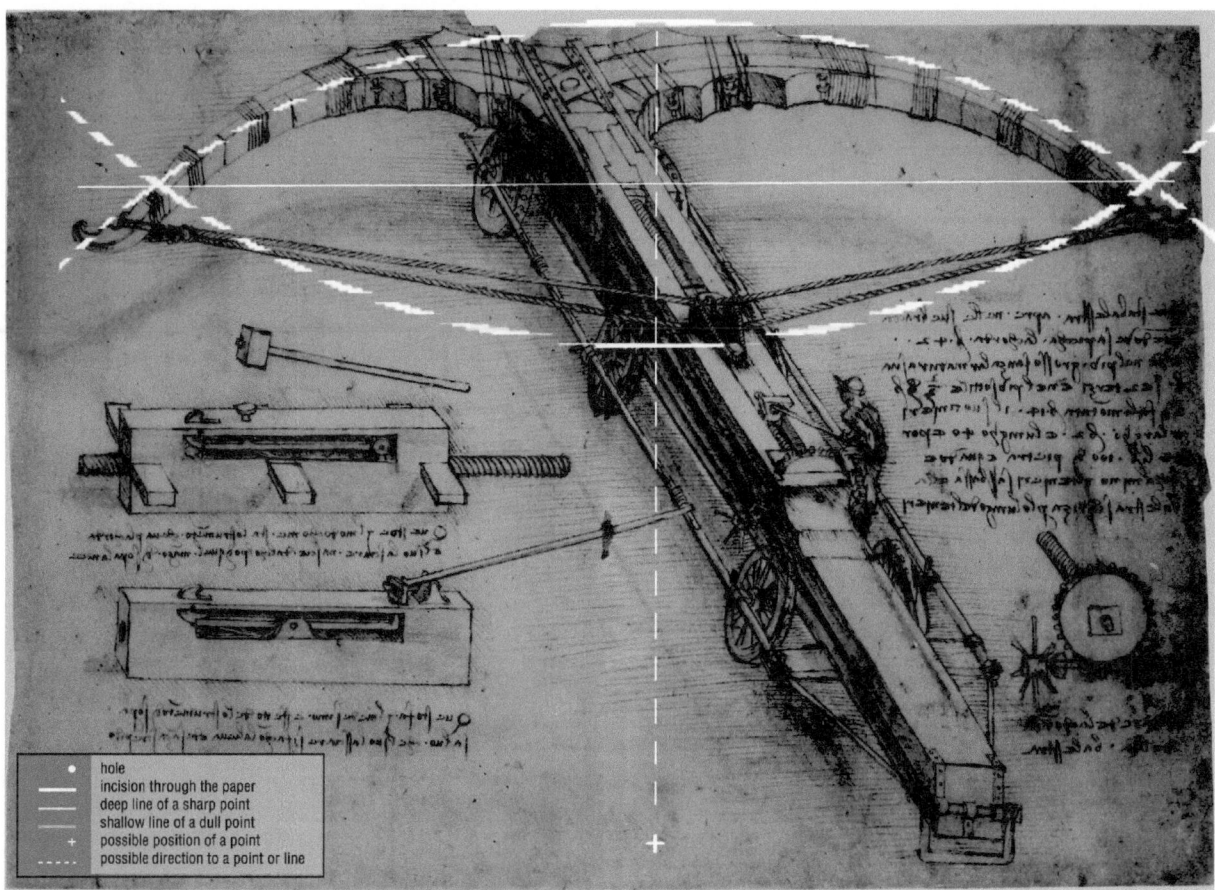

hole
incision through the paper
deep line of a sharp point
shallow line of a dull point
+ possible position of a point
- - - - possible direction to a point or line

Fig. 8.6. Superimposition of white lines over CA 149br, illustrating the position of a metal stylus incision and possible directions of a divider when estimating the position of the armature

According to Angelo Martini and other authors, a late 15th century *braccio romano* (aka. a *passetto romano*) was around 67 cm.[4] Based on this comparison between a 67 cm *braccio* and the room's measurements, it would appear that Pontelli and/or Giovannino de' Dolci considered a long Roman cubit the same length as a Roman *braccio*. Nonetheless, depending on the way that the builder measured the room during construction, its measurements were determined with a minimum margin of error plus-or-minus thirteen centimetres. The accuracy of measurements of the room's dimensions, in relation to the Old Testament dimensions, was therefore not a serious concern for the builder or project manager. Regardless of comparisons between the exact measurements and the Old Testament cubit, however, the room's proportions are remarkably precise. Such comparisons might indicate that the room is not built on a consistent 1 : 1 scale (cubit per measured length), but that it was most important that its proportions were the same as those in 1 Kings 6 : 2. Furthermore, the dimensions are in thirds, as in a number of Leonardo's designs.

This habit of working in thirds dispenses with the need for a direct relationship between the scale of a design and specific measurements of the resulting construction. Without making rigorous measurements, one could produce with divider and stylus a design that is relatively proportional to its life-size construction. It would

---

[4] See the "Lenghts" study further above. Martini (1883) *Manuale di metrologia.* Cavalli (1874) *Tableaux comparatifs des measures* [...]. Chambers (1728) *Measures,* Ciclopædia [...]. Colonel Cotty (1819) *Aide-mémoire* [...]. Diderot and Le Rond d'Alembert (1751–1765) *Pied,* Encyclopédie [...] 7. Horace Doursther (1840) *Dictionnaire universel des poids et mesures anciens et modernes,* Brussels. Eusebio (1899) *Compendio di Metrologia* [...].

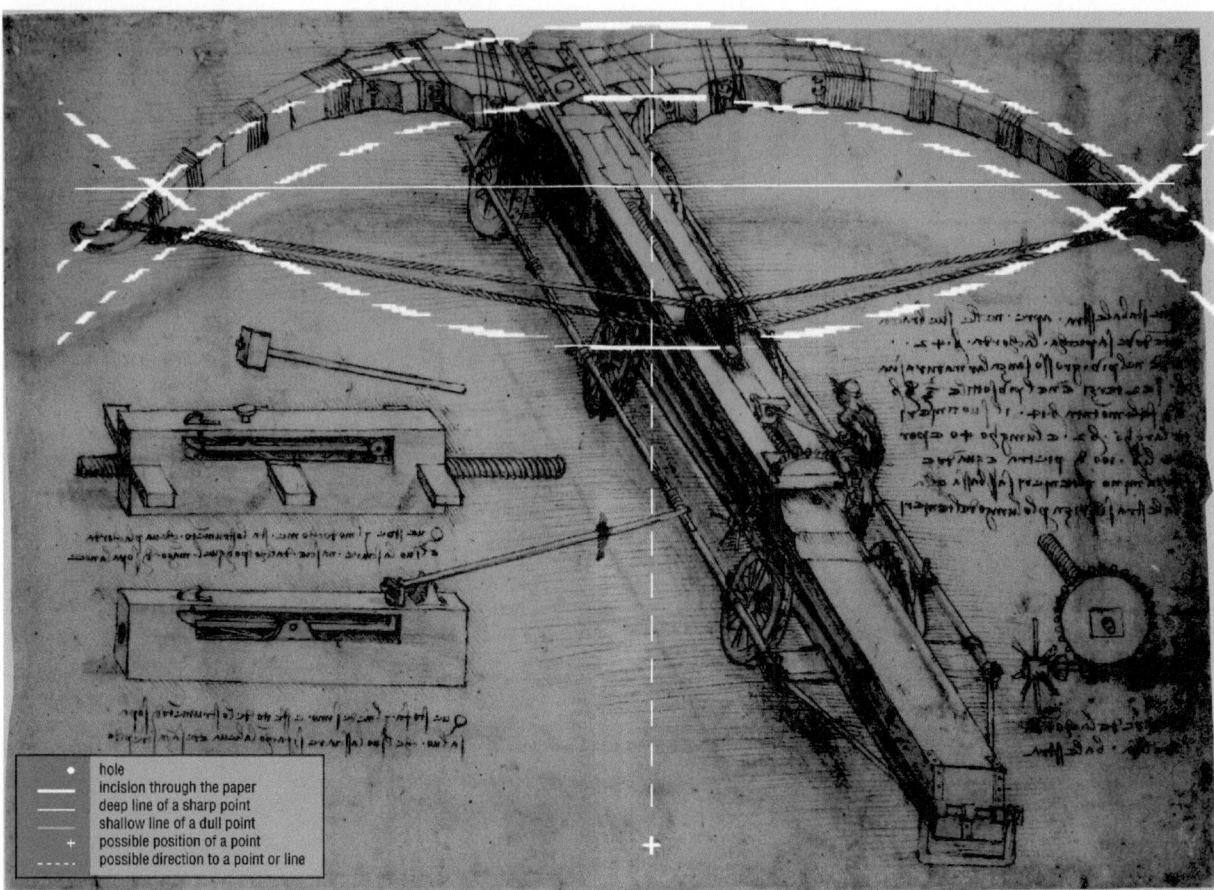

•	hole
————	incision through the paper
————	deep line of a sharp point
————	shallow line of a dull point
+	possible position of a point
- - - - -	possible direction to a point or line

appear as though exact measurements for the *Giant Crossbow* design were not as important to Leonardo as long as estimated measurements could be used to develop proportional components that would make the bow function. For the components to work, they had to be proportionate to one another, an assessment one could determine with estimated measurements at first in the diagram and/or in the textual instructions. Moreover, for the machine to appear functional, its components had to look proportional to the overall design. To give this impression of proportional components throughout the design, Leonardo refers to third portions in the text and the drawing. The stated armature width of 42 *braccia* happens to be three times the specified elevation or draw of fourteen *braccia*; the visible and noted carriage thicknesses are in third portions; the wheel axles are at ⅓ positions along the oblique carriage; the illustrated position of the trigger nut is at the rear third portion of the screw.

In general, there are few scholarly responses to this problem of third portions in Leonardo's mechanical designs. Regarding his general approaches with draughting tools, one could look to the studies of Marco Carpiceci, Carmen Bambach, Edward Ford, and of Klaus Schröer and Klaus Irle.[5] One reason for the relative paucity of scholarship on the methods of layout and composition in Leonardo's mechanical drawings may be due to the difficulties with strictly limited access to sources such

Fig. 8.7. Superimposition of white lines over CA 149br, illustrating the position of a metal stylus incision and possible directions of a divider when estimating the position of the armature

---

[5]   Marco Carpiceci (1986) *Leonardo, La misura e il segno*, L. Chiovini, Rome. Carmen C. Bambach (1999) *Drawing and painting in the Italian Renaissance workshop*, Cambridge Univeersity Press, Cambridge. Edward Ford (2003) *Interpretations of marks from draughting tools in some Italian Renaissance drawings*, Thesis, Oxford University. Klaus Schröer and Klaus Irle (1998) *Ich aber quadriere den Kreis ...*, Waxmann, Münster.

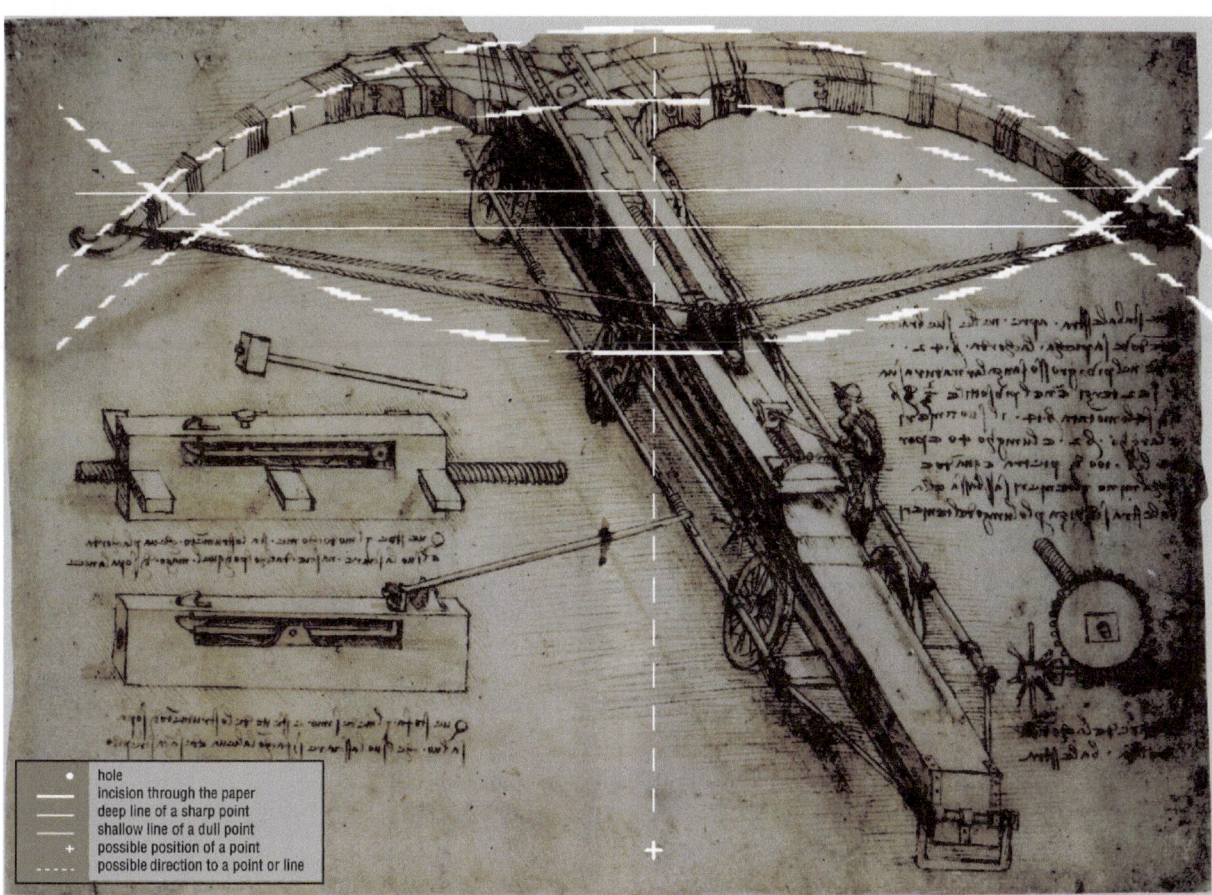

Fig. 8.8. Superimposition of white lines over CA 149br, illustrating the positions of a metal stylus incisions and possible directions of a divider when estimating the position of the armature

as the Codex Atlanticus at the Biblioteca Ambrosiana in Milan. An appropriate study of them requires a direct view of the shallow metal stylus incisions on their surfaces, since the incisions are usually invisible in normal photographs. The best way to see the marks is to direct a light source across the surface of a sheet in such a way that a thin shadow forms on the inside of the incisions or any other shallow indentation on the surface. This type of lighting is "raking light". The process also involves the direction of light beneath the surface (if possible) so that one can identify places, like pinholes, where the light shows through to the surface. Metal stylus marks often show some of the earliest compositional studies of a design, over which Leonardo often traced ink lines. At the Biblioteca Ambrosiana in February 2004, I observed with raking light the metal stylus marks on CA 149br [53vb], 143r [51rc], 145r [51vb], 147av [52va], 148ar [53ra], 148br [53rb], 149ar [53va], 152r [54va], 158r [56vb], 159br [57rb], and 160br [57vb]. These mechanical studies have remarkably similar metal stylus marks, many of them as lines along the longest edges of the designs. Before Leonardo applied ink to certain portions of the page, metal stylus marks determined the optimal format and location of the design. It appears as though his standard approach to this metal stylus formatting was to use a proportional system of third portions. The following discussion addresses this issue of third portion proportional formatting, especially with metal stylus lines and holes, on CA 149br and related drawings.

Page Format

An obvious condition of CA149br is that someone trimmed it down to its current size. Cut away portions include the top and right sides of the armature and the worm

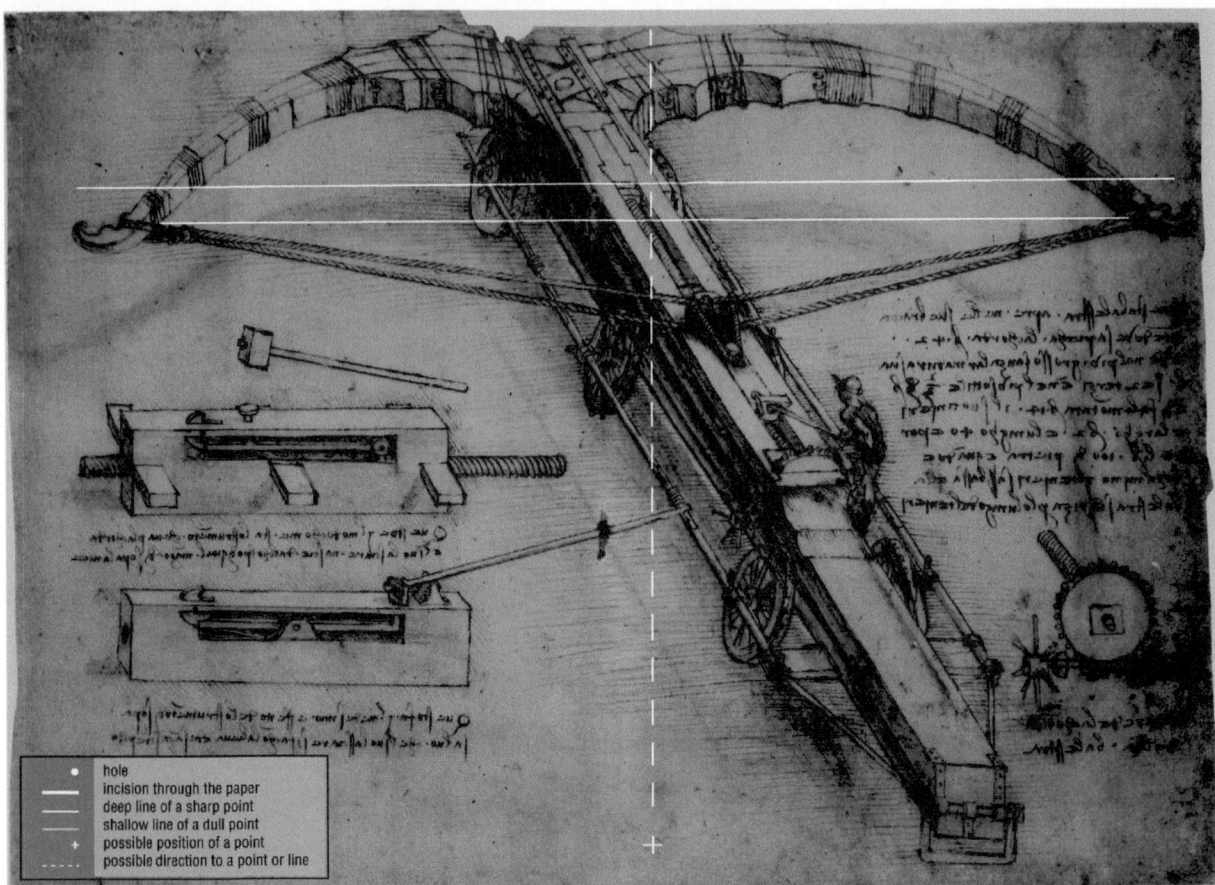

hole
incision through the paper
deep line of a sharp point
shallow line of a dull point
+   possible position of a point
----- possible direction to a point or line

screw winding wheel at the lower right. At first, Leonardo might have removed the drawing's edges, centring the crossbow according to the width of its armature. It is unlikely, however, that he would have trimmed the drawing so closely at the right side. Following a period when the edges could have become damaged or bent, a later editor of his drawings—such as Pompeo Leoni (c. 1590–1608) or Galeazzo Arconati (c. 1632–1637)—might have cropped the edges to their current positions.[6]

The vertical measurements of CA149b are between 209 and 204.1 mm, and horizontal measurements are between 282.1 and 278.6 mm. If one measures from the lower left of the drawing to the outer edges, the drawing is thereby measured at its largest size, whereas measurements between the upper right corner and the corresponding outer edges provide the drawing's shortest measurements.[7] Thus, CA 149b is 209×282.1 mm at its largest.

This sheet is the lower half of folio 149, which included 149a as the upper half. The upper sheet measures 210×278.6 mm. When 149a and 149b were joined, along the former sheet's bottom edge and the latter sheet's top edge, the combination of these two sheets measured at least 419×278.6 mm. Some of the unbisected Codex Atlanticus

Fig. 8.9. Superimposition of white lines over CA 149br, illustrating the positions of a metal stylus incisions and possible direction of a divider

---

[6] There are various discussions on the history of Leonardo's manuscripts. A handy source on this, though not fully representative of the state of the research, is: Carlo Pedretti and Marco Cianchi (1995) *Leonardo, I codici*, Art e Dossier, Giunti, Florence.

[7] I measured the original drawing at the Biblioteca Ambrosiana. For their help with access to the Codex Atlanticus drawings in February 2004, I am grateful to Professor Massimo Rodella, Monsignor Professor Gianfranco Ravasi, Mr. Vittorio Bergnach, Monsignor Professor Pier Francesco Fumagalli, Professor Martin Kemp, Professor Paolo Galluzzi, Professor Pietro Marani, and Professor Carlo Pedretti.

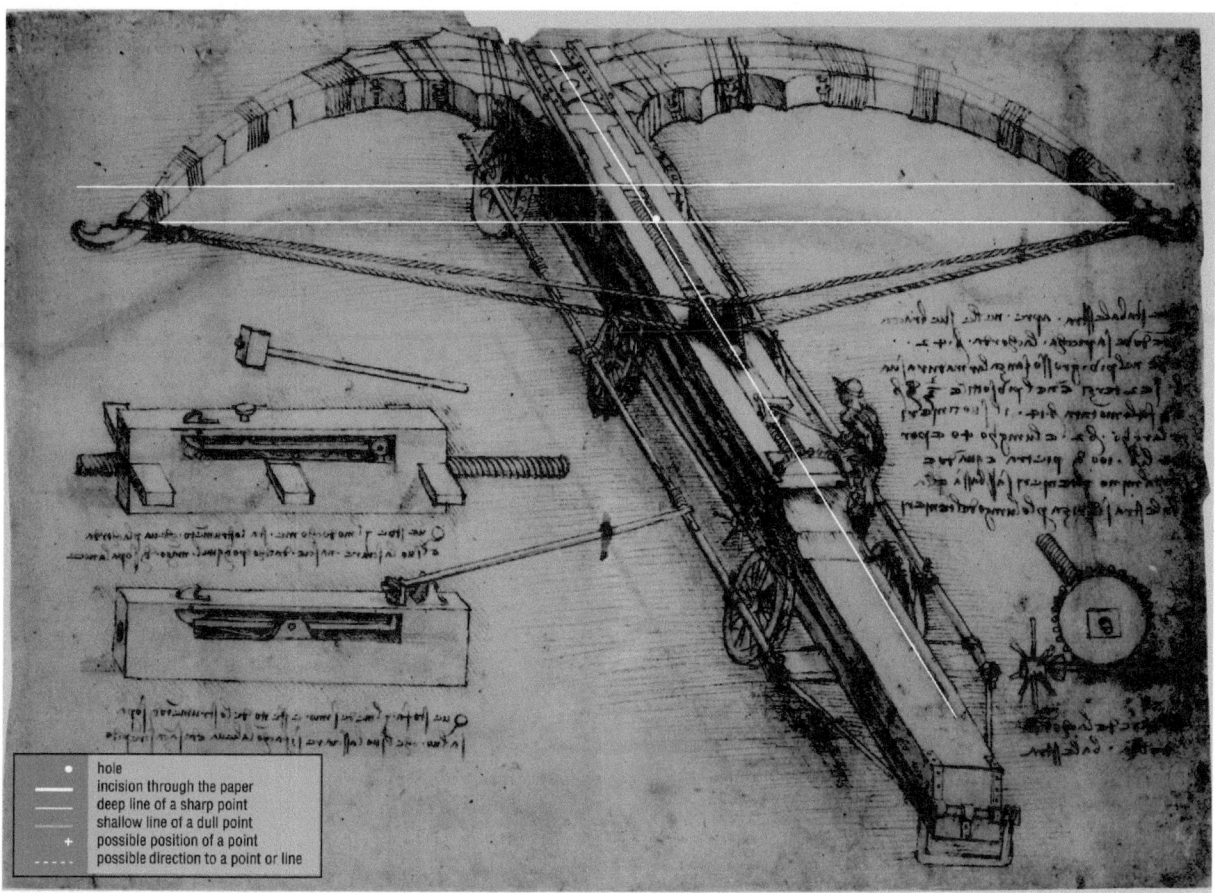

Fig. 8.10. Superimposition of white lines over CA 149br, illustrating the positions of metal stylus incisions

sheets measure an average of 417 mm by 287 mm. This average is determined by comparing the following folios:

- 47   401 × 288 mm
- 77   417 × 286 mm
- 455   434 × 289 mm
- 463   426 × 293 mm
- 899   414 × 279 mm
- 1105   408 × 285 mm

These sheets are the largest among the very few non-bisected folios in the CA. Most of the remaining 1 119 CA sheets are half portions, similar to 149b, of larger sheets measuring around 417×287 mm. Among the dimensions noted in the group above, maximum dimensions for a double-width CA folio would be around 434×293 mm (vert. × horiz.). This is just under the standard sheet size of a Bolognese or Florentine "*rezzuta*" (originally spelled "*reçute*") at 440×312 mm, and just under the normal half portion of a standard *reale* format sheet. Two portions of 440×303 mm paper make a *reale* sheet of 440×606 mm.[8]

Strict rules governed the standards of paper sizes, as specified by Bolognese statutes in 1389 and 1454.[9] For cartoons (large sketches) of the *Battle of Anghiari* (planned

---

[8] Other standard sizes included the *Imperiale*, at 730 × 498 mm and the *Mezzana*, at 494 × 344 mm.
[9] See: Carmen C. Bambach (1999) *Drawing and painting in the Italian Renaissance workshop: Theory and practice, 1300–1600*, Cambridge, Cambridge, UK, pp 36, 364, 388 nos. 26 and 27.

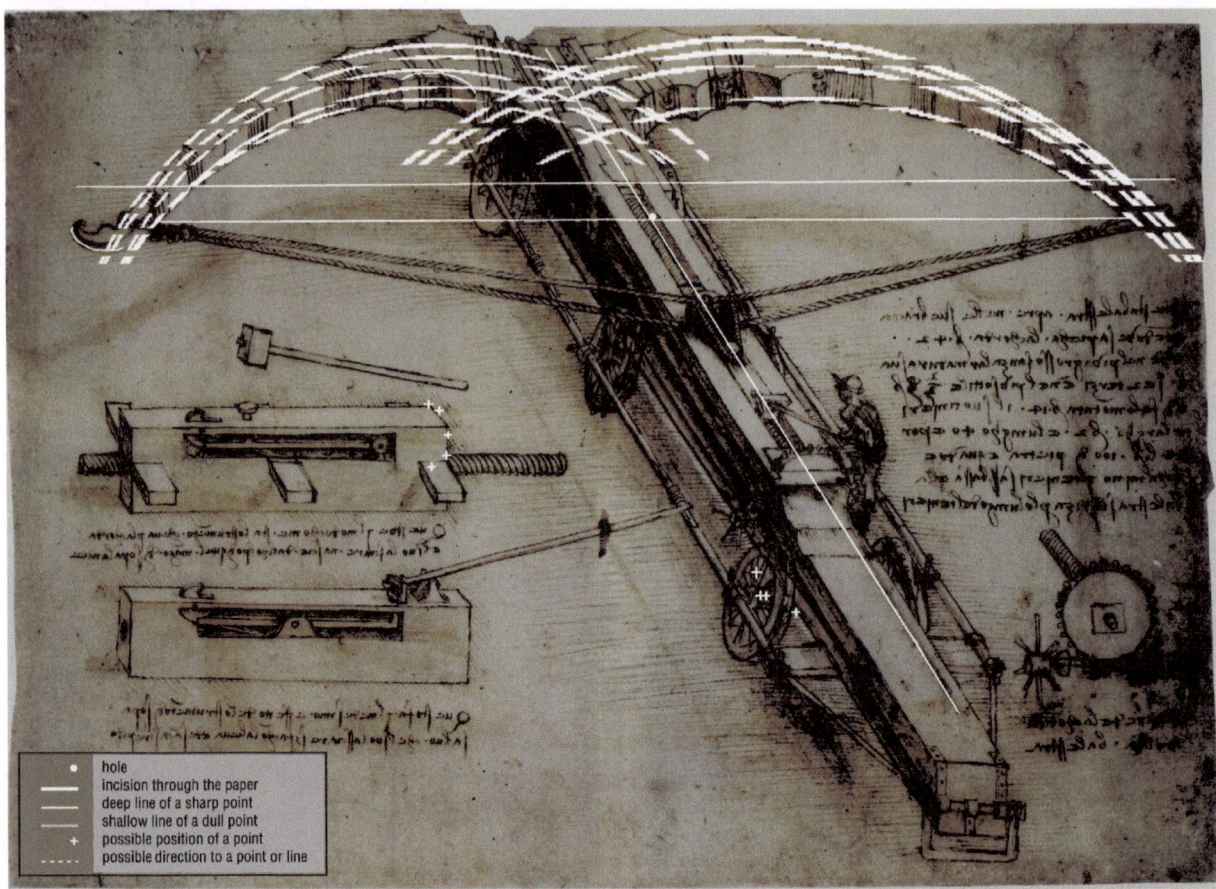

hole
incision through the paper
deep line of a sharp point
shallow line of a dull point
possible position of a point
possible direction to a point or line

1503–1506), Leonardo had acquired several reams (*lisime*) of paper, most probably of the large *reale* format.[10] Like the CA folios, most of the Windsor folios are less than half a *rezzuta*, less than 220×312 mm. Most of the sheets in these CA and Windsor groups happen to be trimmed by at least 10 mm vertically and 10 mm horizontally from the maximum size of a half *rezzuta*. These groups, as well as the Arundel group, consist primarily of centre-folded and cropped *rezzuti*. An untrimmed CA149b could be as much as 220×312 mm, suggesting that the vertical loss of as much as 10 mm and the horizontal loss of as much as 30 mm (down to 209×282.1 mm) was the result of the artist's and/or editor's crop of the drawing one or more times.

Fig. 8.11. Superimposition of white lines over CA 149br, illustrating the positions of a metal stylus incisions and possible directions of a divider when estimating the position of the armature

The Sequence of Marks

It is impossible to know the exact sequence of marks applied to CA 149br, although a good estimate of this process is possible with the help of an analysis of the metal

---

[10] The appropriate documents on this are translated and discussed by the following: Karl Frey (1909) *Studien zu Michelangelo Buonarroti und zur Kunst seiner Zeit*, Jahrbuch der königlich Preussischen Kunstsammlungung, Berlin 30:129, 133, 134 (nos. 137, 161, 164, 193, 198, 219); Luca Beltrami (1919) *Documenti e memorie riguardanti la vita e le opere di Leonardo da Vinci*, Milan, pp 84, 100 (nos. 137, 165); Christian-Adolf Isermeyer (1963) *Die Arbeiten Leonardos und Michelangelos für den grossen Ratsaal in Florenz ...*, Studien zur Toskanischen Kunst: Festschrift für Ludwig Heydenreich, Munich, pp 114–115, 118, 122 (nos. III, VI–VII, XI); and Carmen Bambach (1999) *The purchases of cartoon paper for Leonardo's battle of the Anghiari and Michelangelo's battle of Cascina*, Villa I Tatti Studies, The Harvard University Center for Italian Renaissance Studies, Florence, 8. A 500 sheet *lisime* (ream) consisted of twenty *quaderni* of 25 sheets each.

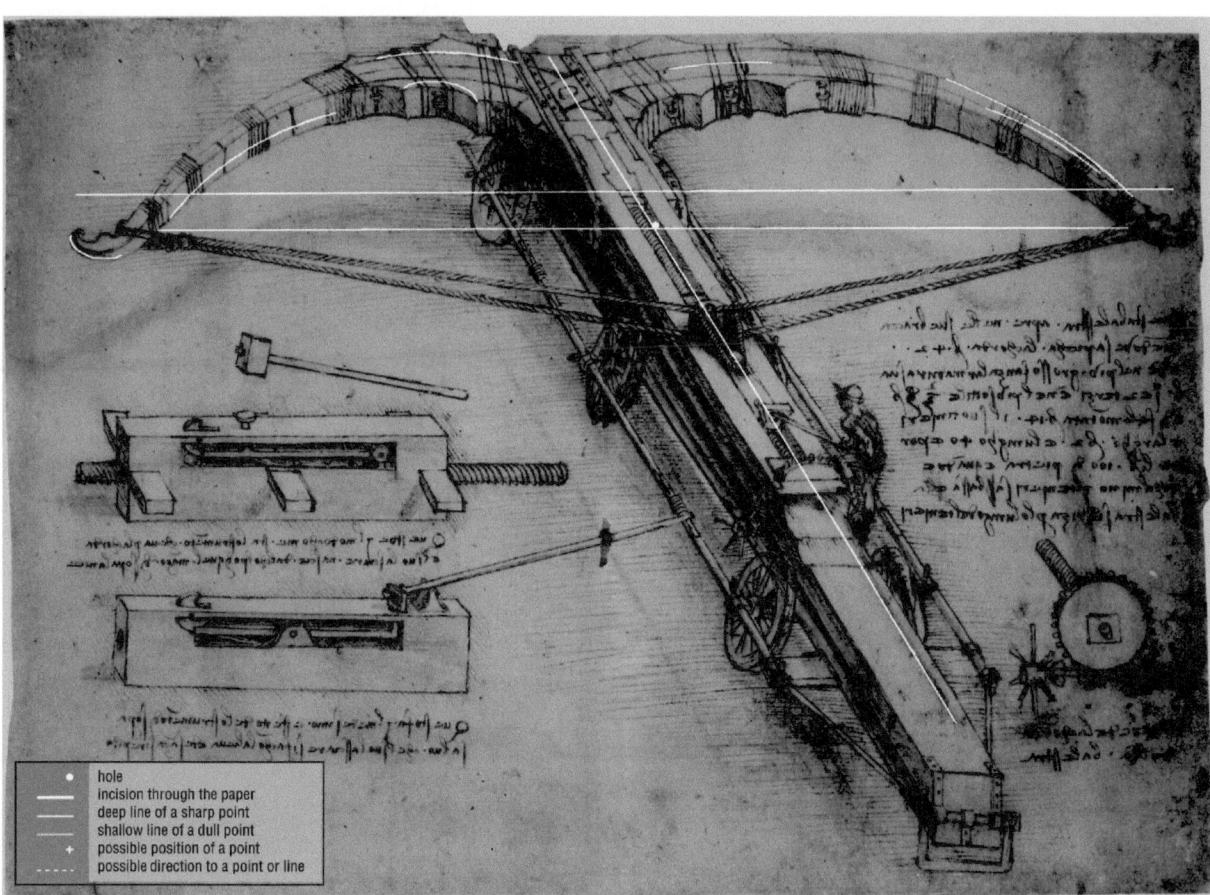

Fig. 8.12. Superimposition of white lines over CA 149br, illustrating the positions of metal stylus incisions

stylus marks on the drawing. Figure 8.20 illustrates with white lines the locations of metal stylus incisions on the sheet. As noted in the key to this figure, located at the lower left of the drawing, I have made five kinds of white marks on this and other illustrations between Figs. 8.8 and 8.20. There are four pinholes in the paper, referred to in the illustrations as white dots. I found these by placing a Mini Maglite® flashlight (torch) beneath the paper and looking for places where the light would shine through.[11] A thick white line represents a place on the upper carriage rear section where a metal stylus tip apparently ripped through the paper. Most of the metal stylus lines are relatively deep incisions, represented by thin white lines. One portion of the drawing, the upper trigger mechanism, has relatively shallow metal stylus lines, which I note on Fig. 8.20 with very thin white and grey lines. To better illustrate the possible methods for making the arcs of the armature, the "possible position of a point" and the "possible direction to a point or line" are noted with cross (+) and dash (– – –) marks, as seen in Figs. 8.8 and 8.11.

According to the prominent positions and thoughtful details of the armatures on CA 149br, 147av, 147bv and 57v-69ar, it would appear that this feature of the crossbow was an early consideration for the formatting of CA 149br. Two straight metal stylus lines extend between each end of the armature, as shown in Fig. 8.8. If one end of an

---

[11] The Mini Maglite® has a peak beam candlepower of 2 200 PBC. An infrared lens is available for this model of Maglite, though it was not used in the present case to study the drawing surface. In order to study CA 149br with infrared light in the Biblioteca Ambrosiana, I would have required a dark room and night vision goggles to view the drawing under the infrared beam. This kind of study would show some of the metal stylus incisions that are covered in ink.

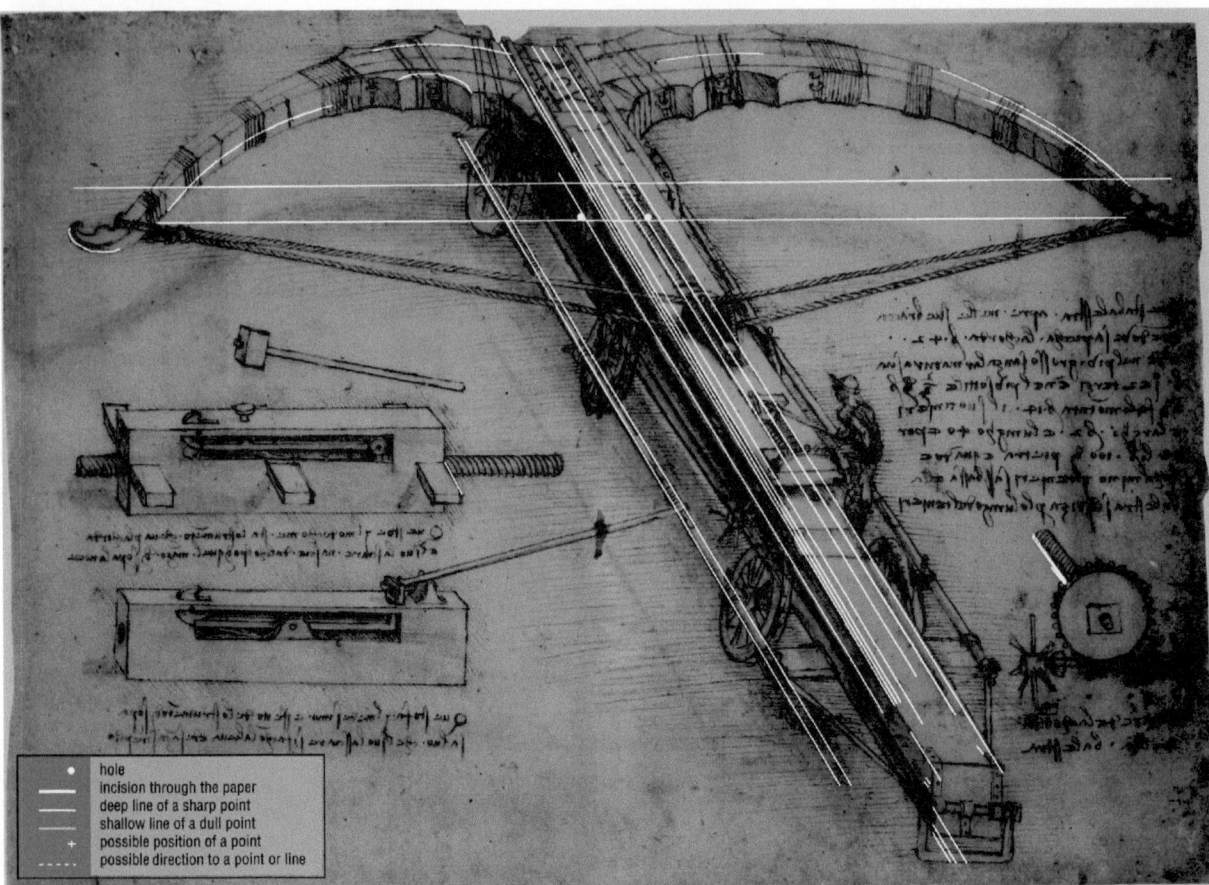

Fig. 8.13. Superimposition of white lines over CA 149br, illustrating the positions of metal stylus incisions

open divider were positioned at the base of CA 149br, as marked with a cross (+) on Fig. 5.1, the other end of the divider would make an arc in the form of the armature that extends to both ends of the upper horizontal metal stylus line. I mark with dashed lines this upper arc of the armature on Fig. 8.8. Before adding any metal stylus lines to CA 149br, it is likely that Leonardo determined the arc of its armature with the help of a divider. This estimated divider arc and its corresponding horizontal metal stylus line fill the upper horizontal portion of the page, an adjustment of which would make a critical change to the drawing's format. The carriage fits onto the design of this arc, with various adjustments to positions of metal stylus lines on the diagonal upper and lower portions. Illustrations of these diagonal adjustments are on Fig. 7.1.

If Leonardo had moved the bottom end of the divider (noted with a cross on Fig. 5.1), up the centre of the sheet onto a portion of 149ar [53va] (Figs. I.4 and 5.1), then it was possible to rotate the front end of that divider so that its lower arc crossed the previous arc. These arcs happen to cross one another at each end of the position of the armature, indicating that he used the crossed arc method to estimate the position of the armature. This is the standard medieval crossed arc method, as discussed by Cennini (*Libro dell'Arte*, ch. 67) for making a straight horizontal line at a right angle to a vertical line.[12] First, (as shown in Fig. 8.1) one would pivot a divider on its back end, in order to mark two sides of an arc with its front end. Second, one would move the divider's back end along the vertical position between the sides of the first arc, shifting that back end from a position below the first arc to a position above that arc. Third, one would pivot the divider on its back end, creating an arc with its front

---

[12] Cennini (1933) *The craftsman's handbook* (c. 1400), pp 42–43.

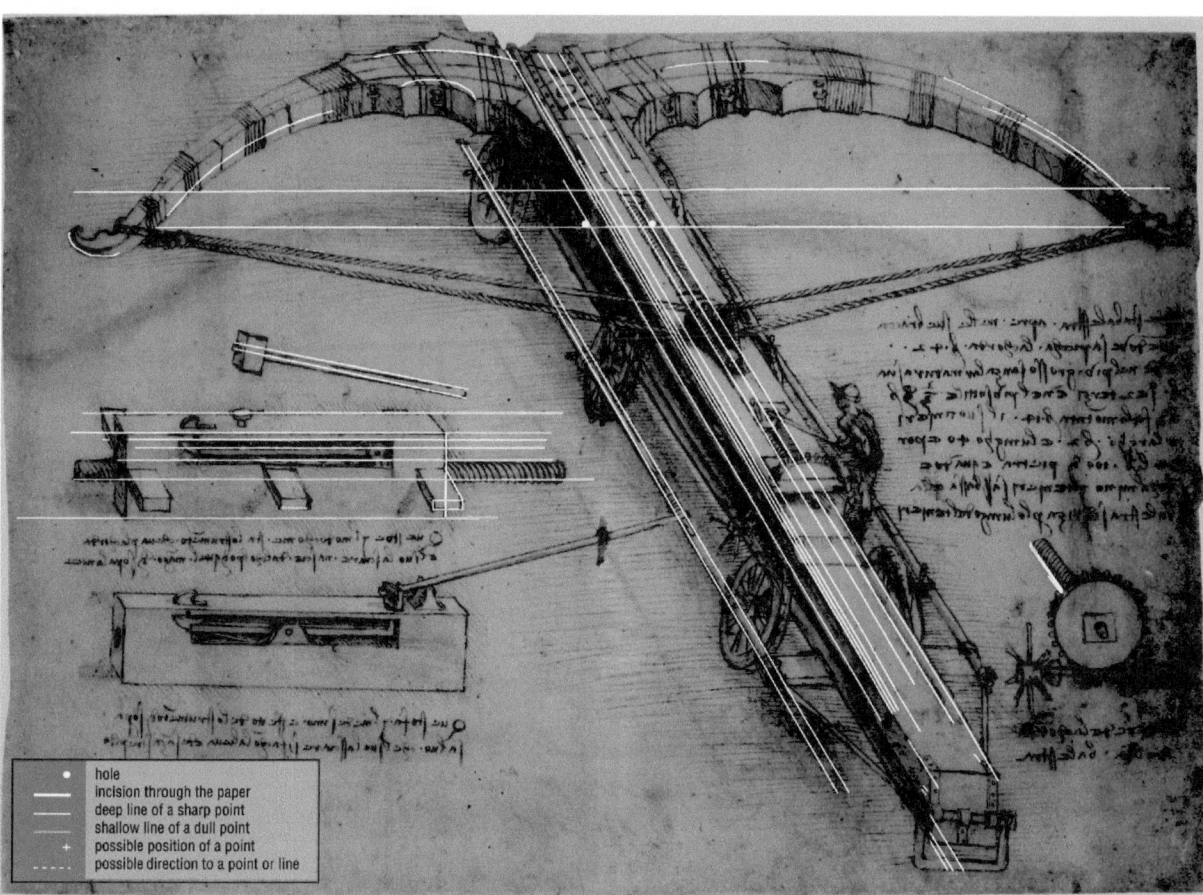

The legend in the image reads:

• hole
incision through the paper
deep line of a sharp point
shallow line of a dull point
+ possible position of a point
possible direction to a point or line

Fig. 8.14. Superimposition of white lines over CA 149br, illustrating the positions of metal stylus incisions

end that crosses the sides of the first arc. If a line is drawn between these two crossed arcs at both sides of the paper, as shown with the upper horizontal metal stylus line on Fig. 8.8, the resulting line would be at 90° to the vertical invisible line between both back ends of the divider. Dashes represent this invisible vertical line on Fig. 8.8.

The upper crossed arcs and the horizontal metal stylus line do not correspond to anything else on the drawing, other than the general positions of the upper edge of the armature and the lower edges of the ropes (Fig. 8.8). Thus it would appear that Leonardo decided to make a second set of crossed arcs approximately one centimetre below the first set. Pivoting the back end of the divider at the position of the cross at the bottom of the sheet (Fig. 8.8), he would have reduced the opening of the divider, making an arc that crosses the edges of the second arc. These crossed arcs are near the ends of a second horizontal metal stylus line, nearly a centimetre below the first horizontal metal stylus line. The ends of the armature meet the ends of the ropes at each end of the second horizontal metal stylus line.

On this lower line, one can see two pinholes, as illustrated with white dots on Fig. 8.20. The pinhole on the right happens to be at the point where a diagonal metal stylus marks the central portion of the front-end upper carriage. Figure 8.11 illustrates the position of this diagonal line, which extends from the front end of the carriage, through the circular post holding the armature beams, through the trigger nut, through the post end of the large gear, and to the rear portion of the upper carriage. Leonardo presumably used some form of protractor or ruler to determine the 60° angle of this line with respect to the horizontal line between rope ends.

The purpose of the two pinholes on the horizontal line is unclear. They would normally mark the previous positions of sharp divider ends, especially in the event of a divider-end pivoting at that point, in order to make crossed arcs with the front

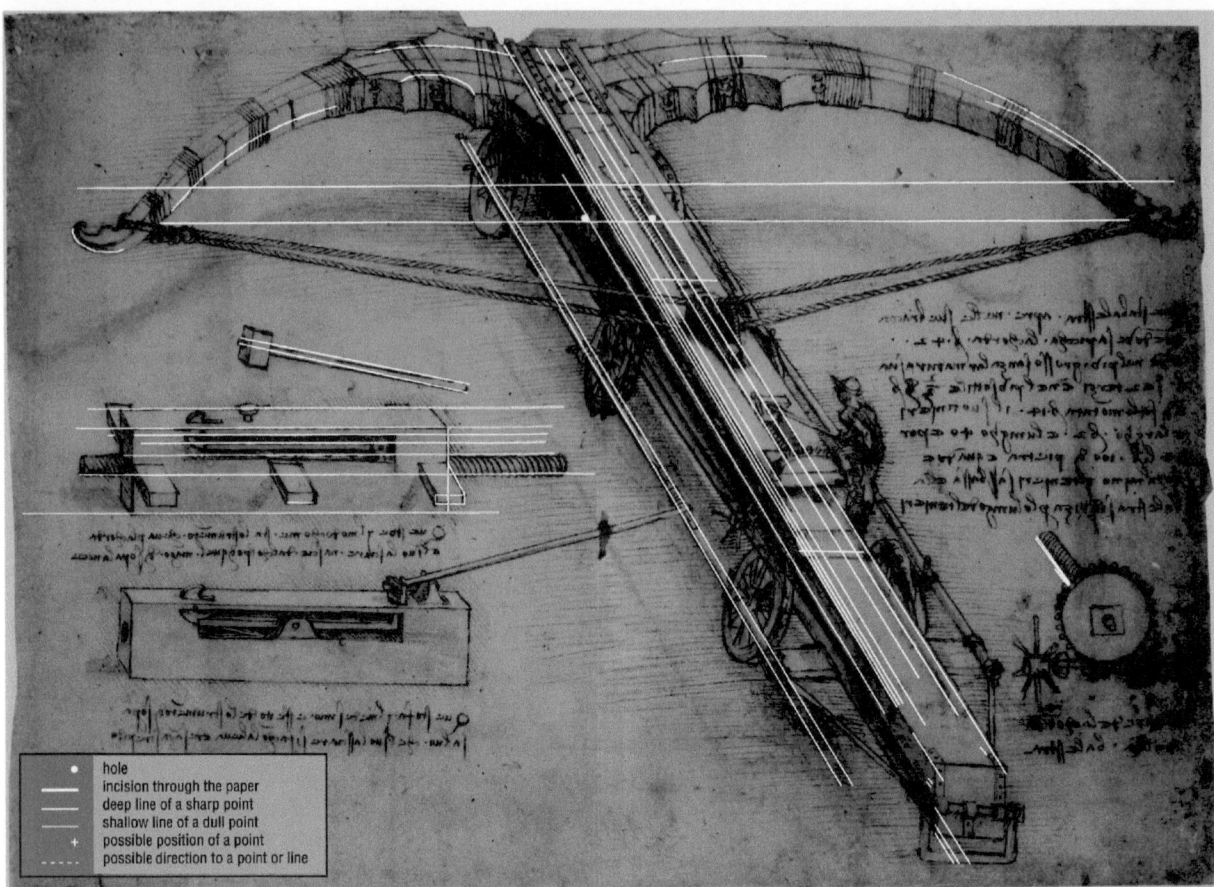

The following legend appears within the figure:

- • hole
- incision through the paper
- deep line of a sharp point
- shallow line of a dull point
- + possible position of a point
- - - - possible direction to a point or line

Fig. 8.15. Superimposition of white lines over CA 149br, illustrating the positions of metal stylus incisions

ends of the divider. Nonetheless, there is no evidence of crossed arcs resulting from the movement of a divider positioned on those two pinholes. Instead, most arcs on the page are the products of dividers that made no holes in the paper. The only exception to this is the hole at the centre of the gear at the bottom right of the page (Fig. 8.20). The circle and outer edge of the gear were probably made with the help of a divider, the other end of which punctured the paper at the gear's centre. Arcs that make the two sides of the armature are also in a circular form, their centres located near the upper trigger mechanism and the left rear wheel of the carriage. Figure 8.8 marks these centres or divider ends with crosses, and the upper arcs of the armature with dashes. As illustrated in Fig. 8.20, only a few minor metal stylus lines are visible on the armature. Other metal stylus lines or spots relating to the armature are presumably covered in ink and therefore impossible to read with the naked eye.

The group of 60° metal stylus lines on CA 149br, as illustrated in Fig. 5.1, mark various estimates of the carriage edges. Leonardo's reason for choosing the position of the front-end upper carriage's central line, as shown on Fig. 8.8, is not obvious. At best, the group of diagonal lines show that he considered several estimates of carriage dimensions. It is possible, for example, that the 60° lines that meet the lower horizontal line at the two pinholes on the upper portion of the sheet were the initial estimates for the carriage width. In any event, this 60° oblique projection of the carriage is a relatively consistent approach in CA 149br, 147av, 147bv, 148ar, and 57v-69ar. This proportional consistency between carriage and armature in each of the drawings might indicate that Leonardo used the edges of a 30°/60°/90° triangle template, as shown at the right of Fig. 7.1, to mark the straight metal stylus lines of the crossbow carriages. Once he had determined the horizontal line between the armature ends, he could then place the narrow bottom edge of the 30°/60°/90° triangle

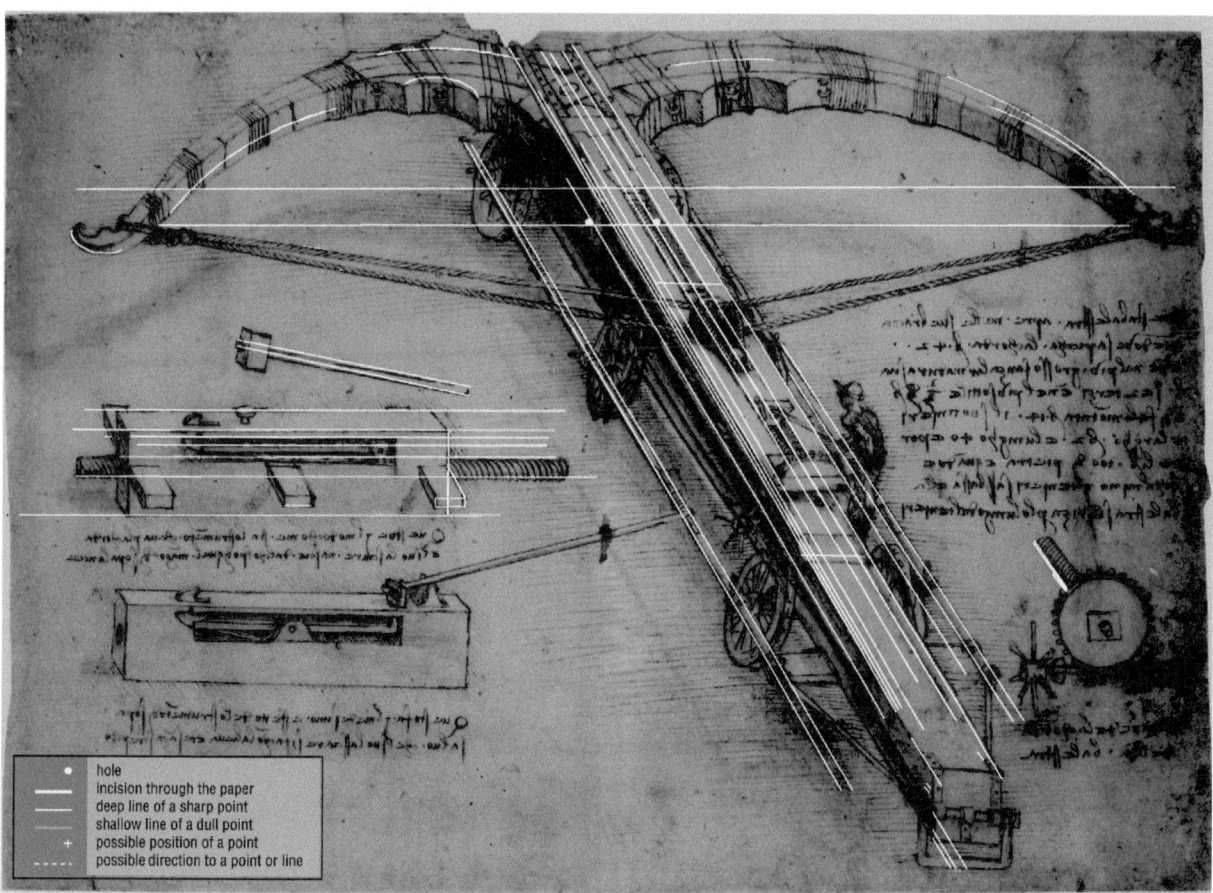

Fig. 8.16. Superimposition of white lines over CA 149br, illustrating the positions of metal stylus incisions

template against this horizontal line and easily draw the 60° angle carriage lines along the longest edge of the triangular template. He also drew these angles when designing the CA 157r cannon and the 175r crossbow trigger mechanism. As noted further above, the closest examples of this kind of drawing technique are Francesco di Giorgio's box enclosures for mills, water pumps and geared pulling devices on *Trattato di Architettura Civile e Militare*, MS Ashburnham 361 ff. 37v, 42v, and 44v though his box angles are relatively random and generally exaggerated, making the boxes look trapezoidal. This is the manuscript with Leonardo's marginalia on twelve of its pages. Both artists favoured the use of straight edge draughting tools and possibly triangle templates, though Leonardo avoided the trapezoid errors, and drew his *Giant Crossbow* to a scale that was relatively proportional to its illustrated features and its written measurements. This kind of precision in mechanical draughting, obviously the result of an elaborate and time-consuming approach, was an unprecedented contribution to engineering. Leonardo could have used other tools to determine the carriage angle, such as a protractor, or especially a sector, which is two rulers attached at one end so that they open at their other ends to various angles. The Florence Cathedral has on its maintenance studio walls two sectors similar to the type used by Brunelleschi (Fig. 8.4). Whatever tool was used to produce the crossbow carriage's oblique projection, it is useful to note in the present study Leonardo's consistent interest in the representation of the carriage at a 60° angle to the armature's 90° horizontal line. The carriage angle happens to be ⅔ the angle of the armature's horizontal (60 ÷ 90 = 0.667). This is consistent with the division of other crossbow portions into thirds.

To draw the right side-pole and the right upper edge of the front-end upper carriage, Leonardo shifted counter-clockwise by 2° the tool used to make the other diagonal angles of the carriage. Figure 8.20 illustrates with white lines the positions of

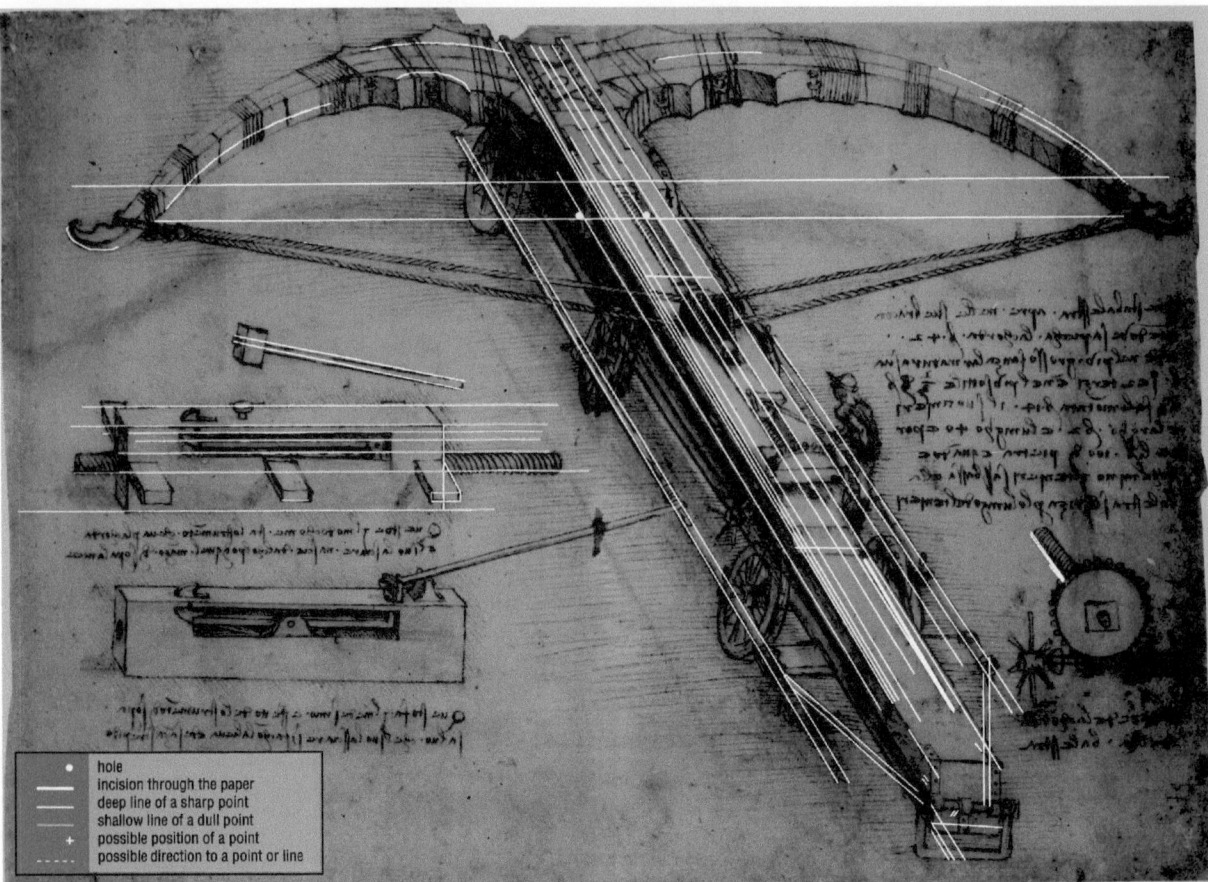

hole
incision through the paper
deep line of a sharp point
shallow line of a dull point
possible position of a point
possible direction to a point or line

Fig. 8.17. Superimposition of white lines over CA 149br, illustrating the positions of metal stylus incisions

these metal stylus lines on the right side of the carriage. The main reason for the alteration of the right side pole and front-end upper edge by 2° is unknown; they appear to be later additions to the other diagonal lines, meant to complete the right side of the carriage design. Sometime after the placement of the carriage's three right-side diagonal metal stylus lines, Leonardo sketched with a metal stylus the outer edges of the crossbowman.

It appears that he added the carriage's three right-side diagonal metal stylus lines after he added two horizontal lines marking the width of the upper carriage (Fig. 8.20). After adding these horizontal lines, he could have drawn with metal stylus the seven lines along the length of the upper trigger mechanism to the left of the sheet. These horizontal lines on the upper carriage and upper trigger are at 90° to the 60° diagonals of the carriage left side. This suggests that the same tool producing the 60° diagonals produced the 90° horizontals. One would only need to rotate by 180° a larger version of the 30°/60°/90° triangle at the right side of Fig. 7.1 and place that next to the 60° carriage diagonals in order to get horizontal lines at 90° to the carriage diagonals. Another way to produce these horizontal lines would have been to position a T-square or draughting square (left side of Fig. 8.4) at the 90° side of the page and shift it up or down, marking lines along its horizontal edge. The widths of both horizontal lines of the upper carriage equal in length and 90° angle the space between the two pinholes on the lower horizontal line between armature ends (Fig. 7.1). These horizontal marks refer to estimates of the upper carriage width. A horizontal metal stylus line of the same width, but at an angle that differs by 2°, is at the back end of the lower carriage, as noted in Fig. 8.20. This line also refers to the carriage width, though it was probably added to the drawing at a different time than the 90° lines on the upper carriage. It is also possible that other metal stylus lines cross the width of the

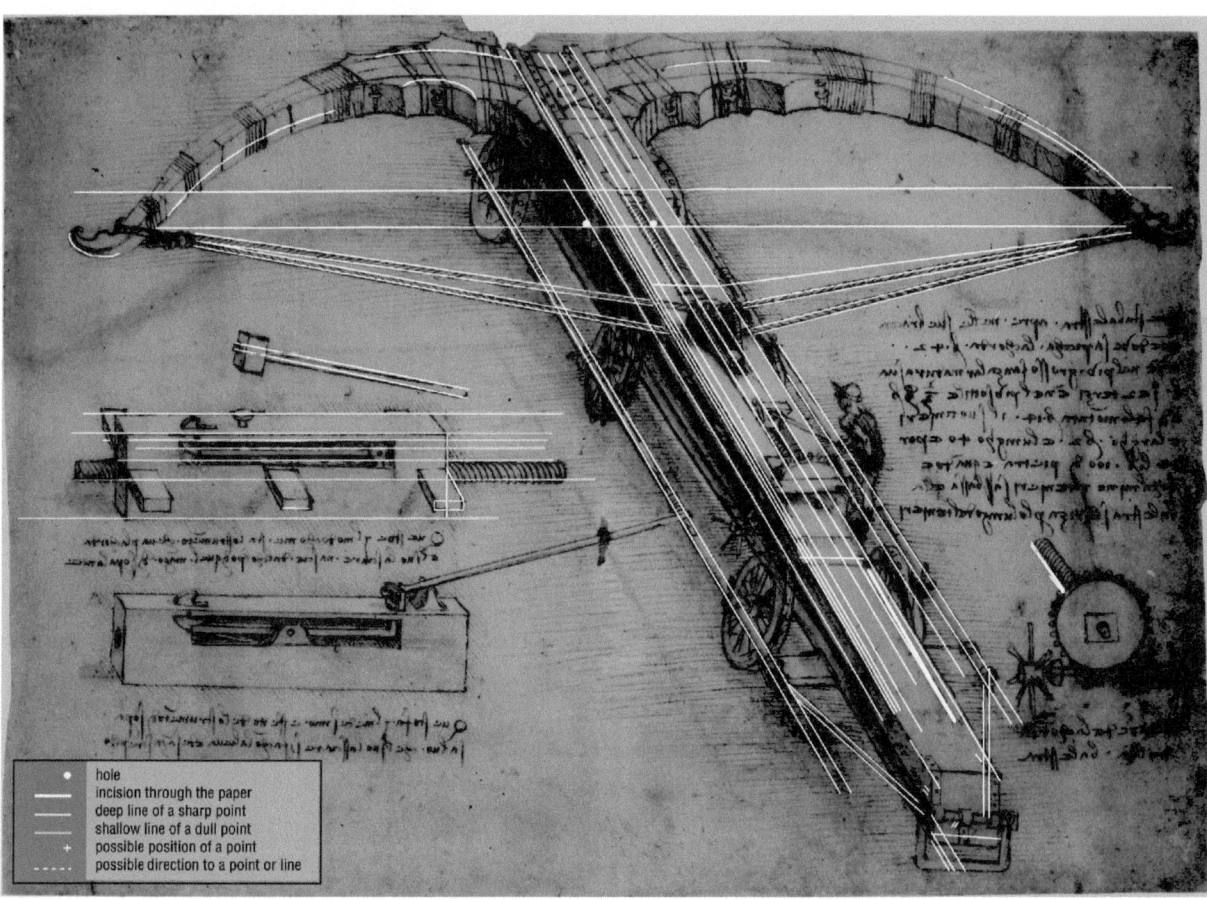

Legend:
- hole
- incision through the paper
- deep line of a sharp point
- shallow line of a dull point
- + possible position of a point
- ----- possible direction to a point or line

Fig. 8.18. Superimposition of white lines over CA 149br, illustrating the positions of metal stylus incisions

carriage, but are invisible to the naked eye due to the ink lines traced over them. After marking the width of the carriage with short horizontal lines, Leonardo could mark the outer edges of its right side and side poles with the three outer diagonal lines, repositioning a straight edge of a draughting tool by 2°.

Because the seven metal stylus lines along the length of the upper trigger mechanism are at 90° to the 60° carriage diagonals, one might expect that this trigger was drawn close to the same time as the upper carriage's two horizontal lines. These nine 90° horizontals are the only visible metal stylus lines on the page that are at a 90° angle to the 60° diagonals. The metal stylus lines on the upper trigger mechanism are unlike other lines on the sheet, however, in that they are relatively shallow marks. Thus, it is possible that this trigger's lines were incised into the sheet before or after all other metal stylus lines were incised. As a concept, this form of trigger, released with the blow of a hammer, did not apparently come before or after the concept of the lower form of trigger, the release of which, "is done with the lever and it is without clamour." This quote is the last portion of Leonardo's statement below the lower trigger mechanism. He considered both types of trigger mechanism on CA 1048br (Fig. 1.6), without an apparent preference for one or the other form of mechanism. According to the formats of CA 1048br and 149br, there appears to be no obvious reason for the positions of each of the triggers in their respective places. In other words, one cannot determine with certainty which trigger design Leonardo drew first or last on each sheet. The relatively shallow metal stylus lines on the upper trigger mechanism therefore bear a relationship to the rest of CA 149br's composition only in terms of their 90° angles to the 60° carriage diagonals.

Metal stylus lines on the upper carriage are also similar in angle to the lines of the upper trigger mechanism's right side, along the diagonal of its right sidebar, and

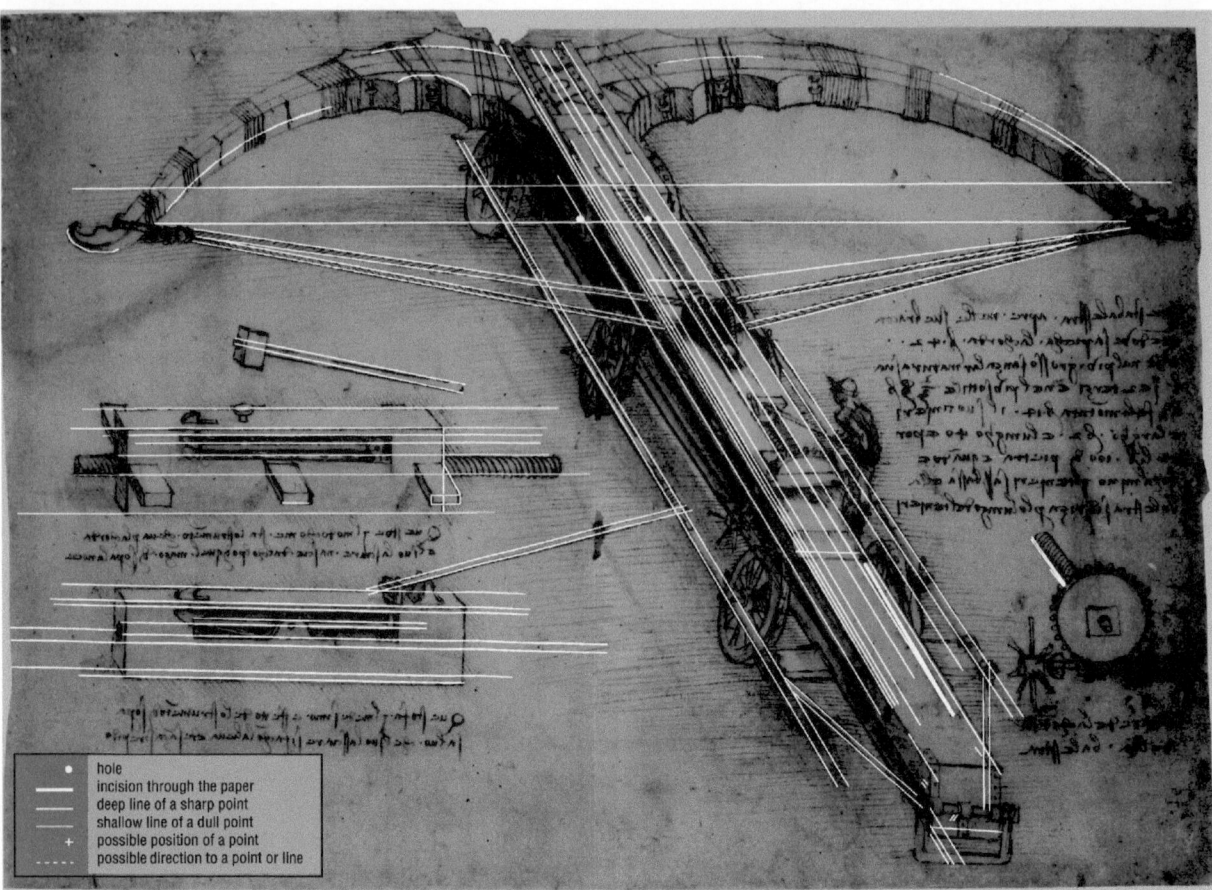

legend:
● hole
— incision through the paper
— deep line of a sharp point
— shallow line of a dull point
+ possible position of a point
----- possible direction to a point or line

Fig. 8.19. Superimposition of white lines over CA 149br, illustrating the positions of metal stylus incisions

around the rectangular end of this right side bar. The sidebars are at a 60° angle to the perpendicular metal stylus lines of the rest of the trigger. Given the similar angles of this design's metal stylus lines and many of the upper carriage's lines, it would appear that the upper trigger mechanism's metal stylus incisions were drawn just before or after some of the upper carriage and armature positions were marked with metal stylus incision. Moreover, as shown in Fig. 8.11, the upper right edge of the upper trigger appears to be the location of the centres of five circles, the circumferences of which make the edges of the left side of the armature. No pinpoint indentations or holes are visible in this area, though the places marked with a white cross (+) could have been the back end locations of a divider making with its front end circular estimates of the left side of the armature.

Extending from the upper horizontal metal stylus line of the upper carriage, just above the ropes, is a metal stylus line that links the carriage right side with the armature's right tip (Fig. 8.20). It would appear that this line was an early estimate for the placement of the ropes. Thereafter, one might have expected Leonardo to add the eight neatly drawn metal stylus lines for ropes of the crossbow. Note that these lines end at the edges of the upper carriage, suggesting that their additions to the drawing follow the additions of the upper carriage diagonals. Furthermore, the left edge of the upper horizontal metal stylus line is directly below the hole at the centre of armature's arc. Figure 8.11 shows with a dashed line this central division of the circular form that marks the general arc and position of the armature. To note the centre of this circle, a white cross marks the possible position of the back end of a divider at the bottom of the sheet. The triangle formed by the positions of the central dashed line, the 60° carriage diagonal line, and the horizontal line across the upper carriage happens to be a 30°/60°/90° triangle. The consistent uses of these

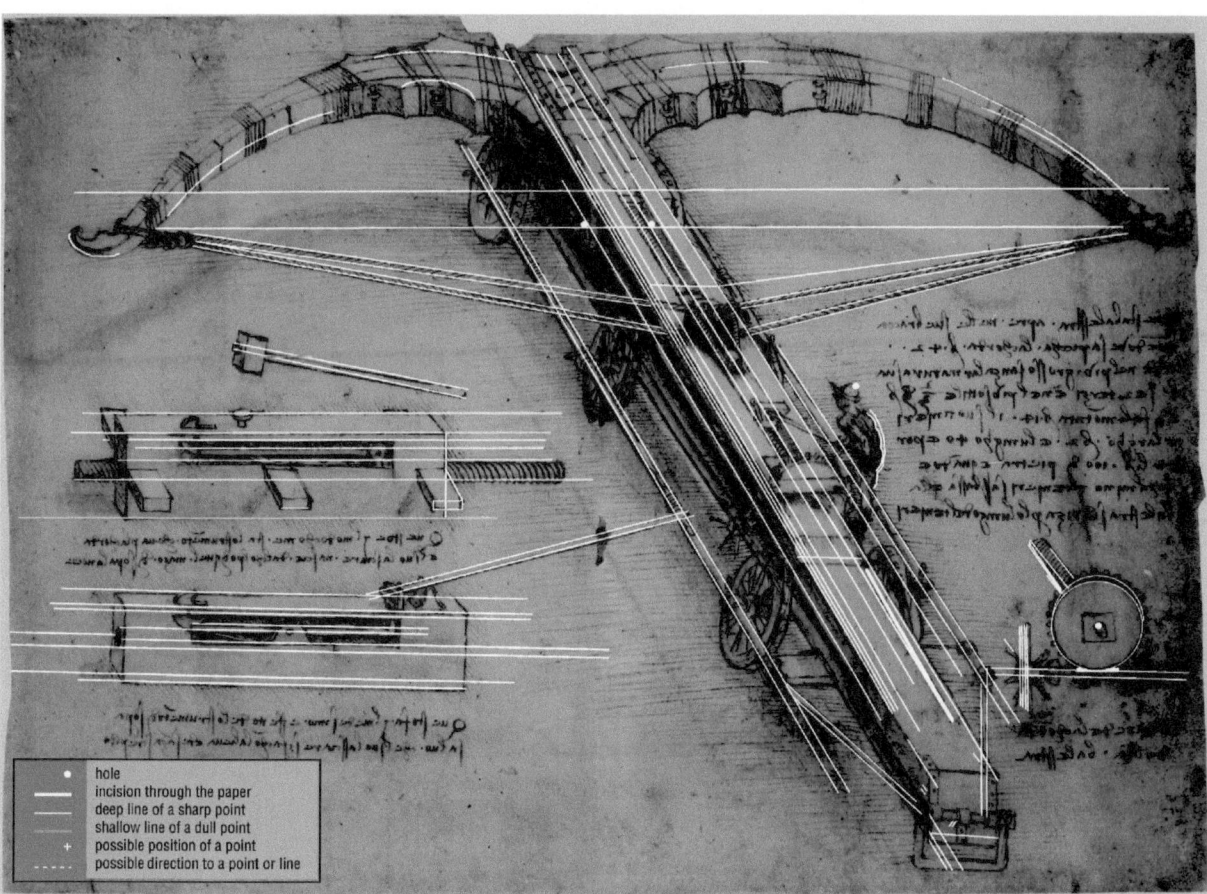

hole
incision through the paper
deep line of a sharp point
shallow line of a dull point
possible position of a point
possible direction to a point or line

Fig. 8.20. Superimposition of white lines over CA 149br, illustrating the positions of metal stylus incisions

angles for parts of the carriage, armature, and upper trigger mechanism reflect Leonardo's meticulous interest in the crossbow design's geometrical and proportional accuracy.

At the bottom of Fig. 8.20, metal stylus lines and some of the ink lines for the lower trigger mechanism and worm screw mechanism are at the same angles to one another. These lines differ from the other 60° and 90° angle lines on the sheet by only 2°. It appears as though Leonardo shifted the upper end of his draughting tool to the right by 2° when making the perpendicular and 60° lines on the lower trigger and the worm screw. In the case of the right side of the carriage, the three outer metal stylus lines express a shift of the upper portion of his draughting tool by 2° to the left. Since he made the lines with his left hand, the right hand shifted the draughting tool or ruler, most probably *without* an interest in changing the angles of the drawing's lines. The slight angular shift of the draughting tool was possibly a natural shift of the hand when moving the metal stylus along the sliding draughting tool. What is remarkable is the exact shift on most of the drawing of lines 2° to the left or to the right of a set of 30°/60°/90° angles. For example, the third horizontal metal stylus line from the bottom of the lower trigger mechanism is in line with the upper horizontal metal stylus line on the worm screw bar at the right. Both of these lines continue along the same trajectory, suggesting that they were drawn along the same straight edge of a draughting tool. It is possible that Leonardo drew the lower trigger mechanism first, at an angle 2° higher on the left, compared to the upper trigger mechanism. He then extended to the right of the page a straight edge so that the third metal stylus line from the bottom of the lower trigger would be in line with an illustration of the worm screw mechanism on the left. In this way, he made the lines of both lower illustrations perpendicular to one another. As for some of the lines on the central wind-

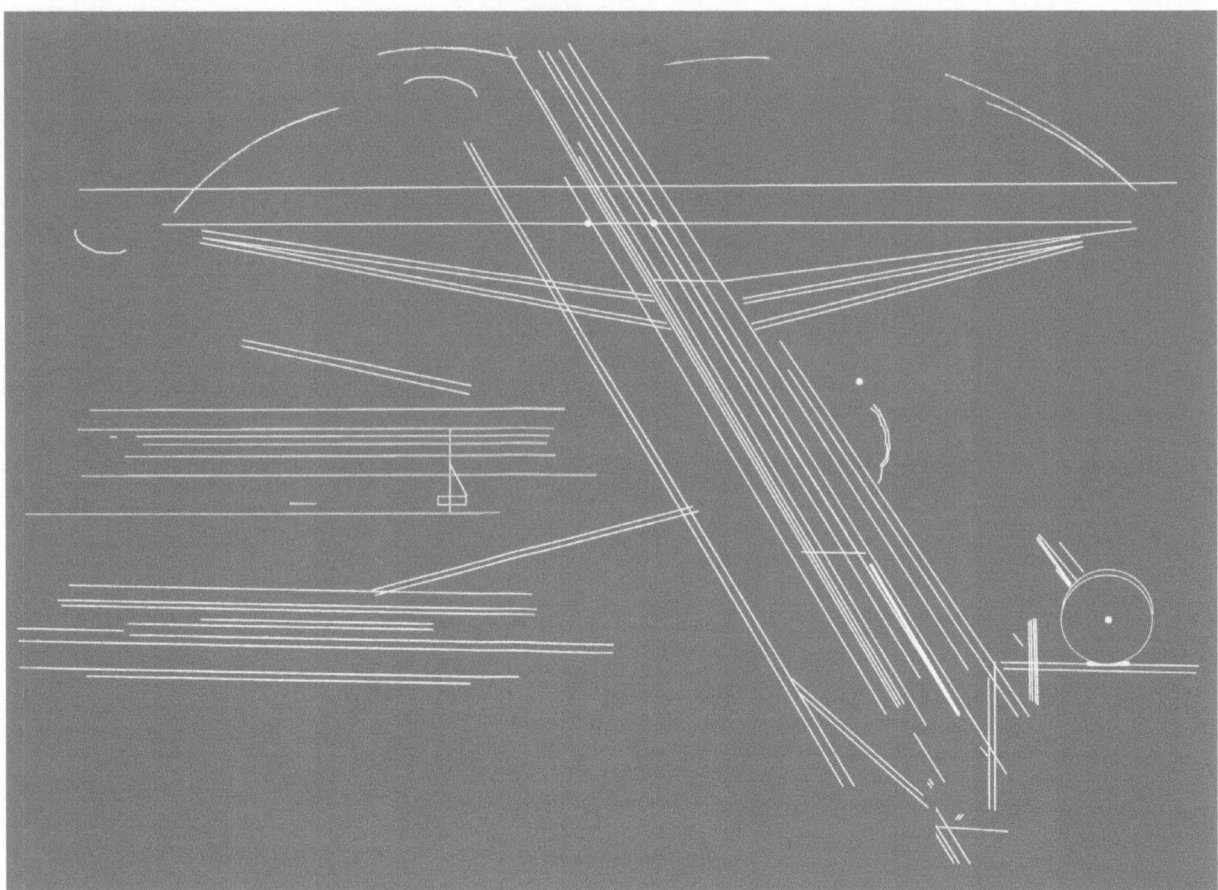

Fig. 8.21. Illustration the positions of metal stylus incisions that originally appeared on CA 149br

ing screw on the worm screw mechanism and ink lines for the lower trigger box's diagonals, these are also 2° higher on the left side than the 60° angle neighbouring lines on the carriage and upper trigger.

Because of its insertion into a very narrow space, it appears as though the worm screw mechanism was one of the last additions to the sheet. Leonardo possibly ran out of room elsewhere on the drawing for this mechanism and therefore tried to fit it into its present spot. By placing the worm screw low on the right, he left room for the text that would be written above it.

This concludes a brief discussion of the metal stylus marks found on CA 149br. A more extensive study of preparatory marks on this drawing and its neighbouring drawings in the Codex Atlanticus would offer much more information about Leonardo's draughting procedures. To continue such work within the framework of the present study, however, would not necessarily assist the present argument about proportion theories in his early work. The presence of numerous metal stylus marks on CA 149br offers ample evidence of his particular interests in proportional preparatory draughting techniques. The drawing's consistently applied 60° and 90° angles reflect the ⅔-to-one (0.667 : 1) ratio of proportions applied elsewhere in the design, as illustrated in the carriage components, and as specified in the text on the right of the folio. Thanks to a reading of the metal stylus marks on CA 149br, one can get a sense of the careful systematic sequence of its graphic construction. Figures 8.3 and 8.5 through 8.20 illustrate the possible sequence of metal stylus incisions on CA149br, as a group of thirteen successive examples. This sequence or process of marks demonstrates a proportional mechanical draughting technique unlike anything that came before it.

Conclusion

As confirmed by the present study, an analysis of the features of the Giant Crossbow presentation drawing, along with its preparatory drawings, historical context, and related documents, show that Leonardo intended to offer in a treatise on military engineering a thorough assessment of his giant crossbow's design, construction and operation. To this end, he used proportional techniques to produce the first known dimetric scale diagram, the measurements and proportions of which military engineers could have understood and could have used to build the machine. Making sure that the geometry of the design was accurate, which would have conceivably convinced Renaissance engineers of its structural integrity, he used Euclidian and Archimedean calculations to determine the crossbow's shape and appearance. Although certain portions of the *Giant Crossbow* are not at the same scale, the general proportional consistency of the design suggests that Leonardo associated the size and proportions of the built project with the scale and proportions of the drawing, to the point of creating a drawing at ⅟₁₀₈ the scale of the proposed project. He might have considered this scale because it is divisible by three, since he also designed each of its components according to ⅓ portion proportions. The present monograph also confirms for the first time an appropriate English translation of the text on CA 149br, and the large number of related crossbow drawings that link the *Giant Crossbow* to the general military engineering programme of Leonardo's at the Sforza Court. Although this programme is generally dated to around 1485–1488, the present study concludes that there is sufficient evidence to suggest that he had been at work on the military engineering programme as early as 1483 and as late as 1492, the possible dates for the *Giant Crossbow*. This drawing is obviously more than the "paper" engineering project that previous scholars have considered it to be. It is a thoughtful contribution to a planned military engineering treatise. As such, this treatise would have offered practical—although very difficult to build—proposals for Ludovico Sforza's military engineers.

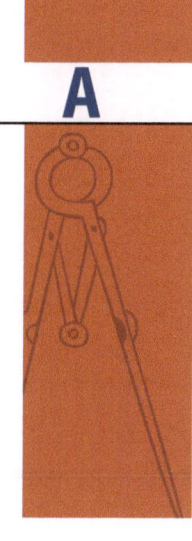

Appendix A
Measurements

The Measurements and Proportions of Leonardo's *Giant Crossbow*

Table A.1 (see p. 142/143) lists the *Giant Crossbow* drawing's actual and proposed measurements and proportions (in terms of scale), with respect to its proposed original size.

I.   Location of a section of the crossbow on the drawing.
II.   Measurement of a section of the crossbow drawing.
III.   Leonardo's written specifications of a section of the crossbow, as written on the drawing.
IV.   The illustrated length of a *braccio*, according to a measurement on the drawing and the proposed length of this section of the crossbow if built to full scale.
V.   The estimated scale of a section of the crossbow with respect to its proposed length if built to full scale.
VI.   The estimated number of *braccia* that a section of the crossbow would measure if built to full scale.
VII.   The estimated number of metres that a section of the crossbow would measure if built to full scale.
VIII. The estimated number of feet that a section of the crossbow would measure if built to full scale.
IX.   Additional notes, particularly with regard to methods for the determination of a measurement or proportion.

Table A.1. *Giant Crossbow* drawing's actual and proposed measurements and proportions

Place on drawing	Orig. cm	Leonardo's written specifications	Ill. br	Est. scale	br	m	ft	Notes
Stylus incision between ends of armature	22.7	where the rope is attached, 42 *braccia*	0.54	108	42	24.51	80.4	2451.12 cm ÷ 22.7 cm = scale
Between outer tips of rope at armature ends	24.52	where the rope is attached …	0.58	100	42	24.52	80.4	A possible armature scale of 100 : 1
Radius of circle outlining the armature perimeter	21.6		0.54	108	40			
Length of armature outer perimeter	34		0.54	108	63			c = (⅔)(2r); 54 = 62.84
Upper carriage "thickness" near the worm screw	1.23	thickest, without armature, 1 and 2 thirds *braccia*	0.74	80	1⅔	0.975	3.19	
Upper carriage top section at worm screw	0.49	at its thinnest, ⅔ of a *braccio*	0.74	80	⅔	0.39	1.28	
Lower carriage at worm screw	0.49	(at its thinnest, ⅔ of a *braccio*)	0.74	80	⅔	0.39	1.28	
(Not illustrated)		an elevation of 14 *braccia*	0.74	80	14	8.17	26.80	58.36 br × 14 = elevation, ground to armature; scale re. "thickness" of carriage
(Not illustrated; distance betw. top of lower carriage and bottom of armature at an elevation of 14 br)			0.74	80	10⅔	6.2	20.4	14 − (2⅓ distance betwe. ground to cart + ⅓ axle thickness + ⅔ cart thickness)
(Not illustrated; 16.5° angle, at the carriage hinge, at an elevation of 14 br)								sin θ = height of 10⅔ br ÷ hypotenuse of 37.5 br = 0.284 sin θ = 16.5°
Carriage width at the rear end	1.62	carriage is 2 *braccia* wide	0.81	72	2	1.167	3.83	Illustrated 33% too wide; orig. would be 66% of illustrated width; ⅓ cr
Carriage width at the front end	1.23	carriage is 2 *braccia* wide	0.74	80	2	1.167	3.83	
(Lower) carriage length	1.23	carriage is … 40 long	0.74	108	40	23.34	76.6	108 scale, because length stylus incision was applied close to the time that the width stylus incision was applied; both are similar sizes
Illustrated upper carriage length	20	carriage is … 40 long	0.54	108	37⅓	21.78	71.47	Upper section is only 37.3 br, lower carriage extends another 2.66 br
Trigger mechanism width	0.54	⅓ of 2, if "carriage is 2 *braccia* wide"	0.54	108	⅔	0.39	1.27	
Illustrated trigger mechanism width	0.54		0.54	108	1	0.5836	1.91	
Illustrated screw dia.	0.19		0.54	108	⅓	0.194	0.64	Various measurements
Illustrated trigger mechanism height	1.23		0.74	80	1⅓	0.97	3.20	Top trigger mechanism, left of drawing

Table A.1. *Continued*

Place on drawing	Orig. cm	Leonardo's written specifications	Ill. br	Est. scale	br	m	ft	Notes
Illustrated trigger mechanism length	2.68		0.54	108	5	2.92	9.57	
Illustrated distance between side poles	4.9		0.74	80	6⅔	3.92	12.86	
Illustrated "thickness" of wheel, felloe, hub, spoke, trigger braces, central side poles, side pole brace, U-bolt, etc.	0.19		0.54	108	⅓	0.19	0.63	7.56"; presumably an approximate expected or recommended thickness for planks and poles
Wheel diameter	2.7		0.54	108	5	2.92	9.57	2⅓ br upper wheel + ⅓ br axle dia. + 2⅓ br lower wheel = 5 br
Illustrated distance of the carriage from the wheel spoke near the hub on the inside of the rear wheel	0.81		0.81	72	1	0.58	23	
Illustrated distance of the carriage from the wheel spoke near the hub on the inside of the central and front wheels	0.54		0.54	108	1	0.58	23	
Wheel height when tilted			0.54	108	4⅔	2.72	8.94	Armature rests on wheel and prop
Est. distance of wheel top edge from carriage					⅓	0.19	0.64	A clearance of at least 7.56" would be reasonable, though this is not illustrated
Illustrated and calculated tilt of wheels	0.54		0.54					An avg. of 72° (tan θ74°, cos θ74°, sin θ69°; ½ wheel = 2.5 br hyp; ½ tilted height = 2⅓ br opposite side; ⅔ br adjacent side)
Distance of lower carriage from ground	1.26		0.54	108	2⅔	1.55	5.1	2⅓ br + ⅓ br axle diameter
Illustrated carriage elevation (estimate)			0.54	108	2⅓	1.36	4.47	Armature rests on wheel and prop; angle of incline is 3.56°
Illustrated height of crossbowman	3.36		0.74	80	4½	2.69	8.82	
Est. distance betw. rear axle and rear carriage	6.77		0.54	108	12⅔	7.39	24.25	20 cm ÷ 37.3 = 0.536 cm / br; approx. ⅓ of 40 br
Est. distance betw. central and rear axles	6.80		0.54	108	12⅔	7.39	24.25	Approx. ⅓ of 40 br
Est. distance betw. front and central exles	5.07		0.54	108	9⅓	5.44	17.86	
Est. distance betw. central axle and front of lower carriage	6.94		0.54	108	12⅔	7.39	24.25	Approx. ⅓ of 40 br
Est. distance betw. front axle and front lower carriage that extends below the armature			0.54	108	2⅔	1.55	5.1	

Lengths[1]

braccio a panno		
fiorentino	58.36	cm = 20 *soldi* = 60 *quattrini* = 240 *denari* = 2 880 *punti*
soldo	2.918	cm = 3 *quattrini* = 12 *denari* = 144 *punti*
quattrino	0.9727	cm = 4 *denari* = 48 *punti*
denaro	0.2432	cm = 12 *punti*
punto	0.0203	cm

braccio mercan-tile milanesi	59 cm (58.97 cm c. 1500, 59.49 cm c. 1600) = 12 *once* = 144 *punti*
trabucco	6 *piede*, 2.61 m (8' 7")
gettata	2 *trabucchi*, 5.22 m (17' 2")

braccio o pas-setto romano	67 cm = 3 *palmi romani* = 36 *once*

large cubit (Roman)	A Roman ell, between 55.5 cm and 59.33 cm; Pliny states that this is 7 palms, or 28 fingers
small cubit (Greek a. Roman)	One forearm, 45 cm; Vitruvius refers to this as a quarter of the body, or 6 palms, or 24 fingers
piede	12 *once* = 144 *punti* = 43.5 cm
oncia	12 *punti* = 144 *atomi* = 36.25 mm
punto	12 *atomi* = 3 mm
atom	0.25 mm

Weights

libbra	339.542	g = 12 *once*
marco milanese	235	g = 8 *once* (half a modern US pound)
oncia	28.295	g = 24 *denari*
denaro	1.179	g = 24 *grani*
grano	0.049	g

---

[1] Angelo Martini (1883/1976) *Manuale di metrologia*, E. Loescher, Turin 1883, and Editrice Edizioni Romane d'Arte, Rome 1976. Giovanni Croci (1860) *Dizionario universale dei pesi e delle misure in uso presso gli antichi e moderni con ragguaglio ai pesi e misure del sistema metrico*, Milan. Other sources on early Milanese as well as Florentine measurement standards include: Hercule Cavalli (1874) *Tableaux comparatifs des mesures, poids et monnaies modernes et anciens. ...*, Paul Dupont, Paris, 2. Ephraim Chambers (1728) *Measures*, in: *Ciclopædia: or an universal dictionary of arts and sciences*, vol. 2, London. Colonel Cotty (1819) *Aide-mémoire a l'usage des officiers d'artillery de France*, 2, Paris, pp 896–897. Denis Diderot and Jean Henri le Rond d'Alembert (1751–1765) *Pied*, in: *Encyclopédie ou dictionnaire raisonnée des arts, sciences et métiers* 7, Paris, pp 562–563. Horace Doursther (1840) *Dictionnaire universel des poids et mesures anciens et modernes*, Brussels. Anonymous (1935–1943) *Piede*, in: *Enciclopedia italiana di scienze, lettere ed arti* 27, Rome, p 168. Ludovico Eusebio (1899/1967) *Compendio di Metrologia Universale e Vocabolario Metrologico*, Unione Tipografico Editrice Torinese, Turin 1899, and Forni Editore, Bologna 1967. Luciana Frangioni (1992) *Milano e le sue misure. Appunti di metrologia Lombarda fra Tre e Quattrocento*, Edizioni Scientifiche Italiani, Naples. Anonymous (1970) *Misure*, in: *Grande dizionario enciclopedico* 12, Unione Tipografico-Editrice Torinese, Turin, p 626. Johann Georg Krüniß (1788) *Öconomische Encyklopädie oder allgemeines System der Staats-, Stadt-, und Landwirtschaft, in alphabetischer Ordnung* 15, Jospeh Georg Traßler, Brünn, pp 519–522. L. Malvasi (1842–1844) *La metrologia italiana ne' suoi cambievoli rapporti desunti dal confronto col sistema metrico-decimale*, Fratelli Malvasi, Modena. Barnaba Oriani (1891) *Istruzione su le misure e su i pesi che si usano nella Repubblica Cisalpina*, Milano. Luigi Pancaldi (1847) *Raccolta ridotta a dizionario di varie misure antiche e moderne coi loro rapporti alle misure metriche ...*, Sassi, Bologna. Anonymous (1803) *Tavole di ragguaglio fra le nuove e le antiche misure ... del Regno d'Italia publicate per ordine del Governo* 1, Stamperia Reale, Milan.

Appendix B
Drawings That Were Closely Associated with the *Giant Crossbow* Project

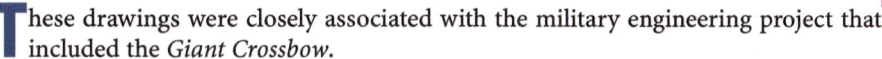

These drawings were closely associated with the military engineering project that included the *Giant Crossbow*.

B 5v through D	(c. 1485–1490)	Assorted military studies, particularly on B ff. 5v-11v, 20r-21r, 23v-25r, 27r-v, 30v-33v, 35r-37r, 39v, 40v-56r, 57v-70r, 73v-93r, 94v-95r, 90r-100v, A.1, A.2, B.1, B.2, D.
B 7v	(c. 1483–1490)	Crossbows and dart throwers
B 27v	(c. 1483–1490)	Toothed laminations
B 50v	(c. 1483–1490)	Dart
B 62v	(c. 1483–1490)	Note about soaking willow in oil
B 74r	(c. 1483–1490)	Bat wing structure
CA 57v [17ra]	(c. 1482–1490)	Giant crossbow sling
CA 90r [33ra]	(c. 1480–1490)	Toothed laminations with linchpins
CA 112ar [39vc]	(c. 1487–1490)	Screw-load crossbows
CA 113r [40ra]	(c. 1485–1490)	Cranequin
CA 142r [51rb]	(c. 1485–1490)	Spring trigger auto drop/load
CA 143r [51rc]	(c. 1485–1490)	Auto drop/load mechanism
CA 147av [52va]	(c. 1483–1490)	Great crossbow
CA 147bv [52vb]	(c. 1485–1490)	Armature design
CA 148ar [53ra]	(c. 1485–1490)	Ratchet pawls
CA 149ar [53va]	(c. 1485–1490)	Screw-loading crossbow lever
CA 149br [53vb]	(c. 1485–1490)	*Giant Crossbow*
CA 153r [54vb]	(c. 1485–1490)	Fold-loading and shooting mechanisms, with circular trigger
CA 154br [55vb]	(c. 1485–1490)	Study for a cannon carriage and wheels
CA 155r [56ra]	(c. 1485–1490)	"This crossbow has the nut that goes with the cord"
CA 175r [61vb]	(c. 1485–1490)	"When the stirrup 'c' will be the entrance of its nut and you take the curved neck 'a' back to its shoulders 'b' 'n', also send the cord with its nut…"
CA 175v [61rb]	(c. 1485–1490)	Calculations of 8, 100, 400, etc.
CA 754r [278rb]	(c. 1490)	Spring-loading arrow launcher, double crossbow
CA 1048br [376rb]	(c. 1485–1490)	*Giant Crossbow* triggers, ratchet and windlass
CA 1054r [379ra]	(c. 1478–1490)	Triggers, armatures, pulleys, trough, worm screw
CA 1054v [379va]	(c. 1478–1490)	Screw sectioning with box clamps
CA 1058r [381ra]	(c. 1485–1490)	Screw tensile strength, visual studies
CA 1063v [384rb]	(c. 1487–1490)	Square frame with inset double armature and windlass loading mechanism, other studies of bows
CA 1070r [387ra]	(c. 1485–1490)	Giant treadmill for rotating and shooting four crossbows in rapid succession, with protective shield, presentation drawing
CA 1071v [387rb]	(c. 1485–1490)	Initial design for 1070r
CA 1094r [394rb]	(c. 1483–1490)	Large armature for a stone-launching catapult

CC 20r&v	(c. 1504–1508)	Crossbow tension strengths
CT p. 99 (54r)	(c. 1485–1409)	Crossbow
UF no. 446ev	(c. 1478)	Screw, windlass, ratchet, crossbow
W 12647r	(c. 1485–1490)	Courtyard of a cannon foundry
W 12649r	(c. 1485–1490)	Boat, paddle whel, folding bow, "*armadura di bar[ca]*" (armature of boat)
W 12469v	(c. 1485–1490)	Boat and paddle-wheel studies, "51 – 42 = 09"
W 12650r	(c. 1483–1490)	Boat studies, calculations, a gear of 60 "*denti*"
W 12650v	(c. 1483–1490)	Paddle-wheel boat and portrait in profile
W 12651r	(c. 1483–1490)	An incendiary dart, tail fin, and lance
W 12652r	(c. 1483–1490)	Studies of cannon, a mortar, and a boat
W 12652v	(c. 1483–1490)	Studies of a town wall being blown up
W 12653r	(c. 1483–1490)	Studies of antique weapons, from MSS illustrations

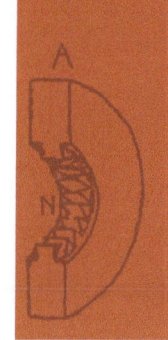

Glossary
Crossbow-Related Terms
Used by Leonardo

antenne motrice	"Engine of poles" that launches a projectile from a catapult (see CA f. 145r [51vb])
armadura	"Armature," the bow, prod, or lath of a crossbow (see CA f. 149br [53vb])
balestriere	"Crossbowman" (see CA f. 149br [53vb])
balestra	"Crossbow," from the Latin *ballista*: catapult (see CA f. 142r [51rb])
bomba	"Bomb" (see MS B f. 31v)
cacciafrusto	"Whip-driver," a medieval mechanism that launches projectiles (see CA 144r [51va], and MS I f. 99v [I² f. 51v])
circumfolgore	A boat with cannon mounted on a large weel at its centre, capable of firing simultaneously in all directions (see MS B f. 82v)
clotonbrot	A type of incendiary bomb that is shot from cannon (see see MS B f. 31v)
dardo	"Dart" (see CA f. 902br [329rb])
falarica	A bent arm engine, spear launching machine (see MS B f. 6r)
frombola	"Sling" (*frommola*) that launches stones or darts (see CA f. 141r [51ra])
fulminario	A trebuchet (large catapult), given a medieval name with reference to *fulminare*: to strike with lightening (see CA f. 113v [40va])
fuoco greco	"Greek fire," tar and/or oil that is set ablaze within a half-shell or bomb casing and then catapulted at, or dropped on, an enemy (see Triv p. 48)
lanciacampi	"Field-launcher" (*lanciasassi*), a catapult that flings projectiles (see MS I f. 99v [I² f. 51v], and CA 140r [50vb])
moscette	"Arrow" (see MS I f. 99v [I² f. 51v])
noche	"Nut," a round catch on a crossbow that holds the string back when the crossbow is spanned (see CA f. 155r [56ra])
rotelle	A type of incendiary bomb that has multiple incendiary bombs within it (see MS B f. 59r)
tranubolo	A large trebuchet (giant catapult, see CA f. 113v [40va])

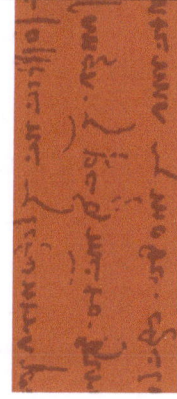

Bibliography

Manuscripts and Archival Sources

Manuscript A (2172 and 2185), Institut de France, Paris
Codex Arundel 263, British Library, London
Ashmolean Museum, Oxford
Manuscript B (2173 and 2184), Institut de France, Paris
Manuscript C (2174), Institut de France, Paris
Codex Atlanticus and assorted sheets, Biblioteca Ambrosiana, Milan
Christ Church College, Oxford
Forster Codices I, II, and III (804.AA.0150 – 0152), Victoria and Albert Museum, London
Codex Leicester, Collection of Bill and Melinda Gates, Seattle
Madrid Codices I and II (8937 and 8936), Biblioteca Nacional, Madrid
Codex Trivulzianus (N 2162), Castello Sforzesco, Milan
Codice Urbinate Latinus 1270, Biblioteca Apostolica Vaticana, Vatican City
Manuscript D (2175), Institut de France, Paris
Manuscript E (2176), Institut de France, Paris
Manuscript F (2177), Institut de France, Paris
Manuscript G (2178), Institut de France, Paris
Manuscript H (2179), Institut de France, Paris
Manuscript I (2180), Institut de France, Paris
Manuscript K (2181), Institut de France, Paris
Manuscript L (2182), Institut de France, Paris
Manuscript M (2183), Institut de France, Paris
Musée du Louvre, Paris
Codex on the Flight of Birds (Cod. Varia 95) and assorted drawings, Biblioteca Reale, Turin
Gabinetto Disegni e Stampe, Uffizi, Florence
Gallerie dell'Accademia, Venice
Windsor Castle Royal Library, Windsor

Printed Primary Sources

Albertini (1510) Memoriale di molte statue et picture sono nella inclyta cipta di Florentia per mano di sculptori et pictori excellenti moderni et antiqui. Antonio Tubini, Firenze
Amoretti C (1804) Memorie storiche su la vita, gli studj, e le opere di Lionardo da Vinci. Dalla Tipografia di Giusti, Ferrario, e C.o, Milano
Beltrami L (ed) (1919) Documenti e memorie riguardanti la vita e le opere di Leonardo da Vinci in ordine cronologico. Treves, Milano
Calvi G (1925) I manoscritti di Leonardo da Vinci dal punto di vista cronologico, storico e biografico. Nicola Zanichelli Editore, Bologna
Clark K, Pedretti C (1968) The drawings of Leonardo da Vinci in the collection of Her Majesty the Queen at Windsor Castle, 2nd ed. Revised 3 vols., Phaidon, London
Dei B (1985) La cronica dall'anno 1400 all'anno 1500. Barducci R (ed), F. Papafava, Firenze
Fabriczy C von (ed) (1893) II Codice dell'Anonimo Gaddiano (Cod. magUabechiano XVIL 17) nella Biblioteca Nazionale di Firenze. tip. M. Cellini e C., Firenze
Farago CJ (1992) Leonardo da Vinci's paragone: A critical interpretation with a new edition of the text in the Codex Urbinas. E. J. Brill, Leiden New York
Frey K (ed) (1892) Il Codice Magliabechiano cl. XVII. 17. G. Grote, Berlin
Kemp M, Walker M (1989) Leonardo da Vinci on painting. Yale University Press, New Haven London
Leonardo da Vinci (1956) Treatise on painting [Codex Urbinas Latinus 1270]. McMahon AP (trans, annot), Heydenreich LH (intro), Princeton University Press, Princeton

Leonardo da Vinci (1970) The notebook of Leonardo da Vinci. Richter JP (ed), Dover Publications, New York

Leonardo da Vinci (1974) I Codici di Madrid / The Madrid Codices. 5 vols., Reti L (ed), McGraw-Hill, New York

Leonardo da Vinci (1995) Libro di Pittura. Codice Urbinate lat. 1270 nella Biblioteca Apostolica Vaticana. 2 vols., Pedretti C (ed), Vecce C (trans), Giunti Gruppo Editoriale, Firenze

Leonardo da Vinci (1999–2007) The manuscripts of Leonardo da Vinci in the Institut de France. Manuscripts A, B,... [through] M. Venerella J (ed), Ente Raccolta Vinciana, Castello Sforzesco, Milano

Leonardo da Vinci (2000) Il Codice Atlantico della Biblioteca Ambrosiana di Milano. Ed. Augusto Marinoni. 3 vols., Giunti, Firenze

Leonardo da Vinci (2006) Il mondo e le acque, vol. XI [manoscritti D, F], Cura di Giovanni Majer. Neri Pozza, Vicenza

McCurdy E (ed) (1938) The notebooks of Leonardo da Vinci. 2 vols., Reynal & Hitchcock, New York

Marinoni A (ed) (1973–1975) Il Codice Atlantico di Leonardo da Vinci. 12 vols., Giunti, Firenze

Marinoni A (ed) (1952) Tutti gli scritti. Scritti letterari. Rizzoli, Milano

Martini, Francesco di Giorgio (1979) Il codice Ashburnham 361 della Biblioteca medicea laurenziana di Firenze: trattato di architettura di Francesco di Giorgio Martini. 2 vols., Giunti Barbera, Firenze

O'Malley D, Saunders JB de CM (1952) Leonardo da Vinci on the human body. Henry Schuman, New York

Pedretti C (c. 1978) Leonardo da Vinci, Codex Atlanticus, A catalogue of its newly restored sheets. 2 vols., Johnson Reprint Corp., New York

Pedretti C (1977) The literary works of Leonardo da Vinci, Compiled and edited from the original manuscripts by Jean Paul Richter. Commentary, 2 vols., University of California Press, Berkeley Los Angeles

Richter IA (ed, trans) (1949) Paragone: A comparison of the arts by Leonardo da Vinci. Oxford University Press, London New York Toronto

Richter JP (ed) (1970) The literary works of Leonardo da Vinci. 2 vols., 1883; Oxford University Press, London; 1939; Phaidon, New York

Shaw JB (1976) Drawings of old masters at Christ Church. 2 vols., Oxford University Press, Oxford

Solmi E (1976) Scritti Vinciani. Le Fonti dei Manoscritti di Leonardo da Vinci e altri studio. Orig. 1904–1924, La Nuova Italia editrice, Firenze

Stites RS (1970) The sublimations of Leonardo da Vinci with a translation of the Codex Trivulzianus. Smithsonian Institution Press, Washington

Valturio R (2006) De Re Militari. Riproduzione facsimilare dell'editio princeps [1472] e saggi critici. Delbianco P (ed) et al., Guaraldi Press, Milano

Vasari G (1980) Le vite de piu eccellenti architetti, pittori, et scultori italiani. 2 vols., Edizione Giuntina, Firenze, 1550; rpt. Broude International Editions, New York

Villata E (ed) (1999) Leonardo da Vinci: i documenti e le testimonianze contemporanee. Ente Raccolta Vinciana, Milano

Printed Secondary Sources

Ackerman JS (1974) Notes on Bramante's bad reputation. In: Comitato Nazionale per le Celebrazioni Bramantesche (ed) Studi Bramanteschi. Atti del Congresso internazionale. Milano, Urbino and Roma, 1970. Italy. De Luca Editore, Roma, pp 339–349

Adorno P (1991) Il Verrocchio: Nuove proposte nella civiltà artistica del tempo di Lorenzo Il Magnifico. Casa Editrice Edam, Firenze

Agghàzy MG (1989) Leonardo's Equestrian statue. Simon Á (trans), Akademiai Kiadó, Budapest

Alberti, Leon Battista (1946) I primi tre libri della Famiglia. Pellegrini FC (trans, annot), Spogano R (rev, intro), Sansoni, Florence

Alberti, Leon Battista (1972) On painting and on sculpture. The Latin texts of De pictura and De statuua. Grayson C (ed), Phaidon Press, London

Alberti, Leon Battista (1988) De re aedificatoria incipit..., Florence 1486. In: Rykwert J, Leach N, Tavernor R (eds, trans) On the art of building in ten books. Cambridge, MA

Alberti, Leon Battista (1991) On painting. Grayson C (trans), Kemp M (intro), Penguin, London

Alberti, Leon Battista (1998) De statua. Collareta M (ed), Sillabe, Livorno

Albertini L (2001) Uomo del Rinascimento, Genio del futuro, No. 1. De Agostini, Novara

Allegri E, Cecchi A (1980) Palazzo Vecchio e i Medici. Guida storica. Studio per edizioni scelte, Firenze

Alm J (1995) European crossbows: A survey by Josef Alm. Wilson GM (ed), Bartlett Wells H (trans) (1st ed, 1947), Royal Armouries Monograph 3, London

Ames-Lewis F (1983) Drawing in Renaissance Italy. Yale, New Haven, CT

Ames-Lewis F (2002) The intellectual life of the Early Renaissance artist. Yale, New Haven, CT

Ames-Lewis F, Rogers M (eds) (1999) Concepts of beauty in Renaissance art. Scolar Press, Aldershot

Anonymous (1803) Tavole di ragguaglio fra le nuove e le antiche misure ... del Regno d'Italia publicate per ordine del Governo, 1, Stamperia Reale, Milan

Anonymous (1935–1943) Piede. Enciclopedia italiana di scienze, lettere ed arti, 27, Rome

Anonymous (1970) Misure. Grande dizionario enciclopedico, 12, Unione Tipografico-Editrice Torinese, Turin

Arano LC (1980) Leonardo: Disegni di Leonardo e delia sua cerchia alle Gallerie dell'Accademia. Exh. Cat. Venezia, Gallerie dell'Accademia, May–June 1980. Ejecta Editrice, Milano

Arano LC (1989) Two angels from the Del Maino Workshop. Achademia Leonardi Vinci 2:177

Arano LC (1992) Leonardo e i De Predis. Studi di storia dell'arte sul medioevo e il rinascimento nel centenario delia nascita di Mario Salmi, vol. n. Atti del Convegno Intemazionale, Arezzo-Firenze, 16–19 November 1989. Edizioni Polistampa, Firenze, pp 729–737

Arasse D (1985) Ritratto di Isabella d'Este. Leonardo. La Pittura. Giunti Martello Editore, Firenze, pp 102–103

Argan GC (1985) 5 daghossto 1473. Leonardo. La Pittura. 1977. Giunti Martello Editore, Firenze, pp 12–15

Aristotle (1984) On the soul, II 412a 20-21. In: Barnes J (ed) The complete works of Aristotle. Bollingen Series LXXI 2, Princeton University Press, Princeton

Audoin E (1913) Essai sur l'armée royale au temps de Philippe Auguste. Champion, Paris

Azarpay G (1987) Proportional guidelines in ancient near eastern art. J Near Eastern Stud 46(3):183–213

Baatz D (1978a) Recent finds of ancient artillery. Britannia IX:1–17

Baatz D (1978b) Das Torsiongeschütz von Hatra. Antike Welt 4:50–57

Baatz D (1980) Ein Katapult der Legio IV Macedonia aus Cremona. Römische Mitteilungen 87:283–299

Baatz D (1994) Bauten und Katapulte des römischen Heeres. Steiner, Stuttgart

Babinger F (1952) Vier Bauvorschläge Leonardo da Vincis an Sultan Bajezid II (1502/1503). Mit einem Beitrag von L. Heydenreich. Akademie der Wissenschaften zu Göttingen, Philologisch-historische Klasse, Nachrichten 1

Bambach C (1999a) Drawing and painting in the Italian Renaissance Workshop. Cambridge University Press, Cambridge

Bambach C (1999b) The purchases of cartoon paper for Leonardo's battle of the Anghiari and Michelangelo's battle of Cascina. Villa I Tatti Studies 8, The Harvard University Center for Italian Renaissance Studies, Florence

Bambach C (ed) (2003) Leonardo da Vinci: Master draftsman. Metropolitan Museum of Art, New York

Bambach Cappel C (1990) Pounced drawings in the Codex Atlanticus. Achademia Leonardi Vinci: Journal of Leonardo Studies and Bibliography of Vinciana 3:129–131

Bambach Cappel C (1991) Foreshortened letters. Achademia Leonardi Vinci IV:99–106

Banadera S (1997) Agostino de' Fondulis e la riscoperta delia terracotta nel Rinascimento lombardo. Bolis, Bergamo; Banco popolare di Crema, Crema

Bandello, Matteo (1999) La prima parte de le Novelle del Bandello, Lucca, Busdrago 1554. Reproduced in: Villata E (ed) [Novella LVIII] Il Bandello a la molto illustre e vertuosa eroina la signora Ginevra Rangona e Gonzaga. Reprint of Beltrami: Documenti e memorie, 1919; Leonardo da Vinci, I Documenti e le Testimonianze Contamporanee, Raccolta Vinciana. Castello Sforzesco, Milano, pp 300–302

Baratta M (1903) Leonardo da Vinci e I problemi della terra. Turino

Baratta M (1905a) Curiosità Vinciane: perché Leonardo da Vinci scriveva a rovescio, Leonardo da Vinci enigmofilo, Leonardo da Vinci nella invenzione dei palombari e degli apparecchà di salvataggio marittimo. Fratelli Bocca, Torino

Baratta M (1905b) Leonardo da Vinci negli studi per la navigazione dell 'Arno. Pres so la Societa Geografica Italiana, Roma

Barolsky P (1991) Why Mona Lisa smiles and other tales by Vasari. University Park, Pennsylvania State University Press, Pennsylvania

Baron H (1943) Towards a more positive evaluation of the fifteenth-century Renaissance. J Hist Ideas IV

Baroni C (1940–1968) Documentz per la storia dell'Architettura a Milano nel Rinascimento e nel Barocco. G. C. Sansoni, Firenze

Bass G (ed) (2002) A history of seafaring based on underwater archaeology, London: Thames & Hudson, 1974. Angus Konstam, Renaissance War Galley, 1470–1590, Osprey, Oxford

Baxandall M (1988) Painting and experience in fifteenth century Italy: A primer in the social history of pictorial style. Oxford University Press, New York, pp 135–151

Beck J (1974) The Medici inventory of 1560. Antichita viva XIII/3, May–June:64–66; XIII/5, Sept–Oct:61–63

Beck J (1988) Leonardo's rapport with his father. Antichita viva XXVIV 5–6, Oct.–Dec.

Beck J (1993) The dream of Leonardo. Artibus et historiae 27:185–198, (also published as 'I Sogni di Leonardo.' Lettura Vinciana 24, Florence)

Beck J (1999) Italian Renaissance painting, 2nd ed. Konemann, Köln

Beer FP, Johnston ER (1960) Mechanics for engineers. McGraw Hill, New York

Bell J (1992) Filippo Gagliardi on Leonardo's Perspective. Achademia Leonardi Vinci 5

Bellincioni B (1876–1878) Le Rime di Bernardo Bellincioni ricontrate sui manoscritti, emendate e annotate da Pietro Fanfani. Pietro Fanfani (ed), G. Romagnoli, Bologna

Belluzi A (1993) Giuliano da Sangallo e la chiesa delia Madonna dell'Umilta a Pistoia. Alinea Editrice, Firenze

Belt EMD (1969) Leonardo the anatomist. Greenwood Press, New York, (original printing: Univ. of Kansas Press, 1955)

Beltrami L (1891a) Le statue funerarie di Lodovico il Mora e di Beatrice d'Este alla Certosa di Pavia. Arch Storico Dell'arte 4:357–362

Beltrami L (ed) (1891b) Il Codice di Leonardo da Vinci nella Biblioteca del Principe Trivulzio. Angelo delia Croce, Milano

Beltrami L (1894) Il Castello di Milano sotto il dominio dei Visconti e degli Sforza. Ulrico Hoepli, Milano

Beltrami L (1902) La sala delle 'Asse' nel Castello di Milano decorata da Leonardo da Vinci nel 1498. Rassegna d'arte II:65–68; 90–93

Beltrami L (1903) Leonardo negli studi per if Tiburio delia Cattedrale di Milano. Alfieri & Lacroix, Milano

Beltrami L (1917) Bramante e Leonardo practicarono l'arte del Bulino? Un incisore sconosciuto: Bernardo Prevedari. Arte Lombarda 17:187–194

Beltrami L (1920) La Vigna di Leonardo. Tipografia Umberto Allegretti, Milano

Beltrami L (1921) La lite di Leonardo con gli altri figli di Ser Piero da Vinci. Nuova Antologia 56:193–207

Beltrami L (1923) Ancora per la madre di Leonardo. Miscellanea Vinciana. Tipografia Umberto Allegretti, Milano, pp 32–34

Beltrami L (1925) Benedetto Dei e il 'gigante del monte Atalante' di Leonardo. Miscellanea Vinciana 3:18–20

Bennett J (1986) 'The mechanics' philosophy and the mechanical philosophy. Hist Sci 24:1–28, 37

Bennett J (1987) The divided circle: A history of instruments for astronomy, navigation and surveying. Oxford

Bennett J (1998) Practical geometry and operative knowledge. Configurations 6(2):195–222

Bennett J, Mandelbrote S (1998) The garden, the ark, the tower, the temple. Biblical metaphors of knowledge in early modern Europe. Oxford

Bertani L (1999) San Miniato al Monte. Giusti di Becocci Saverio, Florence, pp 9–10

Bertelli C (1982) Verso il vero Leonardo. In: Dell'Acqua GA, Bertelli C (eds) Leonardo e Milano. Banca Popolare, Milan, pp 83–88

Bertieri R (1929) Gli studi italiani sull'alfabeto nel rinascimento. Pacioli e Leonardo da Vinci. Gutenberg Jahrbuch, pp 269–286

Bhattacharya AK (1974) Citralaksana: A treatise on Indian painting. Sarasvat Library, Calcutta

Bhattacharya AK (1976) Technique of Indian painting: A study chiefly made on the basis of the Silpa texts. Sarasvat Library, Calcutta

Bhattacharyya T (1986) The canons of Indian art. Firma KLM Private Ltd., Calcutta

Birnbaum MD (1996) The Orb and the Pen: Janus Pannonius, Matthias Corvinus, and the Buda Court. Balassi, Hungary

Biscaro G (1909) La vigna di Leonardo da Vinci fuori de Porta Vercellina. Arch Storico Lombardo XII:363–396

Bishop MC, Coulston JCN (1993) Roman military equipment. B. T. Batsford, London

Blum A (1939) Leonardo da Vinci as engraver. The Print Collector's Chronicle 2/1:1–3; 13–14

Blunt A (1940) Artistic theory in Italy 1450–1600. Clarendon Press, Oxford

Böninger L (1985) Leonardo da Vinci und Benedetto Dei in Mailand. Mitt Kunsthist I Flo XXIX/2–3: 385–388

Bonnaffe E (1884) Sabba da Castiglione, notes sur la curio site italienne a la Renaissance. Gaz Beaux-Arts XXX/July:19–33

Borsi S (1985) Giuliano da Sangallo: i disegni di architettura e dell'antico. Officina Edizioni, Roma

Botero G (1610) Detti memorabili di persone illustri. Brescia

Bovi A (1959) L'opera di Leonardo per il monumento Sforza a Milano. L. S. Olschki, Firenze

Bowie T (1959) The sketchbook of Villard de Honnecourt. Indiana University Press, Bloomington London

Brizio AM (1974) Bramante e Leonardo alla corte di Lodovico il Moro. Studi Bramanteschi-Atti del Congresso Internazionale Milano-Urbino-Roma 1970, De Luca, Roma, pp 1–26

Brizio AM (ed) (1996) Scritti scelti di Leonardo da Vinci. Unione tipografico editrice torinese, Torino, (original printing: 1952)

Brown RC (1967) Observations on the Berkhamstead Bow. Journal of the Society of Archer Antiquaries 10:12–17

Brown CM (1969) Little known and unpublished documents concerning Andrea Mantegna, Bernardino Parentino, Pietro Lombardo, Leonardo da Vinci and Filippo Benintendi, part Two. L'Arte W7–8: 182–214

Brown CM (1982) Isabella d'Este and Lorenzo da Pavia: Documents for the history of art and culture in Renaissance Mantua. Librairie Droz S.A., Geneve

Brown DA (1983) Leonardo and the idealized portrait in Milan. Arte Lombarda 67:102–116

Brown DA (1990) Leonardo and the ladies with the ermine and the book. Artibus et historiae 2200:47–61

Brown DA (1998) Leonardo da Vinci: Origins of a genius. Yale University Press, New Haven London

Brown CM, Lorenzoni AM (1973) Gleanings from the Gonzaga documents in Mantua – Gian Cristoforo Romano and Andrea Mantegna. Mitt Kunsthist I Flo 17:153–159

Bruschi A (1974) Bramante, Leonardo e Francesco di Giorgio a Civitavecchia. Studi Bramanteschi. Atti del Congresso internazionale. Milano, Urbino and Roma, 1970. De Luca Editore, Italy, pp 535–565

Bruschi A (1977) Bramante. Foreword by Peter Murray, Thames and Hudson, London, (1973)

Bruschi A (1978) Luca Pacioli, De Divina Proportione. Nota introduttiva. Scritti rinascimentali di architettura. In: Bruschi A, Maltese C, Tafuri M, Bonelli R (eds) Trattati di architettura 4. Edizioni il Polifilo, Milano, pp 26–52

Budd D (2002) Leonardo da Vinci through Milan: Studies on the documentary evidence. Doctoral dissertation, Columbia University

Buonarroti, Michelangelo (1965–1993) Il carteggio di Michelangelo. 5 vols., Poggi G, Barocchi P, Ristori R (eds), Sansoni, Firenze

Bush V (1978) Leonardo's Sforza monument and Cinquecento sculpture. Arte Lombarda 1:47–68

Busignani A (1969) Pollaiuolo. Edizioni d'Arte il Fiorino, Florence

Butazzi G (1998) Note per un ritratto: vesti e acconciatura della Dama con l'ermellino Leonardo: La dama con l'ermellino. In: Fabjan B, Marani PC (eds) Exh. Cat. Palazzo del Quirinale, Rome, 15 October to 14 November 1998; Pinacoteca di Brera, Milan, 19 November to 13 December 1998; and Palazzo Pitti, Florence, 16 December 1998 to 24 January 1999. Silvana Editorale, Cinisello Balsamo (Milano), pp 67–71

Butterfield A (1997) The sculptures of Verrocchio. Yale University Press, New Haven London

Calvi G (1869) Notizie dei principali professori di belle arti che fiorirono in Milano durante il govemo de' Visconti e degli Sforza, vol. III: Leonardo da Vinci. Tipografia Fratelli Borroni, Milano

Calvi F (1884) Famiglie notabili Milanesi. 4 vols., Vallardi, Milano

Calvi G (1907) Leonardo da Vinci e il Conte di Lignyed altri appunti su personaggi vinciani. Raccolta Vinciana 3:99–110

Calvi G (1916) Contributi alla Biografia di Leonardo da Vinci (Periodo Sforzesco). Arch Storico Lombardo III:417–508

Calvi G (1919a) Il vero nome di un allievo di Leonardo: Gian Giacomo de' Caprotti detto 'Salaj'. Rassegna d'Arte, pp 138–141

Calvi G (1919b) L'adorazione dei Magi di Leonardo da Vinci. Raccolta Vinciana X:1–44

Calvi G (1926–1929) Spigolature vinciane dall'Archivio di Stato di Firenze. Raccolta Vinciana XIII: 35–43

Calvi G (1930–1934a) L'insidia delia falsificazione. Raccolta Vinciana XIV:344–345

Calvi G (1930–1934b) Vecchie e nuove riserve sull'Annunziazione di Monte Oliveto. Raccolta Vinciana XN:201–239

Cambi G (1770–1789) Storie Fiorentine. Delizie degli eruditi toscani. Gaetano Cambiagi, Firenze

Campori MG (1865) Nuovi documenti per la Vita di Leonardo da Vinci. Atti e Memorie delle R. R. Deputazioni di Storia patria per Ie provincie modernosi e parmenesi, vol. III, pp 43–51

Canestrini G (1939) Leonardo costruttore di machine e veicoli. Tumminelli, Milan

Cantù C (1874) Aneddoti di Lodovico il Moro. Arch Storico Lombardo I:483–487

Cardano Girolamo (1663) De vita propria (1576). In: Sponi C (ed) Opera omnia. 10 vols., Lyons

Cardile PJ (1981–1982) Observations on the iconography of Leonardo da Vinci's Uffizi Annunciation. Stud Iconogr 7/8:189–208

Carl D (1982) Zur Goldschmiedefamilie Dei mit neuen Dokumenten zu Antonio Pollaiuolo und Andrea Verrocchio. Mitt Kunsthist I Flo XXVI/2:129–166

Carpiceci A (1982) Il progetto di Leonardo per San Satiro a Milano. Raccolta Vinciana XXIII:121–160

Cartwright J (1908) Beatrice d'Este Duchess of Milan 1475–1497. J. M. Dent & Co., London

Carusi E (1926–1929) Ancora di Salai. Raccolta Vinciana XIII:44–52

Casati C (1876) Vicende Edilizie del Castello di Milano: Libreria Editrice di G. Brigola, Milano

Castelfranco G (1965) The paintings of Leonardo da Vinci. New Random House, York, 1965, (All'insegna del pesce d'oro, Milan, 1956)

Castelfranco G (1968) Drawings by Leonardo da Vinci. Phillips FH (trans), Dover Publications, Inc., New York

Castelfranco G (1996) Studi Vinciani. De Luca Editore, Roma

Catterson-Silver L (2002) Donatello's legacy and the training of Michelangelo: Sculptural practice in Quattrocento Florence. Diss., Columbia University

Cavalli H (1874) Tableaux comparatifs des mesures, poids et monnaies modernes et anciens : cours des changes, usages du commerce : de tous les états du monde : compares avec le système métrique francais et les poids et mesures anglais, 2nd ed. Libr. administrative de Paul Dupont, Paris

Cennini C d'Andrea (1960) The craftsman's handbook "Il Libro dell'Arte. Thompson DV Jr. (trans), Dover Publications, New York, 1960, (reprint of the English translation volume of 1933, Yale University Press, New Haven)

Chambers E (1728) Measures. Ciclopédia: Or an universal dictionary of arts and sciences, vol. 2., London

Chambers DS (1992) A Renaissance cardinal and his worldly goods: The will and inventory of Francesco Gonzaga (1444–1483). Warburg Institute, University of London

Chastel A (1961) The genius of Leonardo da Vinci: Leonardo da Vinci on art and the artist. Callman E (trans), The Orion Press, New York

Chatterjee Sastri A (1971) Visnudharmottarapuranam (Citrasutram). Varanasi

Chevedden PE, Kagay DJ, Padilla PG (eds) (1996) The artillery of King James I the conqueror. Iberia and the Mediterranean world of the Middle Ages: Essays in honor of Robert I. Burns. E. J. Brill, Leiden, pp 47–94

Ciabani R, et al. (1992) Le famiglie di Firenze. 4 vols., Casa Editrice Bonechi, Firenze

Cianchi R (1952) Vinci: Leonardo e la sua famiglia. Con appendice di documenti inediti, publ. no. 10, Museo delia scienze e delIa tecnica, Milano

Cianchi R (1964) Figure nuove del mondo vinciano: Paolo e Vannoccio Biringuccio da Siena. Raccolta Vinciana XX:277–297

Cianchi M (c. 1984) Leonardo da Vinci's Machines. Florence, Becocce Editore

Cinthio GG (1554) Discorsi di M. Giovambattista Giraldi Cinthio nobile Ferrarese, e segretario dell'illustriss. Duca di Ferrara intorno al comporre de i Romanzi, delle Comedie, e delle Tragedie, e di altre maniere di Poesie. Con la tavola delle cose piu notabili in tutti essi discorsi contenute. Con privilegio. G. Giolito de Ferrari, Vinegia

Clark K (1933) The madonna in profile. Burlington Mag March:136–140

Clark K (1935) A catalogue of the drawings of Leonardo da Vinci in the collection of His Majesty the King at Windsor. Phaidon, Cambridge, p ix

Clark K (1969) Leonardo and the antique. In: O'Malley CD (ed) Leonardo's legacy: An international symposium. University of California Press, Berkeley, pp 1–34

Clark K (1988) Leonardo da Vinci. Revised edition with an introduction by Martin Kemp (orig. 1939), Harmondsworth, U.K.

Clark K, Pedretti C (1968) The drawings of Leonardo da Vinci in the collection of Her Majesty the Queen at Windsor Castle, 2nd ed. Revised 3 vols., Phaidon, London

Clayton M (1996) Leonardo da Vinci: A singular vision. New Abbeville Press, York

Clayton M (2003) Leonardo da Vinci, the divine and the grotesque. Royal Collection Enterprises, London

Clephan RC (1903) Notes on Roman and Medieval military engines. Archaeologia Aeliana 24:69–114

Cole-Ahl D (ed) (1995) Leonardo da Vinci's Sforza Monument Horse: The art and the engineering. Assoc. University Press, London

Collins BI (1997) Leonardo, psychoanalysis and art history: A critical study of psychobiographical approaches to Leonardo da Vinci. Northewestern University Press, Evanston, Illinois

Condivi A (1999) The life of Michelangelo, 2nd ed. Sedgwick Wohl A (trans), Wohl H (ed), Pennsylvania State University Press, University Park, Pennsylvania

Convegno internazionale di studi 'Milano nell'eta? di Ludovico il Moro' (1983) Milano nell'eta? di Ludovico il Moro: atti del Convegno internazionale, 28 febbraio–4 marzo 1983. Comune di Milano, Archivio storico civico e Biblioteca trivulziana, Milano

Cooper M (1965) The inventions of Leonardo da Vinci. Macmillan, New York

Corbeau A (1968) Les soutractions de Libri. In: Les manoscrits de Léonard de Vinci: examen ctitique et historique de leurs elements externs; Les manoscrits de Léonard de Vinci de la Bibliothäque nationale de Madrid: description critique et histoire. Centre regional de documentation pédagogique, Caen, pp 187–190

Corti G, Hartt F (1962) New documents concerning Donatello, Luca and Andrea Delia Robbia, Desiderio, Mino, Uccello, Pollaiuolo, Filippo Lippi, Baldovinetti and others. Art Bull XLIV:155–167

Cotty C (1819) Aide-mémoire a l'usage des officiers d'artillerie de France. 2 vols., Paris

Courajod L (1877) Conjectures a propos d'un buste en marbre de Beatrix d'Este au Musee du Louvre. Gaz Beaux-Arts, Oct.:330–344

Covi DA (1981) Verrocchio and the Palla of the Duomo. In: Barasch M, Freeman Sandler L (eds) Art, the ape of nature: Studies in honor of H. W. Janson. H. N. Abrams, New York; Prentice-Hall, Englewood Cliffs, pp 151–169

Covi DA (2005) Andrea del Verrocchio, life and work. Olschki, Florence

Croci G (1860) Dizionario universale dei pesi e delle misure in uso presso gli antichi e moderni con ragguaglio ai pesi e misure del sistema metrico. Milan

Crombie AC (1971) Robert Grosseteste and the origins of experimental science 1100–1700. Carendon, Oxford, (1953)

Cumont F (1936) The excavations at Dura-Europos preliminary report of sixth season of work. New Haven, Yale

Curlee K (1989) The Sforza court. Milan in the Renaissance 1450–1535. Gallery Guide, The Archer M. Huntington Art Gallery/The University of Texas, Austin

d'Ancona A (1971) Origini del Teatro Italiano. Libre Tre. 1891; Bardi Editore, Roma

Dami L (1915) La Basilica di San Miniato al Monte. Boll Arte, August

de Hamel C (1994) A history of illuminated manuscripts. Phaidon Press Ltd., London

de Rinaldis A (1926) Storia dell'Opera Pittorica di Leonardo da Vinci. Castiglione P (trans), Zanichelli, Bologna

de Roover R (1963) The rise and decline of the Medici Bank, 1397–1494. Harvard University Press, Cambridge, MA

De Toni N (1930–1934) Saggio di onomastica Vinciana. Raccolta Vinciana XIV:54–117

De Toni GB (1969) Studio di meccanica. In: Leonardo da Vinci. atti del Simposio Internazionale di Storia della Scienza, Florence

De Toni N (1974) Giovanni Battista Venturi, ed, I manoscritti dell'Ambrosiana a Parigi nel 1797. In: Frammenti Vinciani XXXI, as taken from: Commentari dell'Ateneo di Brescia. Geroldi, Brecia

De Toni GB, Solmi E (1904/1905) Intorno all'andata di Leonardo da Vinci in Francia. Atti del Regio Istituto Veneto di Scienze. Lettere ed Arti LXIV:487–495

Dei B (1985) La cronica dall'anno 1400 all'anno 1500. Barducci R (ed), F. Papafava, Firenze

Della FP (1474–1482) De prospectiva pingendi (On the perspective of painting)

Desrey P (1834) Relation du voyage du Roy Charles VIII pour la conqueste du royaume de Naples. Archives curieuses de l'histoire de France III:215–216

Devries K (2002) A cumulative bibliography of Medieval military history and technology. Brill, Leiden

Di Teodoro FP (1993) '… pare che il cielo ne abbia invidia …'. Fulmini sulla lanterna della cupola di Santa Maria del Fiore: danni e restauri. Arte Lombarda 105/106/107, 2-3–4:181–189

Diderot D, le Rond d'Alembert JH (1751–1765) 'Pied', Encyclopédie ou dictionnaire raisonnée des arts, sciences et métiers. 7 vols., Paris, pp 562–563

Diels H (1924) Antike Technik. Teubner Verlag, Leipzig

Dina A (1921) Isabella d'Aragona, duchessa di Milano e di Bari. Arch Storico Lombardo VIII:269–457

Dodsworth BW (1995) The Arca di San Domenico. Vol. 2 of Intercultural Studies, Peter Lang Publishing, New York

Doglio MM (1983a) Leonardo 'apparatore' di spettacoli a Milano per la corte degli Sforza. Leonardo e gli spettacoli del suo tempo. Electa Editore, Milano, pp 41–76

Doglio MM (1983b) Spettacoli a Milano nel periodo sforzesco. Leonardo e gli spettacoli del suo tempo. Electa Editrice, Milano, pp 20–40

Doursther H (1840) Dictionnaire universel des poids et mesures anciens et modernes. M. Hayez, Brussels

Duhem P (1906) Les origins de la statique. A. Hermann, Paris

Duhem P (1906–1913) Études sur Léonard de Vinci: ceux qu'il a lus et ceux qui l'ont lu. 3 vols., Hermann, Paris

Edgerton SY, Jr. (1985) Pictures and punishments: Art and criminal prosecution during the Florentine Renaissance. Cornell University Press, Ithaca London

Eissler KR (1961) Leonardo da Vinci: Psychoanalytic notes on the enigma. International Universities Press, Inc., New York

Eivind L (1966) Technological studies in ancient metrology. Nordisk, Copenhagen

Elam C (1992) Lorenzo de' Medici's sculpture garden. Mitt Kunsthist I Flo 36:41–83

Elkins J (1991) The case against surface geometry. Art Hist 14(2):143–174

Elkins J (1994) The poetics of perspective. Cornell, Ithaca

Epstein SA (1996) Genoa and the Genoese, 958–1528. University of North Carolina, Chapel Hill

Esche S (1953) Leonardo da Vinci. Das Anatomische Werk, mit kritischem Katalog und hundertfünfundseibzig Abbildungen. Holbein-Verlag, Basel

Eusebio L (1967) Compendio di Metrologia Universale e Vocabolario Metrologico. Unione Tipografico Editrice Torinese Turin 1899; Forni Editore, Bologna 1967

Evans MW (1969) Medieval drawings. Paul Hamlyn, London

Fabriczy C von (ed) (1893) Il Codice dell'Anonimo Gaddiano (Cod. magliabechiano XVIL 17) nella Biblioteca Nazionale di Firenze. tip. M. Cellini e C., Firenze

Fabriczy C von (1895) Andrea del Verrocchio ai servizi de' Medici. Arch Storico Dell'arte 1:163–175

Fabriczy C von (ed) (1969) Il Libro di Antonio Billi e le sue copie nella Biblioteca Nazionale di Firenze. 1891; Gregg, Farnborough 1969

Farago CJ (1992) Leonardo da Vinci's paragone: A critical interpretation with a new edition of the text in the Codex Urbinas. E. J. Brill, Leiden New York

Farago C (1993) Fractal geometry in the organization of Madrid MS II. Achademia Leonardi Vinci VI:47–55, 12 plates

Farago C (1994) Leonardo's battle of Anghiari: A study in the exchange between theory and practice. Art Bull LXXVI/2:301–330

Farago C (ed) (1999) Leonardo. Selected scholarship. 5 vols., Garland Publishing, Inc., New York London

Favaro A (1913–1917) L'abitazione delia famiglia di Leonardo a Firenze in via Ghibellina. Raccolta Vinciana 9:175–176

Favaro G (1917) Il Canone di Leonardo sulle proporzioni del corpo umano. Atti del Reale Istituto Veneto di Scienze, Lettere ed Arti 77(2)

Favaro G (1918) Misure e Proporzione del Corpo Umano Secondo Leonardo. Atti del Reale Istituto Veneto di Scienze, Lettere ed Arti 78(2)

Favaro G (1943) Alberto Magno ed Alberto di Sassonia nei manoscritti di Leonardo. Atti e Memorie della R. Accademia di Scienze, Lettere ed Arti di Modena V(VI)

Fengler CK (1974) Lorenzo Ghiberti's second commentary: The translation and interpretation of a fundamental Renaissance treatise on art. Diss. University of Wisconsin

Fergusson FD (1977) Leonardo da Vinci and the tiburio of Milan Cathedral. Architectura 7/2:175–192

Ferrari D (1999) L'inventario dei beni dei Gonzaga (1540–1542). Quaderni di Palazzo Te 6:85–103

Ficarra A (ed) (1968) L'anonimo magliabechiano. Fiorentino, Naples

Ficino Marsilio (1978) The letters of Marsilio Ficino, vol. I. Shepheard-Walwyn, London

Field JV (1997) The invention of infinity: Mathematics and art in the Renaissance. OUP, Oxford

Field JV (2005) Piero della Francesca: A mathematician's art. Yale, New Haven

Fienga DD (1971) The 'Antiquarie prospetiche romane, composte per prospectivo melanese depictore': A document for the study of the relationship between Bramante and Leonardo da Vinci. Diss. University of California

Fiorio MT (1998) The many faces of Leonardismo. The legacy of Leonardo: Painters in Lombardy 1490–1530. Skira Editore, Milan; Abbeville Publishing Group, New York, pp 39–63

Flamini F (1891) La lirica toscana del rinascimento anteriore ai tempi del Magnifico. Tipografia T. Nistri e C., Pisa

Foot PG, Wilson DM (1970) Viking achievement. Sidgwick & Jackson, London

Ford E (2003) Interpretations of marks from draughting tools in some Italian Renaissance drawings. Thesis, Oxford University

Foster P (1981) Lorenzo de' Medici and the Florence Cathedral Fayade. Art Bull LXIII/3:495–500

Frangioni L (1992) Milano e le sue misure. Appunti di metrologia Lombarda fra Tre e Quattrocento. Edizioni Scientifiche Italiani, Naples

Franklin J (2000) Diagrammatic reasoning and modelling in the imagination: The secret weapons of the scientific revolution. In: Freeland G, Corones A (eds) 1543 and all that: Image and word, change and continuity in the proto-scientific revolution. Kluwer, Dordrecht, pp 53–115

Frey K (ed) (1892) Il Codice Magliabechiano cl. XVII. 17. G. Grote, Berlin

Frey K (1911) Zur Baugeschichte der St. Peter: Mitteilungen aus der Reverendissima Fabbrica di S. Pietro. Jahrbuch der Königlich Preußischen Kunstsammlungen, Beiheft, pp 1–95

Frey K (1923) Der Literarische Nachlass Giorgio Vasaris. Georg Müller, München

Frommel CL (1964) Leonardo fratello della Confraternita delia Pietil dei fiorentini aRoma. Raccolta Vinciana XX:369–373

Fusco L, Corti G (1991) Giovanni Ciampolini (d. 1505), a Renaissance dealer in Rome and his collection of antiquites. Xenia XXl:7–46

Fusco L, Corti G (1992) Lorenzo de' Medici on the Sforza Monument. Achademia Leonardi Vinci V:11–32

Gaffurio, Franchino (1492) Theorica musice. (rev; orig. 1480)

Gaffurio, Franchino (1993) The theory of music. In: Palisca CV (ed) Music Theory Translation Series. Eng. trans. and notes: Walter Kurt Kreyszig, Yale, New Haven

Galluzzi P (ed) (1974) Leonardo da Vinci letto e commentato da Marinoni, Heydenreich, Brizio, Reti, De Toni, Mariani, Salmi, Pedretti, Steinitz, Maccagni, Garin, Vasoli. Lettura Vinciana I–XII (1960–1972), Florence

Galluzzi P (1989) Leonardo e i proporzionanti. Lettura Vinciana 28: the lecture of 16 April 1988. Giunti Barbera, Florence

Galluzzi P (1997) Mechanical marvels. Istituto e Museo di Storia della Scienza, Florence

Galluzzi P (1999) The art of invention: Leonardo and Renaissance engineers. The Science Museum, Exhibition Road, London, 15 October 1999 to 24 April 2000. Istituto e Museo di Storia della Scienza, Florence

Gardiner R, Morrison J (eds) (1995) The age of the galley: Mediterranean oared vessels since pre-classical times. Conway Maritime Press, London

Gay V (1928) Glossaire archéologique du Moyen Age et de la Renaissance, vol. 2. Stein B (ed), Picard, Paris

Gaye G (1968) Carteggio inedito d'artisti dei secoli XIV, XV, XVI. 1840. Bottega d'Erasmo, Torino

Ghinzoni P (1889) Lettera inedita di Bernardo Bellincioni. Arch Storico Lombardo XVI:417–418

Gianani F, Modesti O (1989) Il Duomo di Pavia 1488–1932. Tipografia Editrice Artigianelli, Pavia

Gibbs-Smith C (1978) The inventions of Leonardo da Vinci. Phaidon, London

Gilbert C (1980) Italian art 1400–1500. Sources and documents. Prentice-Hall, Inc., Englewood Cliffs

Gilbert C (1983) Bramante on the road to Rome (with some Leonardo sketches in his pocket). Arte Lombarda 66:5–14

Gille B (1972) Leonardo e gli ingegneri del Rinascimento. Feltrinelli, Milan

Giovio P (1977) Leonardi Vinci Vita. In: Richter JP, Pedretti C (eds) The literary works of Leonardo da Vinci, 1st ed. University of California Press, Berkeley, pp 2–3

Giovio P (1999) Leonardi Vincii Vita. Leonardo. In: Farago C (ed) Selected scholarship, vol. I: Biography and early art criticism of Leonardo da Vinci. Garland Publishing, Inc., New York London, pp 70–72

Giraldi S (1990) The Bacchereto Twelve. Achademia Leonardi Vinci 3:114–115

Giulini A (1912) Bianca Sanseverino Sforza figlia di Lodovico il Moro. Arch Storico Lombardo XVIII: 233–252

Glasser H (1977) Artists' contracts of the Early Renaissance. Garland Publishing, New York

Goethe JW von (1817) Joseph Bossi über Leonardo da Vincis Abendmahl zu Mailand. Über Kunst und Alterthum III, Weimar

Gould C (1981) The newly-discovered documents concerning Leonardo's 'Virgin of the Rocks' and their bearing on the problem of the two versions. Art Hist 3:73–76

Gramatica DL (1919) Le memorie su Leonardo da Vinci di Don Ambrogio Mazenta. Milan

Grant E (ed) (1974) A source book in Medieval science. Harvard University Press, Cambridge

Gravett CR, Hook C (1990) Medieval siege warfare. Osprey, Oxford

Grayson C (1957) The humanism of Alberti. Ital Stud 12(1):37–56

Grayson C (1998) Studi su Leon Battista Alberti. In: Claut P (ed) Ingenium, no. 1. Olschki, Florence

Grierson P (1959) Ercole d'Este and Leonardo da Vinci's equestrian statue of Francesco Sforza. Ital Stud XIV:40–48

Guasti C (1857) La Cupola di S. Maria del Fiore. Barbera, Bianchi e Comp., Firenze

Gudea N, Baatz D (1974) Teile spätrömischer Ballisten aus Gornea und Orsova (Rumänien). Saalburg Jahrbuch XXXI:50–72

Guillaume J (1988) Leonard et Bramante. L'emploi des ordres a Milan a la fin du Xve siecle. Arte Lombarda 3–4:101–106

Gukovskj MA (1964) Leonardo e Galeno. Raccolta Vinciana XX:359–367

Gurrieri F, Berti L, Leonardi C (1988) La Basilica di San Miniato al Monte a Firenze. Gurrieri Associati, Florence

Hall M (1973) Reconsiderations of sculpture by Leonardo da Vinci. A bronze statuette in the J. B. Speed Art Museum. J. B. Speed Art Museum Bulletin XXIX

Harris HA (1974) Lubrication in Antiquity. Greece Rome XXI:32–36

Hart I (1961) The world of Leonardo da Vinci, man of science, engineer and dreamer of flight. MacDonald, London

Hart I (1963) The mechanical investigations of Leonardo da Vinci. University of California Press, Berkeley

Hartt F, Corti G, Kennedy C (1964) The chapel of the Cardinal of Portugal 1434–1459 at San Miniato Florence. University of Pennsylvania Press, Philadelphia

Hatfield R (1996) Giovanni Tornabuoni, i fratelli Ghirlandaio e la cappella maggiore di Santa Maria Novella. Domenico Ghirlandaio, 1449–1494: atti del convegno internazional. Firenze, 16–18 ottobre 1994. Centro Di, Firenze, pp 112–117

Hemsoll D (1995) Giuliano da Sangallo and the new Renaissance of Lorenzo de' Medici. In: Ames-Lewis F (ed) The Early Medici and their artists. Birbeck College, University of London, London, pp 187–205

Herlihy D, Klapisch-Zuber C (1996) Census and property survey of Florentine domains in the Province of Tuscany, 1427–1480. Machine-readable data file. Online Catasto of 1427, Version 1.1. Online Florentine Renaissance Resources: Brown University, Providence, R.I.

Herzfeld M (1920–1922) La rappresentazione della 'Danae' organizzata da Leonardo. Raccolta Vinciana XI:226–228

Herzfeld M (1926–1929a) Noch einmal Leonardo und Ligny: eine Ergänzung zu G. Calvi's Aufsatz. Raccolta Vinciana XIII:53–62

Herzfeld M (1926–1929b) Zur Geschichte des Sforzadenkmals. Raccolta Vinciana XIII:80–86

Heydenreich LH (1929) Die Sakralbau-Studien Leonardo da Vinci's. C. u. M. Vogel, Engelsdorf-Leipzig

Heydenreich LH (1951) Leonardo da Vinci the Scientist. Leonardo da Vinci, International Business Machines Corporation

Heydenreich LH (1954) Leonardo da Vinci. 2 vols., George Allen and Unwin Ltd., London; MacMillan Company, New York; Holbein-Verlag, Basel

Heydenreich LH (1969) Leonardo and Bramante: Genius in architecture. In: O'Malley CD (ed) Leonardo's legacy. An international symposium. University of California Press, Berkeley Los Angeles, pp 125–148

Heydenreich LH (1974) The military architect. In: Reti L (ed) The unknown Leonardo. McGraw Hill Book Company, New York, pp 136–165

Heydenreich LH (1977) Giuliano da Sangallo in Vigevano: ein neues Dokument. In: Ciardi Dupré Dal Poggetto MG, Dal Poggetto P (eds) Scritti di storia dell'arte in onore di Ugo Procacci. Electa, Milano, pp 321–323

Hill G Sir (1978) Medals of the Renaissance. Rev. and enlarged by Graham Pollard. 1920; British Museum Publications Ltd., London

Hill GF, Pollard G (1967) Renaissance medals from the Samuel H. Kress Collection at the National Gallery of Art. Phaidon Press, London

Hochstetler Meyer B (1975) Louis XII, Leonardo and the Burlington House cartoon: Six letters, 1507. Gaz Beaux-Arts Oct:105–109

Holder P (1987) Roman artillery I. Military Illustrated 2:31–37

Hope C (2000) 'Composition' from Cennini and Alberti to Vasari. Taylor P, Quiviger F (eds) Pictorial composition from Medieval to Modern art. Warburg Institute Colloquia 6, London Warburg

Horne HP (1903) A newly discovered 'Libra di Ricordi' of Alesso Baldovinetti. Burlington Magazine II/IV (June 1903):22–32; II/V (July 1903):167–174; II/VI (August):377–390

Howard D (2000) Venice and the East. Yale, New Haven

Huard P (1961) Leonard de Vinci. Dessins anatomiques (anatomie artistique, descriptive etfontionnelle). Choix et presentation par Pierre Huard. Les Editions Roger Dacosta, Paris

Hughes G (1997) Renaissance Cassoni. Masterpieces of early Italian art: Painted marriage chests 1400–1550. Starcity Publishing, Sussex; Art Books International, London

Hui DCK (1996) The science of beauty? Theories of proportion from the 16th to the 20th century. PhD, Cambridge University

Hunt RW (1961) Saint Dunstan's classbook from Glastonbury. Umbrae Codicum Occidentalium 4

Hyman I (1974) Brunelleschi in perspective. Prentice-Hall, Inc., Englewood Cliffs, New Jersey

Isermeyer C-A (1963) Die Arbeiten Leonardos und Michelangelos für den grossen Ratsaal in Florenz. Studien zur Toskanischen Kunst: Festschrift fur Ludwig Heydenreich, Munich

Iverson E (1955) Canons and proportions in Egyptian art. Sidgwick and Jackson, London

Iverson E (1957) The Egyptian origin of the archaic Greek canon. Mitteilungen des Deutschen Archäologischen Instituts, Abteilung Kairo 15, Otto Harrassowitz, Wiesbaden

Iverson E (1990) Metrology and canon. Mitteilungen des Deutschen Archäologischen Instituts, Abteilung Kairo 46

Jardine L (1996) Wordly goods. Macmillan, Kent

Jardine L (2000) Global interests: Renaissance art between east and west. Cornell U., Athens, NY

Jones DE (2000) An instinct for dragons. Routledge, New York

Kahle U (1982) Renaissance-Zentralbauten in Oberaltalien: Santa Maria Presso San Satiro. Das Frühwerk Bramantes in Mailand. Nitz Verlag, Munich

Kalkmann A (1893) Die Proportionen des Gesichts in der Griechischen Kunst. Georg Reimer, Berlin

Katz V (1993) A history of mathematics. Harper Collins, New York

Kecks RG (1995) Ghirlandaio: Catalogo completo. Octavo, Firenze

Kemp M (1971) 'Il concetto dell'anima' in Leonardo's Early Skull Studies. J Warburg Courtauld XXXV:115–134

Kemp M (1972) Dissection and divinity in Leonardo's late anatomies. J Warburg Courtauld XXXV: 200–225

Kemp M (1976) 'Ogni dipintore dipinge sè': A Neoplatonic echo in Leonardo's art theory? In: Clough CH (ed) Cultural aspects of the Italian Renaissance. Essays in Honour of Paul Oskar Kristeller. New York, pp 311–323

Kemp M (1977a) From 'Mimesis' to 'Fantasia'. The quattrocento vocabulary of creation, inspiration and genius in the visual arts. Viator-Mediev Renais 8:347–398

Kemp M (1977b) Leonardo and the perspective pyramid. J Warburg Courtauld 40:128–137

Kemp M (1981) Leonardo da Vinci: The marvelous works of nature and man. Harvard University Press, Cambridge Mass.

Kemp M (1985) Leonardo da Vinci: Science and the poetic impulse. Proceedings, The Royal Society of Arts Journal 5343(133):196–214

Kemp M (1990) The science of art: Optical themes in western art from Brunelleschi to Seurat. Yale University Press, New Haven

Kemp M (1992) In the beholder's eye: Leonardo and the 'errors of sight' in theory and practice. Hammer Prize Lecture, Achademia Leonardi Vinci V:153–162

Kemp M (2004) Leonardo. Oxford University Press, Oxford

Kemp M, Walker M (1989) Leonardo da Vinci on painting. Yale University Press, New Haven London

Kennedy RW (1938) Alesso Baldovinetti: A critical and historical study. Yale University Press, New Haven; Oxford University Press, London

Kiadó C (1981) Bibliotheca Corviniana: The library of King Matthias Corvinus of Hungary. Hom Z (trans), 1967; Magyar Helikon, Budapest

Krautheimer R (1970) Lorenzo Ghiberti, 2nd ed. Princeton University Press, Princeton

Krishna A (1977) Certain Citrasutra traditions and their application in Rajasthani and Mughal painting. Beiträge zur Indienforschung, Museum für Indische Kunst, Berlin

Kristeller P (1913) Die Lombardische Grafik der Renaissance. Bruno Cassirer, Berlin

Kristeller O (1964) Eight philosophers of the Italian Renaissance. Stanford University Press, Stanford, CA

Kurz O (1936) A contribution to the history of the Leonardo drawings. Burlington Mag 69

Kurz O (ed) (1964) La letteratura artistica: Manuale delle fonti della storia dell'arte moderna. La Nuova Italia, Florence

Kwakkelstein MW (1994) Leonardo da Vinci as a physiognomist: Theory and drawing practice. Primavera, Leiden

Landels JG (1980) Engineering in the ancient world. Chatto and Windus, London

Landrus M (1996) Leonardo da Vinci's non-reductive method: Representing chaos. Athanor XIV:9–19

Landrus M (2006a) Treasures of Leonardo da Vinci. Carlton and Harper-Collins, London New York

Landrus M (2006b) Leonardo's Canons: Standards and practices of proportional design in his early work. DPhil Thesis, University of Oxford

Landrus M (2007) The proportions of Leonardo's Last Supper. Raccolta Vinciana 32:303–355

Landrus M (2008) The proportional consistency and geometry of Leonardo's Giant Crossbow. Leonardo 41(1):57–63

Landrus M (2009) Leonardo and theories of beauty. In: Kaniari A, Wallace M (eds) Acts of seeing: Artists, scientists and the history of the visual, A volume dedicated to Martin Kemp. Zidane, London

Landrus M (2010) The proportional geometry of form, balance, force, and motion in Leonardo da Vinci's work. Proportions, Centre d'Etudes Superieures de la Renaissance, Université François-Rabelais. Brepols, Turnhout (Belgium)

Landrus M, Kemp M (2003) One hundred years of powered flight 1903~2003, "Leonardo da Vinci's Flying Machine". Winchester Group, London, p 16

Landucci L (1927) A Florentine diary from 1450 to 1516 by Luca Landucci continued by an anonymous writer till 1542 with notes by Iodoco del Badia. de Rosen Jervis A (trans), J. M. Dent & Sons, Ltd., London; E. P. Dutton & Co., New York

Landucci L (1969) Diario Fiorentino dal 1450 al 1516 di Luca Landucci continuato da un anonimofino al 1542. del Badia I (ed) 1883; Studio Biblos, Firenze

Lane F (1934) Venetian ships and shipbuilders of the Renaissance. John Hopkins University, Baltimore, MD

Lane LD (1987) A note of two Leonardo drawings: Puzzling constructions at Christ Church. Gaz Beaux-Arts 129/March:123–126

Laubenbacher R, Pengelley D (1999) Mathematical expeditions: Chronicles by the explorers. Springer-Verlag, New York

Laurenza D (2001) De Figura Umana: Fisiognomica, anatomia e arte in Leonardo. Olschki, Florence

Laurenza D (2004) Leonardo on flight. Giunti, Florence

Lefäve R (1967–1968) Het Laatste Avondmaal Tongerlo. Koninklijk Instituut voor het Kunspartrimonium, Brussels, Bulletin 10:16–31

Lepper F, Frere SS (1988) Trajan's column. Alan Sutton, Gloucester

Lepsius KR (1884) Die Längenmaße der Alten. Berlin

Liebel J (1998) Springalds and great crossbows. Vale J (trans), Royal Armouries, Leeds

Lindberg D (trans) (1970) John Peckham and the science of optics, Perspectiva communis. WI: University of Wisconsin, Madison

Lindberg D (1992) The beginnings of western science. Chicago U, Chicago

Lodge N (1998) A Renaissance king in Hellenistic disguise: Verrocchio's reliefs for King Matthias Corvinus. Italian echoes in the Rocky Mountains: Papers from the 1998 Annual Conference of the American Association for Italian Studies. Provo, Utah. American Association for Italian Studies and the David M. Kennedy Center for International Studies at Brigham Young University, pp 23–46

Lomazzo GP (1973–1974) Scritti sulle arti. 2 vols., Ciardi RP (ed), Marchi & Bertolli, Firenze

Lotz W (1974) La piazza Ducale di Vigevano. Un foro principesco del tardo Quattrocento. In: Studi bramanteschi. Atti del congresso internazionale (Milano, Urbino, Roma, 1970). De Luca, Rome, pp 205–221

Loumyer G (1920) Les Traditions Techniques de la Peinture Medievale. Bruxelles Paris

Ludwig H (ed) (1882–1885) Lionardo da Vinci. Das Buch von der Malerei, Quellenschriften für Kunstgeschte, vols. 15–17, 3 vols., Vienna

Luzio A (1888) Ancora Leonardo da Vinci e Isabella d'Este. Arch Storico Dell'arte 5:181–184

Luzio A (1890) Delle relazioni di Isabella d'Este Gonzaga con Ludovico e Beatrice Sforza. Arch Storico Lombardo XVII:74–119

Luzio A (1900) I ritratti d'Isabella d'Este. Emporium XI:349–359, 427–442

Lyttkens H (1953) The analogy between God and the World. Universitets èrsskrift, Upsalla

Maccagni C (1971) Riconsiderando il problema delle fonti di Leonardo: l'elenco di libri ai ff. 2v-3r del cod. 8936 delia Biblioteca Nacional di Madrid. Lettura Vinciana X:283–308

Machiavelli N (1823) Delle Istorie Fiorentine. Biblioteca storica di tutte Ie nazioni. Nicolo Bettoni, Milano

Machiavelli N (1845) The Florentine histories. Lester CE (ed), Niccolini GB (trans), Paine & Burgess, New York

Machiavelli N (1989) The chief works and others. 3 vols., Gilbert A (trans), Duke University Press, Durham London

Mack RE (2002) Bazaar to piazza. U. California, Berkeley, CA

Magenta C (1883) I Visconti e gli Sforza nel Castello di Pavia. Fusi, Pavia

Maiocchi R (1996) Codice diplomatico artistico di Pavia. Dall'anno 1330 all'anno 1550. 2 vols., Tipografia giil Cooperativa di B. Bianchi, Pavia, (1937–1949). With index published separately: Cipriani, Renata. Indice del Codice diplomatico artistico di Pavia dall 'anno 1350 al 1550 pubblicato da Mons. R. Maiocchi. Isituto Lombardo di Scienze e Lettere, Milano

Makkai L (1975) A history of Hungary. Paményi E (ed), Collet's, London Wellingborough

Malaguzzi Valeri F (1901) II Duomo di Milano nel Quattrocento. Note storiche su nuovi documenti. Repertorium für Kunstwissenschaft I:87–102; II:230–240

Malaguzzi Valeri F (1915) La corte di Ludovico il Moro, vol. II: Bramante e Leonardo da Vinci. Ulrico Hoepli, Milano

Malaguzzi Valeri F (1917) La corte di Ludovico il Moro, vol. III: Gli artisti Lombardi. Ulrico Hoepli, Milano

Malaspina di Sannazzaro G (1816) Memorie storiche della fabbrica della cattedrale di Pavia. Milano

Malvasi L (1842–1844) La metrologia italiana ne' suoi cambievoli rapporti desunti dal confronto col sistema metrico-decimale. Fratelli Malvasi, Modena

Marani PC (1988) Bramante e Leonardo architetti militari. Arte Lombarda 3–4:107–113

Marani PC (1992a) La 'Crocifissione' di Bramantino: progetto, pentimenti, cronologia. In: Marani PC (ed) La Crocifissione di Bramantino, storia e restauro. Quaderni di Brera 7 (Nov):13–46

Marani PC (1992b) Leonardo a Venezia e nel Veneto: documenti e testimonianze. Leonardo e Venezia. Bompiani, Milano, pp 23–36

Marani PC (1998) The question of Leonardo's Bottega: Practices and the transmission of Leonardo's ideas on art and painting. The Legacy of Leonardo: Painters in Lombardy 1490–1530. Sidra Editore, Milan; Abbeville Publishing Group, New York, pp 9–37

Marani PC (2000) Leonardo: una carriera dipittore. F. Motta, Milano, 1999; Eng. trans.: Leonardo da Vinci: The complete paintings. Harry N. Abrams, Inc., New York

Marani PC (2002) Von der Natur zum Symbol. In: Fehrenbach F (ed) Leonardo da Vinci, Natur im Übergang. Wilhelm Fink Verlag, Paderborn, pp 371–390

Marchini G, Micheletti E, Ciardi Dupre Dal Poggetto MG (eds) (1987) La chiesa di Santa Trinita a Firenze. Cassa di Risparmio di Firenze, Firenze

Marcolongo R (1937) Memorie sulla geometria e la meccanica di Leonardo da Vinci. Studi Vinciani, Naples

Marinelli S (1981) The author of the Codex Huygens. J Warburg Courtauld 44:214–220

Marinoni A (ed) (1952) Tutti gli scritti. Scritti letterari. Rizzoli, Milano

Marinoni A (1964) La teoria dei numeri frazionari nei manoscritti vinciani. Leonardo e Luca Pacioli. Raccolta Vinciana XXI:111–196

Marinoni A (1982) Leonardo ingegnere militare. (Exh. Cat.) Shell Italia, Milan

Marinoni A (1989) Le proporzioni secondo Leonardo. Raccolta Vinciana XXIII:259–273

Marrucci RA (ed) (1987) La 'Prospettiva' Bramantesca di Santa Maria presso San Satiro. Storia, restauri e intervento conservativo. Banca Agricola Milanese, Milano

Marsden EW (1969) Greek and Roman artillery: Historical development. Oxford U., Oxford

Marsden EW (1971) Greek and Roman artillery: Technical treatises. Oxford U., Oxford

Martini A (1976) Manuale di metrologia. E. Loescher, Turin, 1883; Editrice Edizioni Romane d'Arte, Rome 1976

Martini, Francesco di Giorgio (1979) Il codice Ashburnham 361 delia Biblioteca medicea laurenziana di Firenze: trattato di architettura di Francesco di Giorgio Martini. 2 vols., Giunti Barbera, Firenze

Masters RD (1998) Fortune is a river. Leonardo da Vinci and Niccolo Machiavelli's magnificent dream to change the course of Florentine history. Free Press, New York

Mazzotta AB (1986) Cesare Cesariano. II Duomo di Milano. Dizionario storico artistico e religioso, Milano, NED, pp 169–170

McCurdy E (1958) The notebooks of Leonardo da Vinci. Eng. trans., Reynal & Hitchcock, New York (1938); Braziller, New York

McMullen R (1975) Mona Lisa. The picture and the myth. Houghton Mifflin Co., Boston

McMurrich JP (1930) Leonardo da Vinci the anatomist (1452–1519). Wilkins & Wilkins Co., Baltimore, for Carnegie Institute of Washington

Milanesi G (1872) Documenti inediti riguardanti Leonardo da Vinci. Arch Storico Ital XVI(III), 71–72, 5-6:219–230

Minar EL (1969) Plutarch's Moralia. 16 vols., trans., Loeb Classical Library, Harvard, Cambridge MA, (1961)

Moffatt CJ (1990) 'Duca di Bari'. Achademia Leonardi Vinci III:125–128

Moffitt JF (1990) Leonardo's 'Sala delle Asse' and the primordial origins of architecture. Arte Lombarda 92–93:76–90

Molinari C (1961) Spettacolifiorentini del Quattrocento. Contributi allo studio delle Sacre Rappresentazioni. Neri Pozza Editore, Venezia

Möller E (1912) Leonardo da Vincis Entwurf eines Madonnenbildes für S. Francesco in Brescia (1497). Repertorium für Kunstwissenschaft XXXV:241–261

Möller E (1916) Leonardos Bildnis der Cecilia Gallerani in der Galerie des Fürsten Czartoryski in Krakau. Monatshefte für Kunstwissenschaft IX:313–326

Möller E (1934) Ser Giuliano di Ser Piero da Vinci e le sue relazioni con Leonardo. Rivista d'Arte, pp 387–399

Möller E (1939) Der Geburtstag des Lionardo da Vinci. Jahrbuch preussischer Kulturbesitz LX:71–75

Monstadt B (1995) Judas beim Abendmahl: Figurenkonstellation und Bedeutung in Darstellungen von Giotto bis Andrea del Sarto. Scaneg, München

Montebelli V (1992) Piero, la matematica e i poliedri. In: Dal Poggetto P (ed) Piero e Urbino, Piero e Ie Corti rinascimentali. Exh. Cat. Urbino, Palazzo Ducale e Oratorio di San Giovarmi Battista, 24 July to 31 October 1992. Marsilio Editore, Venezia, pp 479–485

Monti S (1914) Curiosità Vinciana: 2. I diversi codici di Leonardo da Vinci nella Biblioteca Ambrosiana descritti da Anton Giuseppe delle Torre di Rezzonico circa l'anno 1779. Periodico delia societa storica delia Provincia e antica Diocesi di Como XXI:69–79

Moore JM (1965) The manuscript tradition of Polybius. Cambridge

Morpurgo ST (1895) Ricordi in Firenze. A Leonardo da Vinci a Paolo Toscanelli Le Armi delia Famiglia da Vinci e del Comune di Vinci – Un Fratello di Leonardo Lanaiolo in Firenze e il suo 'Confessionale'. Stabilimento Tipografico Fiorentino, Firenze

Mosteller JF (1990) The problem of proportion and style in Indian art history: Or why all Buddhas in fact do not look alike. The Art J 49(4):388–394

Mosteller JF (1991) The measure of form: A new approach for the study of Indian sculpture. Abhinav Publications, New Delhi

Mostra di disegni: manoscritti e documenti (1952) Quinto Centenario della Nascita di Leonardo da Vinci. Biblioteca Medicea Laurenziana, Firenze

Motta E (1893) Ambrogio Preda e Leonardo da Vinci (Nuovi documenti). Arch Storico Lombardo XX: 972–996

Mukherji DP (2001) The Citrasutra of the Visnudharmottara Purana. IGNCA and Motilal Banarsidass, Delhi

Müller-Walde P (1889) Leonardo da Vinci. Lebensskizze und Forschungen über sein Verhältnis zur Florentiner Kunst und zu Rafael. G. Hirth, Müchen

Müller-Walde P (1897a) Beiträge zur Kenntnis des Leonardo da Vinci. I. Ein neues Dokument zur Geschichte des Reiterdenkmals für Francesco Sforza: Das erste Modell Leonardo's. Jahrbuch der Königlich-Preussischen Kunstsammlungen XVIII:92–136

Müller-Walde P (1897b) Beiträge zur Kenntnis des Leonardo da Vinci. II. Eine Skizze Leonardo's zur stehenden Leda; Eine Skizze nach Praxiteles und der Merkur im Kastell von Mailand. Jahrbuch der Königlich-Preussischen Kunstsammlungen XVIII:137–169

Müntz E (1893) Leonardo da Vinci and the study of the Antique. Portfolio 24:153–160

Müntz E (1898) Leonardo da Vinci: Artist, thinker, and man of science. 2 vols., english. trans. William Heinemann, London; Charles Scribner's Sons, New York

Murray P (1963) Leonardo's approach to anatomy and architecture, and its effect on Bramante. Archit Rev XCCCN:346–351

Mussini M (1991) Il Trattato di Francesco di Giorgio Martini e Leonardo: il Codice Estense restituito. Universita di Parma, Istituto di Storia dell'Arte, Parma

Nardi I (2003) The theory of Indian painting: A critical re-evaluation of the Citrasutras, their uses and interpretations. Thesis, School of Oriental and African Studies, University of London

Natali A (ed) (2000) L'Annunciazione di Leonardo: La montagna sui mare. Cinisello Balsamo, Milano

Newbigin N (1996) Feste d'Oltrarno. Plays in Churches in Fifteenth-century Florence. 2 vols., Leo S. Olschki Editore, Firenze

Nicodemi G (1996a) The life and works of Leonardo. Leonardo da Vinci. Barnes & Noble Books, New York, pp 19–88

Nicodemi G (1996a) The portrait of Leonardo. Leonardo da Vinci. Barnes and Noble Books, New York, pp 9–18

Nicolle D (1999) Arms and armour of the Crusading Era 1050–1350: Islam, Eastern Europe and Asia. Greenhill, London

Nicolle D, Thompson S (illust) (2002) Medieval siege weapons (1), Western Europe AD 585–1385. Osprey, Oxford

O'Malley D, Saunders JB de CM (1952) Leonardo da Vinci on the human body. Henry Schuman, New York

Onians J (1988) Bearers of meaning, The classical orders in Antiquity, the Middle Ages, and the Renaissance. Princeton University Press, Princeton, 176 p

Oriani B (1891) Istruzione su le misure e su i pesi che si usano nella Repubblica Cisalpina. Milano

Pacciani R (1993) 'Tum pro honore publico turn pro commoditate privata'. Un documento del 1490 per l'edificazione di palazzo Gondi a Firenze. Arte Lombarda 105/106/107, 2-3-4: 202–205

Pacetti D (1952) Lorenzo di Ser Piero da Vinci: fratello di Leonardo e il suo 'confessionario' autografo nel Cod. 1420 delia Biblioteca Riccardiana di Firenze. Quaracchi, Firenze

Pacioli FL (1889) Divina proportione, Venice: Paganinus de Paganinus, 1509, folio 1a. Reprint, Verlag von Carl Graeser, Vienna

Palestra A (1969) Cronologia e documentazione riguardante la costruzione della chiesa di S. Maria presso S. Satiro del Bramante. Arte Lombarda 14/2:154–160

Pancaldi L (1847) Raccolta ridotta a dizionario di varie misure antiche, e moderne coi loro rapporti alle misure metriche, ed a quelle di Bologna : compilata dall'ingegnere Luigi Pancaldi. tipografia Sassi, Bologna

Panofsky E (1940) The Codex Huygens and Leonardo da Vinci's art theory; The Pierpont Morgan Library Codex M.A. 1139. Studies of the Warburg Institute 13, Warburg, London

Panofsky E (1962) Artist, scientist, genius: Notes on the 'Renaissance-Dämmerung'. In: The Renaissance: Six Essays, Harper, New York, 1962 (1953, Metropolitan Museum)

Panofsky E (1987) The history of the theory of human proportions as a reflection of the history of styles. In: Meaning in the visual arts. Penguin, New York (Doubleday, 1955)

Parsons W (1968) Engineers and engineering in the Renaissance. Harvard, Cambridge, Mass

Passavant G (1959) Andrea del Verrocchio als Maler. Verlag L. Schwann, Düsseldorf

Passavant G (1969) Verrocchio: sculptures, paintings and drawings. Phaidon, London

Patetta L (1987) S. Maria presso S. Satiro. L'architettura del quattrocento a Milano. Clup, Milano, pp 176–189

Patterson WF (1990) A guide to the crossbow. Society of Archer-Antiquaries, London

Payne-Gallwey R (1903) The crossbow. Mediaeval and modern military and sporting; Its construction history and managment. With a treatise on the balista and catapult of the acients. Longman, Green and Co., London

Peckham J (1482–1483) Archbishop of Canterbury, Prospectiua cois ... ad ungue castigata [per] ... d. Faciu cardanu. [Fazio Cardano]. P. de Corneuo, Milan

Pedretti C (1953) Documenti e memorie riguardanti Leonardo da Vinci a Bologna e in Emilia. Editoriale Fiammenghi, Bolgona

Pedretti C (1957a) Indagine con gli 'infrarossi'in una pagina di Leonardo. Studi vinciani: Documenti, analisi e inediti leonardeschi. Librairie E. Droz, Geneva

Pedretti C (1957b) Studi vinciani: documenti, analisi, e inediti leonardeschi. Librairie E. Droz, Geneva

Pedretti C (1960) Spigolature nel Codice Atlantico. Bibliotheque d'Humanisme et Renaissance. Travaux et Documents XXII:531–533

Pedretti C (1964) The missing folio 3 of MS B. Raccolta Vinciana XX:211–224

Pedretti C (1968) II Nuovi documenti riguardanti la 'Battaglia d'Anghiari. Leonardo da Vinci. Tre Saggi. G. Barbera Editore, Firenze, pp 53–78

Pedretti C (1970) Leonardo da Vinci: Manuscripts and drawings of the French period, 1517–1518. Gaz Beaux-Arts LXXVI:285–318

Pedretti C (1972) Leonardo da Vinci. The Royal Palace at Romorantin. Belknap Press of Harvard University Press, Cambridge

Pedretti C (1973) Leonardo: A study in chronology and style. Thames & Hudson, London

Pedretti C (1977) The Sforza Sepulchre. Gaz Beaux-Arts April:121–131

Pedretti C (1978) Leonardo architetto. Electa, Milano

Pedretti C (1988) The 'libro dj medjcina dj cavallj'. Achademia Leonardi Vinci 1:91–95

Pedretti C (1989) A proem to sculpture. Achademia Leonardi Vinci 2:11–39

Pedretti C (1990) La 'Dama con l'ermellino' come allegoria politica. In: Rota Ghibaudi S, Barcia F (eds) Studi politici in onore di Luigi Firpo, vol. I. F. Angeli, Milano, pp 161–181

Pedretti C (1991) Leonardo and the Antique: A bibliography. Achademia Leonardi Vinci N:214–244

Pedretti C (1992) Paolo di Leonardo. Achademia Leonardi Vinci 5:120–121

Pedretti C (1993) Li Medici mi crearono e desstrussono. Achademia Leonardi Vinci 6:173–184

Pedretti C (1994) Leonardo a Venezia. In: Maschio R (ed) I tempi di Giorgione. Gangemi Editore, Rome

Pedretti C (1995) The Sforza Horse in context. In: Cole Ahl D (ed) Leonardo da Vinci's Sforza monument horse: The art and the engineering. Assoc. University Press, London, pp 27–39

Pedretti C (1999) Leonardo, the machines. Giunti, Florence

Pedretti C, Roberts J (1977) Drawings by Leonardo da Vinci at Windsor newly revealed by Ultra-Violet Light. Burlington Mag June:396–408

Perrin B (1968) Plutarch's lives. 11 vols., trans., Loeb Classical Library, Harvard, Cambridge, MA, (1917)

Pisani M (1923) Un avventuriero del Quattrocento: La vita e Ie opere di Benedetto Dei. Vol. V of Biblioteca delia 'Rassegna'. Societa anonima editrice Francesco Perrella, Genova

Plotinus (1580) Plotini Platonicorum facile coryphaei Operum Philosophicorum Omnium Libri LIV in Sex Enneades distrieuti Ex antiquiss. Codicum fide nunc primum Graece editi, cum Latina Marsilii Ficini interpretatione and commentatione. Perneam Lecythum, Basel

Pochat G (1978) Brunelleschi and the 'Ascension' of 1422. Art Bull 60:232–234

Poggi G (ed) (1909) I ricordi di Alesso Baldovinetti nuovamente pubblicati e illustrati. Frammenti inediti di vitafiorentina. Pirenze, Florence: Libreria editrice fiorentina

Poggi G (1910) Note su Filippino Lippi: La tavola per san Donato di Scopeto e l'Adorazione dei Magi di Leonardo da Vinci. Rivista d'Arte VII:1–44

Polzer J (1980) The perspective of Leonardo considered as a painter. In: Emiliani MD (ed) La Prospettiva rinascimentale: Codificazioni e trasgressioni, no. 1. Centro Di, Florence, pp 233–247

Pope-Hennessy J (1958) Italian Renaissance sculpture. Phaidon Press, London

Popham A (1994) The drawings of Leonardo da Vinci. Kemp M (intro) 1945; Pimlico, London

Popp AE (1928) Leonardo da Vinci, Zeichnungen. R. Riper & Co., München

Prager PD, Scaglia G (1970) Brunelleschi. Studies of his technology and inventions. The MIT Press, Cambridge London

Pratesi F (1995) La splendida basilica di San Miniato al Monte a Firenze. II Rinascimento inizia da qui. Octavo, Firenze

Prou V (1877) La Chirobaliste d'Heron d'Alexandrie. Notices et Extraits des manuscrits de la Bibliothäque nationale et autres bibliothäques 26:1–319

Pudor H (1902) Das Leonardische Abendmahl. Laokoon: Kunsttheoretische Essays. Seemann, Leipzig

Pulci L (1998) Morgante: The epic adventures of Orlando and his giant friend Morgante. Lebano EA (ed) Tunsiani J (trans), Indiana University Press, Bloomington

Radice A (1976) Il Cronaca – A fifteenth-century Florentine architect. Diss. University of North Carolina at Chapel Hill

Ragghianti CL (1954) Inizio di Leonardo. Critica d'Arte I/1954:4–18; I/1954:102–118; IV:302–316

Raphael M (1989) Ein Anmerkung zum Abendmahl Leonardos, (1938). Posthumously retitled 'Fläche und Raum im Abendmahl Leonardos' and published in: Bild-Beschreibung: Natur, Raum und Geschichte in der Kunst, vol. 7, Suhrkamp, Frankfurt

Raw B (1955) The drawing of an angel in MS 28, St. John's College, Oxford. J Warburg Courtauld XVIII: 318–319

Reti L (1968) The two unpublished manuscripts of Leonardo Da Vinci in the Biblioteca Nacional of Madrid – I. Burlington Magazine CX/Jan 1968:10–22; The two unpublished manuscripts of Leonardo Da Vinci in the Biblioteca Nacional of Madrid – II. Burlington Magazine CX/Feb:81–89

Reti L (ed) (1974) The unknown Leonardo. Bührer EM (designer), McGraw-Hill, New York

Reti L (1999) Leonardo da Vinci and the graphic arts: The early invention of relief-etching. In: Farago C (ed) Leonardo's science and technology: Essential readings for the non-scientist. Garland Publishing, Inc., New York London, pp 379–385, originally in Burlington Magazine April 1971, pp 189–195

Richmond IA (1936) Trajan's army on Trajan's column. Papers of the British School at Rome XIII:1–40

Richter JP (1881) Leonardo. Scribner and Welford, New York; Sampson Low, London; Searle & Rivington, Marston

Richter IA (ed, trans) (1949) Paragone: A comparison of the arts by Leonardo da Vinci. Oxford University Press, London New York Toronto

Risner F (1991) Opticae Thesaurus Alhazeni Arabis libri septem: nunc primum editi; Eivsdem liber De Crepvscvlis [et] Nubium ascensionibus; Item Vitellonis Thvringopoloni libri X./Omnes instaurati, figuris illustrati [et] aucti, adiectis etiam in Alhazenum commentarijs. Episcopius. Basileae 1572. Unguru S (trans) Witelonis Perspectivae Liber Secundus et Liber Tertius. Studia Copernicana XXVIII, Polish Academy of Sciences, Wroclow

Robins G (1991) Composition and the artist's squared grid. Journal of the American Research Centre in Eqypt 28:41–54

Robins G (1994) Proportion and style in Ancient Egyptian Art. Fowler AS (illust), Thames and Hudson, London; (U. Texas, Austin)

Rodgers W (1940) Navel warfare under oars, 4th to the sixteenth centuries: A study of strategy, tactics and ship design. Naval Institute, Annapolis, MD

Rosand D (1988) The meaning of the mark: Leonardo and Titian. No. VIII of Franklin D. Murphy Lectures. Spencer Museum of Art, University of Kansas, Lawrence, Kansas

Rossi F (1956) La lanterna della cupola di Santa Maria del Fiore e i suoi restauri. Boll Arte XLI:128–143

Rossi P (1971) I filosofi e le machine (1400–1700). Feltrinelli, Milan

Rubenstein N (1995) The Palazzo Vecchio, 1298–1532: Government, architecture, and imagery in the civic Palace of the Florentine Republic. Clarendon Press, Oxford

Rubin PL (1995) Giorgio Vasari. Art and history. Yale University Press, New Haven London

Ruggeri U (1978) Carlo Urbini e il codice Huygens. Critica Arte 43:167–176

Rzepinska M (1985) La dama dell'ermellino. Leonardo. La Pittura. Giunti Martello Editore, Firenze, pp 66–70

Rzepinska M (1993) The 'Lady with the Ermine' revisited. Achademia Leonardi Vinci 6:191–199

Salvini R, Camesesca E (1965) La Cappela Sistina in Vaticano. Rizzoli Editore, Milan

Sangpo P (1996) The clear mirror depicting the pearl rosaries of thangka painting of the Tsang-pa tradition of Tibet. Tibetan Refugee Self Help Centre, Darjeeling, West Bengal

Sannazzaro G (1816) Malaspina di. Memorie storiche delia fabbrica delia cattedrale di Pavia. Milano

Sannazzaro GB (1993) Per S. Maria presso S. Satiro e Leonardo: nuovi documenti. Raccolta Vinciana XXV:63–85

Saslow JM (1986) Ganymede in the Renaissance: Homosexuality in art and society. Yale University Press, New Haven London

Saviotti A (1920) Una rappresentazione allegorica in Urbino nel 1474. Atti e Memorie delia R. Accademia Petrarca di Scienze, Lettere ed Arti in Arezzo. L'Accademia, Arezzo

Saxonia A de (1482/1487) Tractatus proportionum, Padua, 1482, and Venice 1487

Saxonia A de (1482/1492) Questiones in Aristotelis libros De coelo et mundo, Pavia, 1482, and Venice 1492

Scarpati C (1982) Leonardo e i linguaggi. Studi sul Cinquecento italiano, Milan, pp 3–26

Scarpati C (ed) (1993) Leonardo da Vinci. Il paragone delle arti. Milan

Scarpati C (2001) Leonardo scrittore. Milan

Schiaparelli A (1921) Leonardo retrattista. Fratelli Treves Editori, Milano

Schneider R (1906) Heron's Cheiroballistra. Römische Mitteilungen 21:142–168

Schofield RV (1976) A drawing for Santa Maria presso San Satiro. J Warburg Courtauld 39:246–253

Schofield R (1982) Ludovico il Moro and Vigevano. Arte Lombarda 62:93–140

Schofield R (1989) Amadeo, Bramante and Leonardo. Achademia Leonardi Vinci 2:68–100

Schofield R (1991) Leonardo's Milanese architecture: Career, sources and graphic techniques. Achademia Leonardi Vinci IV:111–157

Schofield R, Shell J, Sironi G (eds) (1989) Giovanni Antonio Amadeo: Documents. Edizioni New Press, Como

Schoot A van der (1999) De ontstelling van Pythagoras. Kok Agora, Baarn (Netherlands)

Schramm E (1904) Zu der Rekonstruktion griechisch-römischer Geschütze. Jahrbuch der Gesellschaft fur lothringische Geschichte und Altertumskunde xvi:142–160

Schramm E (1906) Zu der Rekonstruktion griechisch-römischer Geschütze. Jahrbuch der Gesellschaft fur lothringische Geschichte und Altertumskunde xviii:276–283

Schröer K, Irle K (1998) Ich aber quadriere den Kreis ...: Leonardo da Vincis Proportionsstudie. Waxmann, Münster

Schuster F (1910) Zur Mechanik Leonardo da Vincis. Erlangen

Sciolla GC (1996) Leonardo e Pavia. Lettura Vinciana XXV/1995:1–46, Giunti, Firenze

Scott-Elliot AH (1956) The Pompeo Leoni Volume of Leonardo drawings at Windsor. Burlington Mag 98(634):11–17

Scrivano R (1965) Bernardo Bellincioni. Dizionario Biografico degli Italiani. Instituto della Enciclopedia Italiana, Roma VII:687–689

Seidlitz W v (1911) Regesten zum Leben Leonardos da Vinci. Repertorium für Kunstwissenschaft XXXIV:448–458

Seymour C Jr (1971) The sculpture of Verrocchio. New York Graphic Society, Ltd., Greenwich, CT

Seymour C Jr (1974) Michelangelo's David. A search for identity. W. W. Norton & Company, Inc., New York

Shaw JB (1976) Drawings of old masters at Christ Church. 2 vols., Oxford University Press, Oxford

Sheard WS (1992) Verrocchio's Medici tomb and the language of materials; with a postscript on his legacy in Venice. In: Bule S, Darr AP, Superbi F (eds) GioffrediVerrocchio and Late Quattrocento Italian sculpture. Acts of two conferences at Brigham Young University and Villa I Tatti. Casa Editrice Le Lettere, Firenze, pp 63–90

Shell J (1987) Painters in Milan, 1490–1530. A resource of newly discovered documents. Diss. New York University

Shell J (1993) The Scuola di San Luca, or Universitas Pictorum, in Renaissance Milan. Arte Lombarda 104(1):78–99

Shell J (1995) Pittori in bottega. Milano nel Rinascimento. U. Allemandi, Torino

Shell J (1998a) Ambrogio de Predis. I Leonardeschi: L'eredita di Leonardo in Lombardia. Skira, Milano, pp 123–130; In: The Legacy of Leonardo: Painters in Lombardy 1490–1530. Abbeville Publishing Group, New York, pp 123–130

Shell J (1998b) Cecilia Gallerani: una biografia. Fabjan B, Marani PC (eds) Leonardo: La dama con l'ermellino. Exh. Cat. Palazzo del Quirinale, Rome, 15 October–14 November 1998; Pinacoteca di Brera, Milan, 19 November–13 December 1998; and Palazzo Pitti, Florence, 16 December 1998 to 24 January 1999. Cinisello Balsamo, Silvana Editorale, Milano, pp 51–65

Shell J (1998c) Marco d'Oggiono. The legacy of Leonardo: Painters in Lombardy 1490–1530. Skira Editore, Milan; Abbeville Publishing Group, New York, pp 163–178

Shell J, Sironi G (1989) Giovanni Antonio Boltraffio and Marco d'Oggiono: The Berlin resurrection of Christ with Sts. Leonard and Lucy. Raccolta Vinciana XXIII:119–154

Shell J, Sironi G (1992a) Cecilia Gallerani: Leonardo's lady with an ermine. Artibus et historiae 25: 47–66

Shell J, Sironi G (1992b) Salai and the inventory of his estate. Raccolta Vinciana XXN:109–153

Shukla DN (1957) Hindu canons of painting or Citra-laksanam. Lucknow

Siculus D (1976) Library of history XIV:41, 3–6, Loeb Classical Library, vol. VI. Oldfather CH (trans), Harvard, Cambridge, MA

Simione A (1904) Un' umanista milanese, Piattino Piatti. Arch Storico Lombardo II:5–50; 227–301

Simons P (1987) Patronage in the Tornaquinci Chapel, Santa Maria Novella, Florence. In: Kent FW, Simons P (eds) Patronage, art, and society in Renaissance Italy. Oxford University Press, New York, pp 221–250

Sirén O (1911) Leonardo da Vinci. Aktiebolaget Ljus, Stockholm

Sirén O (1916) Leonardo da Vinci, the artist and the man. Yale University Press, New Haven

Sironi G (1981) Nuovi documenti riguardanti la 'Vergine delle Rocce' di Leonardo da Vinci. de Vecchi PL (intro), Giunti Barbera, Firenze

Sivaramamurti C (1978) Chitrasutra of the Vishnudharmottara. Kanak Publications, New Delhi

Smiraglia Scognamiglio N (1896) Nuovi documenti su Leonardo da Vinci. Arch Storico Dell'arte II: 313–315

Smiraglia Scognamiglio N (1900) Ricerche e documenti sulla giovinezza di Leonardo da Vinci. Marghieri, Napoli

Smith RD, Brown RR (1981) Bombards: Mons Meg and her sisters. Royal Armouries Monograph 1, London

Solmi E (1904) Documenti inediti sulla dimora di Leonardo da Vinci in Francia nel 1517 e 1518. Arch Storico Lombardo II:389–410

Solmi E (1907) Ricordi dell a vita e delle opere di Leonardo da Vinci raccolti dagli scritti di Gio. Paolo Lomazzo. Arch Storico Lombardo VIII:290–331

Solmi E (1910) Su di una probabile gita di Leonardo da Vinci in Genova il 17.3.1498. Arch Storico Lombardo XIV:435–450

Solmi E (1912) Leonardo e Machiavelli. Tipografia Editrice L. F. Cogliati, Milano

Solmi E (1923) Leonardo (1452–1519). 1900; Giunti Barbera, Firenze

Solmi E (1924a) Leonardo da Vinci e la Cattedrale di Piacenza al tempo del Vescovo Fabrizio Marliani. In: Solmi A (ed) Scritti Vinciani. Soc. Anon. Editrice 'La Voce'. Firenze, pp 99–109, (originally published in Bolletino Storico Piacentino 1911)

Solmi E (1924b) Leonardo da Vinci e Papa Giulio II. In: Solmi A (ed) Scritti Vinciani. Soc. Anon. Editrice 'La Voce.', Firenze, pp 241–263, (originally published 1911 in Arch Storico Lombardo XVI:390)

Solmi E (1924c) Leonardo da Vinci nel Castello e nella Sforzesca di Vigevano. In: Solmi A (ed) Scritti Vinciani. Soc. Anon. Editrice 'La Voce', Firenze, pp 77–95, (originally in Viglevanum, vol. V, fasc. V, 1911)

Solmi E (1924d) Leonardo da Vinci: il Duomo, il Castello e l'Universita di Pavia. In: Solmi A (ed) Scritti Vinciani. Soc. Anon. Editrice 'La Voce', Firenze, pp 17–74

Solmi E (1976a) La Festa del 'Paradiso' di Leonardo da Vinci e Bernardo Bellincione. In: Scritti Vinciani. Le fonti dei Manoscritti di Leonardo da Vinci e altri studio Firenze: La Nuova editrice, pp 407–418

Solmi E (1976b) Nuovi contributi alle fonti di Leonardo da Vinci. Scritti Vinciani. Le Fonti dei Manoscritti di Leonardo da Vinci e altri studio. La Nuova Italia editrice, Firenze, pp 345–405, (originally in Giornale storico delia letteratura italiana LVIII/1911)

Spencer J (1973) Il progetto per il cavallo di bronzo per Francesco Sforza. Arte Lombarda 38/39:23–35

Spencer JR (1987) Speculations on the origins of the Italian Renaissance medal. In: Pollard JG (ed) Studies in the history of art 21, 'Italian Medals'. National Gallery of Art, Washington, DC, pp 197–203

Steinitz K (1948) Manuscripts of Leonardo da Vinci, their history, with a description of the manuscript editions in facsimile. Los Angeles

Steinitz KT (1949) A reconstruction of Leonardo da Vinci's revolving stage. Art Quart XII(4):325–336

Steinitz KT (1952) Leonardo da Vinci signs his name: Seven signatures of Leonardo da Vinci in Codex Atlanticus. Raccolta Vinciana XVII:151–156, (originally published in Autograph Collector's Journal VII, fall)

Steinitz KT (1969) Leonardo architetto teatrale e organizzatore di feste Quando s'apre Il Paradiso di Plutone. Lettura Vinciana IX:249–271

Steinitz KT (1970) Leonardo Architetto Teatrale e Organizzatore di Feste. Lettura Vinciana IX, Milan, 1969; G. Barbera Editore, Florence

Steinitz KT, Feinblatt E (trans) (1999) Leonardo da Vinci, c. 1540. Leonardo. In: Farago C (ed) Selected scholarship, vol. I: Biography and early art criticism of Leonardo da Vinci. Garland Publishing, Inc., New York and London, pp 73–75

Steinmann E (c. 1987–c. 1990) Amadori, Giovanni di Zanobi. Lexikon der Kunst: Malerei, Architektur, Bildhauerkunst. Herder, Freiburg

Stites RS (1970) The sublimations of Leonardo da Vinci with a translation of the Codex Trivulzianus. Smithsonian Institution Press, Washington

Streeter EC (1916) The role of certain Florentines in the history of anatomy, artistic and practical. B Johns Hopkins Hosp XXXVII:113–118

Sudhoff KFJ (1913) Die Lehre von den Hirnventrikeln in textlicher und graphischer Tradition des Altertums und Mittelalters. Archiv für Geschichte der Medizin, vol. VII

Sudhoff K (1914–1918) Beiträge zur Geschichte der Chirurgie im Mittelalter. Leipzig

Summers D (1987a) The judgment of sense: Renaissance naturalism and the rise of aesthetics. Cambridge University Press, Cambridge

Summers D (1987b) The stylistics of color. In: Hall MB (ed) Color and technique in Renaissance painting. Italy and the North. Locust Valley, pp 205–220

Taccola M (1449) De machinis (aka. De rebus militaribus), Ms. Codex Latinus Monacensis 28800. Bayerische Staatsbibliothek, München

Taddei M, Zanon E (eds) Laurenza D (text) (2005) Leonardo's machines. Giunti, Firenze

Taddei M, Zanon E, Lisa M (eds) (2005) Leonardo da Vinci, Codice Atlantico. Leonardo3, Milano

Tanaka H (1995) The process of the Nagoya City Sforza reconstruction. In: Cole Ahl D (ed) Leonardo da Vinci's Sforza Monument Horse: The Art and the Engineering. Assoc. University Press, London, pp 129–135

Tatarkiewicz W (1970–1974) History of aesthetics. 3 vols., Harrell J, Barrett C, Petsch D (eds), De Gruyter, Paris The Hague

Themistius (1996) On Aristotle: On the soul. In: Todd RB (trans) Ancient commentators on Aristotle. Series, gen. ed. Richard Sorabji, Duckworth, London

Theophilus (1963) On divers arts. Hawthorne JG, Smith CS (trans, notes), U. Chicago, Chicago

Thévenot M (1693) Veterum mathematicorum Athenaei, Apollodori, Philonis, Bitonis, Heronis, et aliorum opera: graece et latine pleraque nunc primum edita / ex manuscriptis codicibus Bibliothecae Regiae. Ex Typographia Regia, Parisiis

Thiis J (1913) Leonardo da Vinci: The Florentine years of Leonardo and Verrocchio. Muir J (trans), Herbert Jenkins Limited Publishers, London

Thomas I (1939/1980) Selections illustrating the history of Greek mathematics. 2 vols., trans., Harvard, Cambridge, MA

Thomas I (1939/1998) From Thales to Euclid. Trans., Loeb Classical Library 335, Harvard, Cambridge, MA

Thomas A (1976) Workshop procedure of fifteenth century Florentine artists. Diss. Courtauld Institute of Art, University of London

Thomas A (1995) The painter's practice in Renaissance Tuscany. Cambridge University Press, Cambridge

Thompson DV (1933/1960) The craftsman's handbook. Trans., Dover, New York, 1960, (Yale 1933)

Thompson EA (trans) (1952) A Roman reformer and inventor: Being a new text of the treatise, De Rebus Bellicis. Clarendon, Oxford

Thorndike L (1922–1958) History of magic and experimental science. Columbia University Press, New York

Tognetti S (1997) L'attivita di banca locale di una grande compagnia fiorentina del XV secolo. Arch Storico Ital CLV:595–649

Tognetti S (1999) Il banco Cambini: Affari e mercati di una compagnia mercantile-bancaria nella Firenze del XV secolo. Biblioteca storica toscana 27. Leo S. Olschki Editore, Firenze

Tory G (1529) Champfleury, auquel est contenu l'art et Science de la deue et vraie proportion des lettres attiques. Paris

Travers NH (1983) Leonardo da Vinci as mural painter: Some observations on his materials and working methods. Arte Lombarda 66:71–88

Tursini L (1951) Leonardo e l'arte militare. Rivista d'ingegneria 10, Oct.

Tursini L (1954) Navi e scafandri negli studi di Leonardo. In: Leonardo, Saggi e Ricerche. Rome, pp 67–84

Uccelli A (1934) Sopra due presunte carte vinciane esistenti nella raccolta C. L. Ricketts di Chicago. Raccolta Vinciana XV-XVV-36:185–190

Uccelli A (ed) (1940) I libri di meccanica di Leonardo da Vinci. Milan

Unguru S (trans) (1991) Witelonis Perspectivae Liber Secundus et Liber Tertius, Studia Copernicana XXVIII. Polish Academy of Sciences, Wroclow

Usher AP (1929) A history of mechanical inventions. Harvard, Cambridge, MA

Utz H (1973) Giuliano da Sangallo und Andrea Sansovino. Storia dell'arte 19/Sept-Dec:209–216

Uzielli G (1890) Leonardo da Vinci e tre gentildonne Milanesi. Tipografia sociale, Pinerolo

Uzielli G (1896) Ricerche intorno a Leonardo da Vinci, vol. 1. 1872; Ermanno Loescher, Turin, 1896 Vol. II. Salviucci, Rome, 1884

Valentiner WR (1930) Leonardo as Verrocchio's co-worker. Art Bull XII Mar:43–89

Valentiner WR (1937) Leonardo's portrait of Beatrice d'Este. Art Am Elsew XXVII Jan:3–23

Valentiner WR (1950) On Leonardo's relation to Verrocchio. Studies of Italian Renaissance Sculpture. Phaidon Press, London, pp 113–133, 134–177

Valery P (1929) La Corte di Ludovico il Moro. Milan

Valturius R (1472) De re militari. Verona

Varese R (1994) Giovanni Santi. Zampetti P (intro), Nardini Editore, Fiesole

Varese R (ed) (1999) Giovanni Santi: Atti del convengo internazionale di studi Urbino, Convento di Santa Chiara, 17/18/19 marzo 1995. Electa, Milano

Vasari G (1906) Le Vite de' piu eccellenti Pittori, Scultori ed Architettori, con nuovi annotazioni e commenti di Gaetano Milanesi. 9 vols., Milanesi G (trans), 1878–1885; G. C. Sansoni, Florence

Vasari G (1912–1915) Lives of the most eminent painters, sculptors and architects. 10 vols., G du C de Vere (ed, trans), Macmillan and Co. & The Medici Society, London

Vatsyayan K (1983) The square and the circle of Indian arts. Humanities Press, New Jersey

Vecce C (1990) La Gualanda. Achademia Leonardi Vinci TI1:51–72

Vecce C (1998) Leonardo. Salerno Editrice, Roma

Venturelli P (1999) Gioielli e oggetti preziosi nell'Inventario Stivini. Alcune note. Quaderni di Palazzo Te 6:75–80

Venturi A (1885) Relazioni artistiche tra le corti di Milano e Ferrara nel secolo XV. Arch Storico Lombardo XII:225–280

Venturi A (1888a) Gian Cristoforo Romano. Arch Storico Dell'arte I:49–59, 107–118, 148–158

Venturi A (1888b) Nuovi documenti su Leonardo da Vinci. Arch Storico Dell'arte I:45–46

Venturoli P (1983) Interventi sui passato. Un gruppo ligneo ridato aHa luce. Insieme cultura June

Verga E (1912a) Gli epigrammi latini di Francesco Arrigoni per la statua equestre a Francesco Sforza. Raccolta Vinciana 8:155–165

Verga E (1912b) Regesti Vinciani. Terza Serie. Raccolta Vinciana 8:110–151

Verga E (1923–1925) Elenco e analisi delle pubblicazioni pervenute alla Raccolta 1923–1925. Raccolta Vinciana XII:1–159

Verga E (1931) Bibliografica Vinciana, 1493–1930. Zanichelli, Bologna

Vezzosi A (ed) (1982) Leonardo dopo Milano. La Madonna dei fusi (1501). Exh. Cat. Vinci, Castello dei Conti Guido, 16 May to 30 September 1982, Pedretti C (intro), Giunti Barbera, Firenze

Vezzosi A (ed) (1983) Leonardo e il leonardismo a Napoli e a Roma. Exh. Cat. Napoli, Museo Nazionale di Capodimonte; Palazzo Venezia, Roma; Giunti Barbera Editore, Firenze

Vezzosi A (ed) (2000) 'Parleransi li omini di remotissimi paesi l'uno all'altro e risponderansi', Leonardo e l'Europa dal disegno delle idee alia profezia telematica. Exh. Cat. Assisi, Napoli, San Benedetto del Tronto, e Milano, Apr–Dec 2000. Relitalia, Perugia

Villata E (ed) (1999) Leonardo da Vinci: i documenti e le testimonianze contemporanee. Ente Raccolta Vinciana, Milano

Vitruvius (1998) On architecture, books I–X. Granger F (trans), Loeb Classical Library 251, Harvard, Cambridge MA, (1931)

Vitzthum G Graf (1903) Bernardo Daddi. Verlag von Karl W. Hiersemann, Leipzig

Waldman LA (1996) Florence Cathedral: The facade competition of 1476. Note Hist Art XVIII,fall:1–6

Warner P (1968) Sieges of the Middle Ages. Barnes and Noble, New York

Welch ES (1995) Art and authority in Renaissance Milan. Yale University Press, New Haven London

Wilde J (1944) The hall of the Great Council of Florence. J Warburg Courtauld VII:65–81

Wilinski S (1969) Cesare Cesariano elogia la geometria architettonica della Cattedrale di Milano. In: Gatti Perer ML (ed) II Duomo di Milano. Congresso Internazionale Milano, vol. I. Museo della Scienza e delia Tecnica, 8 & 12 September 1968. Edizioni la rete, Milano, pp 132–143

Williams K (1996) Verrocchio's tombslab for Cosimo de' Medici: Designing with a mathematical vocabulary. in: Williams K (ed) NEXUS. Architecture and mathematics. Edizioni dell'Erba, Fucecchio Firenze, pp 193–205

Wintemitz E (1964a) Leonardo's invention of the viola organista. Raccolta Vinciana XXI:1–46

Wintemitz E (1964b) Melodic, chordal, and other drums invented by Leonardo da Vinci. Raccolta Vinciana XX:47–67

Winternitz E (1964c) Leonardo's invention of key-mechanisms for wind instruments. Raccolta Vinciana XX:69–82

Winternitz E (1982) Leonardo da Vinci as a musician. Yale University Press, New Haven

Wittkower R (1977) Hieroglyphics in the Early Renaissance. Allegory and the migration of symbols. Westview Press, Boulder, CO, pp 113–128, (originally in: Levy (ed) developments of the Early Renaissance, 1972)

Wormald F (1952) English drawings of the tenth and eleventh centuries. London

Wright A (1998) Dancing nudes in the Lanfredini Villa at Arcetri. In: Marchand E, Wright A (eds) With and without the Medici: Studies in Tuscan art and patronage 1434–1530. Ashgate Publishing, Hants Vermont, pp 47–77

Yates FA (1947) The French academies of the sixteenth century. Studies of the Warburg Institute XV, London

Young MS (1984) 'T'ai vu Léonard peindre la cene' vers 1490, témoignage de Bandello. Gaz Beaux-Arts Sept:49–51

Yriarte C (1888) Les Relations d'Isabelle d'Este avec Leonardo de Vinci d'apres des documents reunis par Armand Baschet. Gaz Beaux-Arts I:118–131

Zenale e Leonardo (1982) Tradizione e rinnovamento delia pittura Lombarda. Exh. Cat. Museo Poldi Pezzoli, 4 Dec 1982 to 28 Feb 1983. Gruppo Editoriale Electa, Milano

Zöllner F (1985) Agrippa, Leonardo and the codex Huygens. J Warburg Courtauld 48

Zöllner F (1987) Vitruvs Proportionsfigur: Quellenkritische Studien zur Kunstliteratur im 15. und 16. Jahrhundert. Worms

Zöllner F (1989) Die Bedeutung von Codex Huygens und Codex Urbinas für die Proportions- und Bewegungsstudien Leonardo da Vincis. Z Kunstgeschichte 52

Zöllner F, Nathan J (2003) Leonardo da Vinci, the complete paintings and drawings. Taschen, Köln

Zorzi EG, Sperenzi M (eds) (2001) Teatro e spettacolo nella Firenze dei Medici. Modelli dei luoghi teatrali. Leo S. Olschki, Firenze

Zubov V (1968) Leonardo da Vinci. Kraus D (trans), Harvard University Press, Cambridge

Zwijnenberg R (1999) The writings and drawings of Leonardo da Vinci: Order and chaos in early modern thought. van Eck CA (trans), Cambridge

Index

A

A Horseman in Combat with a Griffin (drawing by Leonardo da Vinci) 70
Abbo Cernuus 52
acacia wood 73
Accademia Galleries Venice, military drawings 32
Adoration of the Magi (painting by Leonardo da Vinci) 1, 21, 37, 121
  –, perspective 118
air
  –, flow dynamics 35
  –, resistance 91
Alberti, Leon Battista 73
alder 73
Alexander the Great (reliefs of) 69
alloy 63
ammunition 107
anatomy 93
angle of force 90, 91
animal
  –, form 70
  –, proportions 93
Annunciation (altarpiece by Cosmè Tura) 34
Annunciation (painting by Domenico Veneziano) 102
antenne motrice (engine of poles) 147
approach
  –, proportional 1
  –, proto-scientific 1
arbutus 73
Archimedes 71, 85
  –, axioms 85
  –, steam-powered canon (*Architronito*) 86
architecture 93
Architronito (steam-powered canon by Archimedes) 86
Arconati, Galeazzo 126
Aristotle 82
arithmetic, Aristotelian 81
arm
  –, lenght (see also *braccia*) 9, 44, 61, 94
  –, manuscript (arm and military engineering) 106
armadura (armature) 9, 65, 147
armature (see also *armadura*) 65, 95, 145
  –, arc 130
  –, bat wing shaped 31
  –, Berkhamstead type 53
  –, construction 65–77
  –, design 65, 145
  –, draw 5, 9-10, 94, 95, 102, 107, 116, 124
  –, elevation 102
  –, eyelet 72
  –, force 88
  –, form 71
  –, initial extent 89, 90
  –, length 142
  –, metal 56
    –, illustration 56
  –, operation 79–82, 85–88, 90, 91, 93, 94
  –, plotting 113
  –, position 121–125, 128
  –, proportional design 79–82, 85–88, 90, 91, 93, 94
  –, size 11
  –, spanning 73, 86, 88
  –, study 65
  –, stylus incision 142
  –, toothed lamination 75
  –, width 13, 75, 98, 124
    –, spanned 101
arrow 147
  –, flying 91
  –, launcher 145
  –, platform frame 51
  –, speed 46
arrowhead
  –, bronze 46
  –, first 45
Art of War (treatise by Sun Tzu) 45
Arundel Codex (see *Codex Arundel*)
Atlanticus (see *Codex Atlanticus*)
atom (length unit) 144
axle
  –, distances 104
  –, lubrication 105

B

balance, arm 61
balestra (crossbow) 147
balestriere (crossbowman) 111, 147
balista de torno vel pesarola (great crossbow) 54
balistam
  –, *ad duos pedes* (two-foot crossbow) 53
  –, *ad estrif* (see also *crossbow, stirrup*) 53
  –, *ad tornum* (windlass crossbow) 53
ball, explosive 39
ballast 107
ballista (Latin, stone thrower) 1, 47, 52
  –, bolts 52
  –, *de torno* (see also *balista de torno vel pesarola*) 52
  –, *fulminalis* 51, 52
    –, draw tension 54, 55
    –, operation 51, 52
  –, operation 46, 47

ballista (*continued*)
-, *quadrirotis* 52
-, U-spring or torsion screw equipped 46
ballistics 61, 63
ballistra (Greek, stone thrower; see also *balestra, balista, balistam*)
46, 47, 52
-, mark III 48
-, reproduction, incorrect 50
Bambach, Carmen 15, 124
Bandino Baroncelli, Bernardo di (portrait) 27
Baptism of Christ (painting by Piero della Francesca) 102
Baratta, Mario 37
bat 67
-, study 35
-, wing 31, 68, 71
-, design 66
-, form 71
-, structure 145
-, study 31, 37
battering ram 52
battle
-, *of Anghiari* (cartoons) 127
-, of Ma-Ling, China 45
-, with Venice, preparations 35
beam, laminated 71
beech 53
Bellifortis (treatise by Conrad Kyeser) 53, 54
belly shooter (see also *gastraphetes*) 46
Belopoeica (catapult manual) 48, 50
Beltrami, Luca 15
Bergamo 19
Berkhamstead Castle 53
Biblioteca Ambrosiana, Milan 3
bird 67
-, study 35
-, wing 71
Biton (Roman author) 48, 49
block 71
boat 146
-, armature 146
-, paddle-wheel 146
Bodleian Library, Oxford 53
Bolognese
-, *rezzuta* (standard paper format) 127, 128
-, statutes, paper size rules 127
bolt
-, head 107
-, iron, Roman 48
bomb, incendiary 147
bomba (bomb) 147
bombard 61, 93
Bona of Savoy 34
Böninger, Lorenz 23
Borri, Gentile de' 40
Botticelli, Sandro 27, 29, 120
bow 146
-, armature, metal 56
-, capabilities 88
-, first painting 45
-, folding 146
-, skein 55
-, stave 53
-, tensioning bench (tool) 53
bowstring 53
bowyer (crossbow carpenter) 53

box 74
braccio (length unit) 117, 144
-, *a panno fiorentino* 44, 144
-, Florentine 9, 13, 44
-, illustrated length 141–143
-, *mercantile milanesi* 9, 44, 144
-, Milanese 44
-, *o passetto romano* 144
-, *romano* 123
Bradwardine, Thomas 82
Bramante, Donato 97
brazil 74
Brescia 19
bridge 35
-, design 40
-, dating 40
-, portable 17
British Museum, London, military drawings 32
bronze 92
Brunelleschi, Filippo 97, 121, 133
-, iron screw 38
buckler 68
Buda (present-day Budapest) 49
Budd, Denise 23, 27
Bull, Greorge 68
Bust of a Warrior in Profile (painting by Leonardo da Vinci) 69
butterfly 69

C

cacciafrusto (whip-driver) 147
Caesar 47
calculation 52, 63
-, proportional 92
Calvi, Gerolamo 15, 38, 42
cam 61
Canestrini, Giovanni 15
cannon 8–10, 17, 19, 35, 43, 50, 53, 57, 66, 92, 97, 146, 147
-, 18 pounder 92
-, carriage 42, 145
-, design 57
-, dragon 68
-, factory 37
-, foundry 146
-, illustration 53
-, multi-barrelled 42
-, multiple 37
-, steam-powered 86
-, study 39
-, wheel 145
cantilever principle 106
capstan 13, 97, 115
Cardano, Fazio 71, 84, 86
Carnesecchi Tabernacle (painting by Domenico Veneziano) 102
carpenter 119
Carpiceci, Marco 124
carriage 97
-, elevation 5, 9, 10, 94, 95, 102, 116, 124, 142, 143
-, length 13, 142
-, lifting 46, 102, 106
-, lower 95, 98, 101–104
-, front extension 101–104
-, hinged tail-piece 106
-, measures 98–104
-, metal stylus lines 110

–, part 95
–, proportions 98, 99
–, size 11
–, thickness 109, 110
–, upper 66, 109, 110, 112–118
    –, angle to lower carriage 102
    –, elevation 102
    –, lowering 93
    –, measures 11, 111, 142
    –, spanning screw 73
    –, thickness 142
–, width 13, 142
cartography 93
Castiglione, Sabba 25
Castle Cornet, Guernsey 53
catapult 17, 32, 147
    –, giant 147
    –, large 76
    –, steel spring 56
    –, stone-launching 145
Cathedral of Sta. Maria del Fiore, Florence 121
Cennini, Cennino 130
chariot
    –, battle 32
    –, scythed 32
        –, stylus lines 32
chassi tilting 116
cheiroballistra (Latin for crossbow) 46–48, 50–52, 54, 55
    –, diagram
        –, Byzantine 48
    –, illustration 48, 49
    –, Romanian excavation 48
    –, spanning 48
    –, textual evidence 48
cherry 74
Chinery, Pansy 116
Christ Church College, 20r&v 146
church, architecture 33
circle, circumference 85
circumfolgore (boat with 360° turnable cannon) 147
City Charter of Carlisle 107
Clark, Kenneth 39
clotonbrot (incendiary bomb) 147
Codex Arundel
    –, 54r 87
    –, 263 80
    –, page formats 128
Codex Atlanticus 3, 63, 80, 92, 93
    –, 21r [5va], dating 42
    –, 24r [6rb], dating 42
    –, 26r [7r] 37, 39
        –, dating 37
    –, 31r [9ra] 39
        –, dating 40
    –, 32rb [9rb] 39
    –, 33r [9va] 39
        –, dating 40
    –, 47v [14va] 39
    –, 55r [16va] 40
        –, dating 40
    –, 57v [17ra] 40, 41, 71, 75, 76, 105, 129, 132, 145
    –, 59bv [18rb] 39, 47
    –, 69ar [22ra] 40, 71, 75, 76, 105, 129, 132
    –, 70br [22vb] 66
    –, 71r [23ra] 40

–, 71v [23va] 40, 145
    –, dating 40
–, 76vb [26vb] 39
–, 78r [27rb] 86, 87
–, 85ar [31ra] 15
–, 89r [32va] 42, 73
    –, dating 42
–, 90r [33ra] 42, 73, 75, 145
    –, dating 42, 73
–, 91v [33vb] 42
–, 94r [34va] 39
–, 94v [34ra], dating 42
–, 112ar [39vc] 145
–, 113r [40ra] 145
–, 113v [40va] 147
    –, military drawings 32
–, 114v [40vb] 39
–, 139r [49vb] 42, 75
    –, dating 42, 73
–, 140ar [50va] 15
    –, dating 56
–, 140br [50vb] 147
    –, dating 56
–, 141r [51ra] 15, 147
–, 142r [51rb] 65, 66, 145, 147
–, 143r [51rc] 32, 125, 145
–, 144r [51va] 147
–, 145r [51vb] 15, 16, 37, 56, 125, 147
–, 147av [52va] 6, 12, 13, 15, 40, 53, 66, 71, 74–77, 85, 98, 103, 105, 110, 115, 125, 129, 132, 145
–, 147bv [52vb] 15, 18, 19, 40, 42, 53, 66, 71–77, 85, 98, 102, 103, 105, 110, 129, 132, 145
–, 148ar [53ra] 15, 20, 21, 115, 125, 132, 145
–, 148br [53rb] 15, 125
–, 149ar [53va] 8, 9, 37, 39, 53, 55, 56, 68, 82, 116, 125, 130, 145
–, 149br [53vb] 3–6, 9, 10, 15, 33, 39, 42, 43, 59, 61, 65, 66, 70–73, 75–77, 82–86, 88–90, 92–95, 97–105, 107, 109, 110, 112, 113, 115–139, 145, 147
    –, description 3
    –, front end 95
    –, measures 70
    –, metal stylus incisions 70, 129
    –, notes 65
    –, page format 3, 125–128
    –, pinholes 129
    –, sequence of marks 128, 129
    –, translation 5, 9–11
    –, trigger mechanism 55
–, 150r [54ra] 15
–, 151r [54rb] 15
–, 152r [54va] 15, 125
–, 153r [54vb] 145
–, 154br [55vb] 95, 97, 105, 145
–, 155r [56ra] 145, 147
–, 157r [56va] 42
–, 157v [56vb] 39
–, 158r [56vb] 125
–, 159ar [57ra] 15
–, 159br [57rb] 15, 22, 23, 37, 125
–, 160br [57vb] 15, 125
–, 166r [59vb] 32
–, 172r [61ar] 37, 39
    –, dating 39
–, 175r [61vb] 145
    –, crossbow trigger mechanism 133

Codex Atlanticus (*continued*)
  –, 175v [61rb] 145
  –, 181r [64r] 6, 37, 56
  –, 182ar [64va] 36, 37
  –, 182br [64vb] 15, 25, 37
  –, 581r [217br] 37, 39
    –, dating 39
  –, 638dv [234vc] 38
  –, 693r [257cr] 37, 39
    –, dating 39
  –, 747r [276br] 37
    –, dating 37
  –, 748r [276va] 37, 39
  –, 754r [278rb] 145
  –, 846v [309av] 37, 66
  –, 848r [309bv] 37
    –, dating 37
  –, 860r [313va]
    –, dating 37
  –, 876va [319v] 37
    –, dating 39
  –, 881r [320vb] 37–39
    –, dating 38
  –, 887r [323rb] 19
  –, 888r [324r] 21, 27
  –, 888v [324v] 21, 27, 40, 42, 71, 73, 75
    –, dating 40, 42, 73
  –, 902br [329rb] 40, 147
  –, 902bv [329vb] 40
    –, dating 40
  –, 909v [333v] 37, 38
    –, dating 38
  –, 946r [344va] 42, 73, 75
  –, 1048br [376rb] 15, 26, 27, 55, 114, 115, 135, 145
  –, 1051r [377rb], dating 37
  –, 1052ar [378ra] 37, 39
    –, dating 39
  –, 1054r [379ra] 145
  –, 1054v [379va] 145
  –, 1058r [381ra] 145
  –, 1058v [381va] 37
    –, dating 37
  –, 1063v [384rb] 145
  –, 1069r [386rb] 37
    –, dating 37
  –, 1070r [387ra] 15, 28, 29, 37, 145
  –, 1071v [387rb] 145
  –, 1082r [391ra] 15, 17, 65
  –, 1094r [394rb] 145
  –, access 125
  –, cannon study 39
  –, military drawings 15, 32
  –, notes 34
  –, page formats 126–128
  –, springald 40
Codex Madrid 63
  –, 50v 86
Codex Madrid I 87
  –, 51r 86, 87
Codex Magliabechiano II 333
  –, 51r 21, 23, 29
Codex Magliabechiano XVII
  –, 17 21, 23
Codex on the Flight of Birds (Cod. Varia 95), military drawings 32

Codex Parisinus graec. 2442 49
  –, 13v 56
  –, 76v 50, 51, 54
Codex Parisinus graec. 607 49, 54
  –, 56v 51
Codex Trivulziano 74
  –, 2r 23, 85
  –, 99 [54r] 146
    –, stylus lines 32
Codex Vaticanus 1164 49
Colleoni, Bartolomeo 25, 121
Column of Trajan 47, 48
Commentaries (treatise by Ctesibius) 48
compound pulley 51, 52
consistency, proportional 109, 111–113, 119
Constantinople 27
construction
  –, perspective 93, 98, 109
  –, principles, medieval 63
Construction of War Machines (treatise by Biton) 48
Corsica 34, 35, 39
Corvinus, Matthias (King) 49
  –, Court 49
Cosimo, Piero di 29, 120
Cossa, Francesco del 56
Council of Seventy, Florence 35
craftsmanship 52
cranequin 145
cropping of folios 128
crossbow (see also *Giant Crossbow*) 8, 9, 43, 53, 146, 147
  –, armature, span width 86
  –, box format springald 53
  –, carpenter 53
  –, Chinese 45
    –, technical parameter 46
  –, construction 43
  –, design 43
  –, double 145
  –, drawing 1, 32
  –, evidence 53
  –, France 52
  –, great 1, 145
    –, *balista de torno vel pesarola* 54
    –, illustration 55
  –, handheld (*manuballista*) 45, 52
  –, handling 46
  –, invention 45
  –, large, built 116
  –, Latin terms 46
  –, lever, screw-loading 145
  –, movement, lateral 105
  –, power 91
  –, proportions 88
  –, repeating 28, 29, 37
  –, screw-load 145
  –, size 5
  –, stirrup 52
  –, stone throwing 76
  –, technology 45
  –, tension, strength 146
  –, two-foot 53
  –, usage 43
  –, windlass 53
crossbowman (*balestriere*) 66, 112, 117, 118, 120, 147

-, height 111, 143
-, illustrated size 117
-, position 112, 113
-, proportions 112
crossed arc method 82, 119, 130
Ctesibius (Roman author) 48
cubit (length unit) 122, 123
-, large 144
-, Roman 117
-, small 144
Cupid Gem (Roman jewellery) 48

D

danaro (coin) 71
Dardanelles gun 92
dardo (dart) 147
dart (*dardo*) 145, 147
-, explosive 77
-, incendiary 39, 146
-, thrower 46, 145
De Architectura (treatise by Vitruvius) 48
De Bello Civili (treatise by Caesar) 74
Dolci, Giovannino de' 122, 123
De proportionibus (treatise by Albert of Saxony) 80
De re aedificatoria (treatise by Leon Battista Alberti) 73
De re militari (treatise by Roberto Valturio) 5, 9, 15, 19, 33, 40, 54, 56, 66, 68, 70, 74, 79
-, woodcuts 5, 9–11
De rebus bellicus (treatise by an anonymous author) 51
De Roover, Raymond 23
Dei, Benedetto 21, 25, 29
denaro (length unit) 44, 144
design 119
-, mechanical, third portions 124
-, preparatory 110
-, proportional 79–82, 85–88, 90, 91, 93, 94
device
-, dart smacking 56
-, truss-bending 71
Dibner, Bern 9
dinaro, Milanese (coin) 44
Dionysius of Syracuse 46
discovery, recent 1
Disputation of St. Catherine (painting by Masolino Da Panicale) 102
divident 40 71
divider directions 113, 121–126, 128
diving gear 35, 37
division
-, proportional 13
-, third-part 13
Donatello, (Donato di Niccolò di Betto Bardi) 25, 97, 102
double crossbow 145
drafting strategy, technical 1
dragon cannon 68
draughting technique
-, mechanical 119–138
draw strength, proportional 87
drawing 93
-, naval 39
Duke of Calabria 19, 33, 35
Duke of Milan 21, 23, 25
Dürer, Albrecht 103, 106, 107
dynamics 1

E

Ebolus (late ninth century Abbot of Saint-Germain des Prés) 52
École Nationale Supérieure des Beaux-Arts, Paris, military drawings 32
Edinburgh Castle 107
Elements (treatise by Euclid) 84
ell (Roman length unit) 9, 144
elm 53
engineer 101, 119
-, military 5, 50, 79, 80
-, Renaissance 2
engineering 5
-, military 15, 35
equestrian statue 19, 25
-, Francesco Sforza 34
espringal, French 48
Este, Ercole d' 35, 92
Euclid 71, 85
Eudoxus 85
euthytonon (dart and stone launching ballista) 46
eyelet, looped 72

F

falarica (spear launching machine) 147
Feast of Harod (painting by Donatello) 102
feature, aesthetic 63
felloe, thickness 143
Ferrara 19, 34, 35, 39
Ferreri, Ambrogio 19
field-launcher 53, 147
figure, human 42
Filarete, Antonio Averlino detto 97
finger (length unit) 144
Flagellation (painting by Piero della Francesca) 102
Florence 19, 21, 23, 25, 27, 53
-, Cathedral 38, 82, 133
-, political climate c. 1480 25
Florentine
-, *braccio* 9, 13, 44
-, merchants 23
-, *rezzuta* 127
-, *soldo* 44
flotation device 37
fluid motion 61
flying machine, design 66
Flying Zedoras (circus acrobats) 116
folding bow 146
folio cropping 128
Fontana, Giovanni 97
force 81, 87
-, crossbow, movement 79
-, disproportionately 82
-, movement 79
-, of the mover 88
-, relationship to distance 80
-, relationship to object 80
Ford, Edward 124
forearm (length unit) 144
form, proportional 82, 84, 85
formula, Archimedean 1
Forster Codex I, II 80
fortification 19, 34, 42, 107, 116

fortification (*continued*)
- –, canon fire 47
- –, defence 42, 47, 53

fortress crossbow 107
fraction, Archimedian 82
Fragmenta Vindobonensia 120 49
France, military engineering 106
frombola (sling, also *frommola*) 147
fulminario (trebuchet) 147
fuoco greco (Greek fire) 147

G

Gaddiano, Anonimo 21, 23, 25
Gadio, Bartolommeo 19
galley
- –, Turkish 72
- –, Venetian 72
  - –, armature 54

Galluzzi, Paolo 73
gastraphetes (dart thrower) 46
Gatta, Bartolomeo della 29, 120
Gattamelata Monument, Padua 25
gear 115
- –, locking 115
- –, mechanism 42

gem 48
Genoa 34, 35
geometry
- –, Archimedean 1, 84
- –, Euclidean 1

Germanicus 47
Germany, military engineering 106
gettata (length unit) 144
Ghiberti, Buonaccorso 97
Ghiberti, Lorenzo 97
Ghirlandaio, Domenico 29, 120
Giant Crossbow 1, 35, 145
- –, ammunition 76
- –, analysis 1, 3
- –, application 116
- –, armature 42, 65
  - –, size 11
- –, associated drawings 145
- –, basic features 67
- –, capabilities 66
- –, capstan 13
- –, carriage (see *carriage*)
- –, components 3
- –, concept 6
- –, dating 15, 29, 31, 139
- –, design 1, 31, 124
  - –, functionality 59
- –, development 57, 59
- –, dimensions 13, 70, 82, 84, 100, 101
  - –, proposed 120
- –, draw 5, 9-10, 94, 95, 102, 107, 116, 124
- –, drawing (see also *Codex Atlanticus 149br*) 1, 3–5, 43
  - –, measurements 142
  - –, relationships 44
  - –, stylus lines 32
- –, dynamics 61
- –, elevation 5, 9–11, 94, 95, 102, 116, 124, 142, 143
- –, force 1
- –, gear 13
- –, history 45

- –, inaccuracies 61
- –, inconsistencies 71
- –, interpretation 10
- –, Leonardo da Vinci's intention 43
- –, lifting 103
- –, locating 3
- –, lower carriage (see *carriage, lower*)
- –, material 72
- –, measures 44, 141
  - –, calculated 80–82, 84, 85
- –, movement 1
- –, noise 92
- –, notes 120
- –, portions 44
- –, preliminary study 74
- –, preparatory sketches 2, 12, 13, 18, 19, 40, 41, 61
- –, presentation drawing 139
- –, primary dimensions 1
- –, proportionality 61
- –, proportions 82, 84
  - –, inconsistencies 101
- –, reconstruction 60, 62
- –, scale 70, 101, 111, 120, 139
- –, section 141–143
  - –, measures 141–143
  - –, scale 141–143
- –, side poles 105
- –, size 9, 11
- –, span 86
- –, stabilisation 105
- –, trajectory 66
- –, transport 103
- –, trigger mechanism (see also *trigger mechanism*) 13, 55
- –, upper carriage (see *carriage, upper*)
- –, written specifications 141–143

Gibbs-Smith, Charles 106
Giorgio Martini, Francesco di 33, 39, 56, 97, 133
goat horn 53
grano (weight unit) 144
great crossbow 1, 2, 5, 39, 45, 52, 54, 55, 65, 66, 77, 84, 88, 106, 107, 116, 118, 145
Greek fire 147
guesswork, intuitive 1
Guido da Vigevano, 97

H

hair 46
halberd 32
Hart, Ivor 19
hemp 46, 72
Heron (Roman author) 48–50
Heron of Alexandria 85
hinge 95
hoist 106
horn 53, 65, 73
hub, thickness 143
human proportions 93

I

illustration
- –, architectural 117
- –, orthogonal 98
- –, technical 5

impetus theory 88
incision (see also *metal stylus incision*)
    –, preparatory 1
    –, visualisation 125
insect form 70
Irle, Klaus 124
iron 72
    –, screw 38
iron bolt, Roman 48
Italo-Byzantine human proportions 118
ITN Factual Television 60–62, 92

K

katapeltes (dart thrower) 46
Kemp, Martin 21, 91
King Louis 55
King Matthias Corvinus 49
Kyeser, Conrad 53, 54, 97

L

l'arte de l'abbacho (treatise by Treviso) 85
lamination 71
    –, grooved 42
    –, modular 72
    –, study 75
    –, toothed 42, 75, 145
lance 32, 146
lanciacampi (field-launcher) 53, 147
Landscape with Cannon (engraving by Albrecht Dürer) 103, 106, 107
Landucci, Luca 27
large cubit (Roman length unit) 144
Last Supper (painting by Leonardo da Vinci) 1
lath 65
laurel 74
law
    –, physical 1
    –, pyramidal 86, 91
League against Venice 19
length units 144
Leonardo da Vinci 97
    –, design skills 119
    –, left-hand writing 33
    –, locating c. 1480–1484 19, 21–23, 25
    –, organisation of notes 33
    –, paper formats 128
    –, sources 45
    –, trigonometry 102
Leonardo's Dream Machines (documentary) 2
Leoni, Pompeo 3, 126
libbra (weight unit) 144
Libri, Guglielmo (see *Timoleone, Brutus Icilio*)
Libro dell'Arte (treatise by Cennini) 130
Liebel, Jean 5
light, raking 125
Ligurian Sea 38
linchpin 72
loading mechanism 145
Löffelholz manuscript, MS lat. 599, 34v 55
Lord Ashburnham 33
Lord of the Rings, Two Towers (movie) 61
lower carriage (see *carriage, lower*)
lynch pin 107

M

Machiavelli, Niccolò 33
machine
    –, automatic file-making 42
    –, bat-like 66
    –, flying 31, 35, 37
        –, design 66
        –, study 37
    –, spear launching 147
Madrid I 80, 92, 93
    –, 50v 91
    –, 50v-51r 93
    –, 51r 87, 88, 91, 92
Madrid II
    –, 2v-3r 85
mangonel 17, 50, 66, 76
Mantua 19
manuballista (see *crossbow, handheld*)
Manuscript 63, folio 80v, Universitätsbibliothek, Göttingen 54
Manuscript 3069, folio 22v 53
Manuscript A 80, 87, 92, 93
    –, 29v 80
    –, 30r 87
    –, 35r 87, 88
Manuscript Arundel (see *Codex Arundel*)
Manuscript Ashburnham 361
    –, 37v 97, 133
    –, 42v 97, 133
    –, 44v 97, 133
Manuscript B 32, 35, 92, 93
    –, 4v 79
    –, 5v 42, 73, 145
    –, 6r 42, 73, 147
    –, 7v 68, 145
    –, 8r 68
    –, 11r 37, 38
    –, 11v 145
    –, 20r 145
    –, 21r 145
    –, 23r 40
    –, 23v 145
    –, 24v 39
    –, 25r 145
    –, 27r 145
    –, 27v 11, 42, 73, 75, 145
    –, 30v 39, 145
    –, 31r 39
    –, 31v 39, 147
    –, 32r 39
    –, 33r 86
    –, 33v 145
    –, 35r 145
    –, 37v 39, 145
    –, 39v 145
    –, 40v 145
    –, 43 40
    –, 45v 39
    –, 50 40
    –, 50v 39, 71, 77, 145
        –, explosive dart 77
    –, 54r 39
    –, 55v 39
    –, 56r 145
    –, 57v 145

Manuscript B (*continued*)
 –, 59r 147
 –, 62v 71, 74, 145
 –, 70r 145
 –, 73v 35, 37, 66, 145
 –, 74r 31, 35, 37, 47, 66
 –, 74v 145
 –, 75r 37
 –, 75v 35
 –, 76r 35, 37
 –, 79r-v 35
 –, 80r 37
 –, 80v 40, 66
 –, 81v 37, 39
 –, 82v 147
 –, 84 33
 –, 87v 33, 37
 –, 88r 66
 –, 88v 37
 –, 89v 31, 67, 71
 –, 90r 66, 145
 –, 90v 35, 37
 –, 91v 35
 –, 93r 145
 –, 94v 145
 –, 95r 145
 –, 99v 35
 –, 100v 31, 69, 145
 –, A.1 145
 –, A.2 145
 –, B.1 145
 –, B.2 145
 –, cannon study 39
 –, D 145
 –, dating 38, 39, 42
 –, N. It. 2037, f.74r 31
 –, notes 34
 –, quires 32, 33
 –, springald 40
 –, stolen folios 33
Manuscript B Ashburnham A.2 32
Manuscript Douce 289 9–11
Manuscript E 33
 –, stolen folios 33
Manuscript G
 –, 82r 82
 –, 84v 82
Manuscript I 80, 92, 93, 147
 –, 99v 147
Manuscript M 80
Manuscript Salluzziano 148, 61v 56
Manuscript Sul Volo 68
Marcellinus, Ammianus 50
marco milanese (weight unit) 144
Marcus Aurelius monument, Rome 25
Marinoni, Augusto 39
marks, sequence 119, 128–138
Marsden, Eric 47
Martini, Angelo 123
Martini, Simone 50
Martyrdom of St. Sebastian (painting) 56
Masaccio (Tommaso di Ser Giovanni di Simone) 102
Masolino (Tommaso di Cristoforo Fini) 102
measurement, exact 124
Measurement of a Circle (treatise by Archimede) 85

mechanical draughting 119–121, 123–138
mechanics 61
mechanism
 –, auto drop/load 145
 –, loading 145
 –, shooting 145
 –, spanning 53
 –, trigger, measures 44
 –, worm-screw 97
Medici, Giuliano de' 27
Medici, Lorenzo de' 21, 23, 25, 27
Medusa 69
Melzi, Francesco 3
metal
 –, alloy 63
 –, alloys 57
 –, spike 72
 –, spring
  –, design 63
 –, stylus 32, 98, 101
  –, holes 125
  –, incision 70, 71, 101, 109, 110, 114, 123–138
  –, mark 125, 129
method
 –, crossed arc 82, 119, 130
 –, proportional 1, 2, 63, 93
Migliorotti, Atalante 21, 23
Milan 19, 21, 25, 39, 53, 97, 106
 –, city wall 116
Milanese Court 49
Milemete, Walter de 53
military
 –, engineer 5, 50, 79, 80
 –, engineering 1, 5, 15
  –, manuscripts 106
 –, study 31, 33
mine 17
missile
 –, basket 76
 –, explosive 39
 –, Greek 46
 –, incendiary 35, 55
 –, motion 86, 87
 –, trajectory 57
 –, weight 57
moat 92
model, dynamical 61
Monastery of Vatopedi, Mt. Athos 49
Mons Meg cannon 107
Morgan, Len 48
mortar 17, 61, 146
moscette (arrow) 147
motion of missile 86, 87
movement of force 79
Müller-Walde, Paul 37
Musée Bonnat, Bayonne 27
music 61
Muwahhid Caliph Abu Yaqub 52

N

Naples 19
natural philosophy 93
Natural History (treatise by Pliny) 74
nature design 68

naval fleet, Venetian 34
navy, Carthaginian 46
noche (nut, a crossbow device) 147

O

oak 53, 74
object power 81
oblique projection 118
of Saxony, Albert 80
Old Testament 122
    –, cubit 123
olive 74
    –, wood 73
onagari (catapult) 52
onager ("wild ass", stone-throwing engine) 50
once (length unit) 44, 144
Opera of Santa Liberata 38
optics 61
ordinance 17
    –, light 61
Oresme, Nicholas 82
ornithopter 35, 37
    –, wings 35
Otranto 35
ounce, Milanese (weight unit) 44, 71
oxybeles (dart thrower) 46, 47
    –, parameters 46

P

paddle-wheel 146
page
    –, format 125–128
    –, formatting, proportional 98
painting 93
Palazzo del Bargello, Florence 27, 29
palintonon (dart and stone launching ballista) 46
palla (gilt copper ball) 82
palm (length unit) 44, 144
palo di ferro (spike) 72
Panofsky, Erwin 117
Papal State 19
paper size, standards 127
Parsons, William Barclay 15
passage, winding 17
Pasti, Matteo 9–11, 54
patron 5, 44, 66, 68, 94, 114, 119
Paul III 48
Pavia 106
pawl mechanism 115
Pazzi conspiracy 27
pear 74
Pedretti, Carlo 15, 37–39, 42
Pergamum 47
Perge, Turkey 47
perspective
    –, *Adoration* 118
    –, construction 93
Perugino, Pietro 29, 56, 120
Pharsalia (treatise by Lucan) 74
Philip II Augustus 52
Philon (Roman author) 48, 49
philosophy, natural 93
piede (length unit) 144

Piero della Francesca 101, 118
pine 74
pinhole 112
Pinturicchio, (Bernardino di Betto) 29, 120
pirates 38
pivot 61
Po Delta 34
    –, defence 40
Po River, mills 39
pole, wooden 52
Pollaiuolo, Antonio del 56
Pollaiuolo, Piero del 56
Pontelli, Baccio 122
Pope Sixtus IV 19, 35, 122
Popham, Arthur E. 15, 37
pound, Milanese 71
power 81
    –, quality 81
precision 52
preparatory design 110
presentation, formal 43
principle
    –, proportional 61, 80
    –, Vitruvian 117
problem
    –, arithmetical 80
    –, dynamical 63
    –, of proportion 80
projection
    –, oblique 98, 109, 118
    –, orthographic 59, 97
    –, perspective 109
proportion theory 1, 79, 85
    –, pyramidal 86
proportional division 13
proportions, Vitruvian 118
proposal to Ludovico Sforza 15
    –, dating 19, 21
    –, text 17
Prou, Victor 48
Ptolemy 85
Pulci, Luigi 23
pulley 61, 145
    –, capabilities 35
    –, compound 51, 52
punto (length unit) 44, 144
pyramid, perspective 87

Q

qaws al-lawab (large crossbow) 52, 55
qaws al-ziyar (Arab skein bow) 55
quadratura del circulo (squaring of the circle) 85
quality 81
    –, proportional 81
quattrino (length unit) 44, 144

R

ram battering 52
Ramusio, Paolo 74
ratchet 61, 115, 145, 146
    –, mechanism, inner 115
    –, pawl 145
reale (paper format) 127, 128

rear U-bolt 102
Regisole Monument, Pavia 25
relationship, proportional 80
requirement, mechanical 120
rezzuta (Bolognese/Florentine standard paper format) 127, 128
Roman *braccio* (length unit) 123
Roman ell (length unit) 144
rope
   –, tension 57
   –, treatment 63
Rosselli, Cosimo 29, 120
rotelle (incendiary bomb) 147
Royal Armouries, Fort Nelson 60, 62
rule
   –, proportional 82
   –, of three 70, 79–81

S

Sangallo, Antonio da 97
Sangallo, Giuliano da 97
Santa Reparata 38
Santarem, Portugal 52
Santini, Paolo 97
sarcophagi 48
scale 1, 59
   –, consistency 111
Schofield, Richard 21
Schröer, Klaus 124
science 1
Scipio 69
Scorpio 47
scorpionum catapultarum (scorpio) 47
Scottish siege of Carlisle 50
screw 61, 146
   –, central 116
   –, diameter 142
   –, sectioning 145
   –, spanning 73
   –, tensile strength 145
   –, thread technology 63
   –, winding 88
sculpture 17, 93
sectioning, modular 117
sector 121
self-bow
   –, draw tension 46
   –, technical parameter 46
sequence of marks 119, 128–138
Sforza
   –, Castle, walls 116
   –, Court 139
      –, measuring standards 44
   –, Francesco 25
      –, equestrian statue 25, 34
   –, Galeazzo Maria 19, 25, 34
   –, Horse 25, 34
   –, letter 15
   –, Ludovico 5, 15, 21, 25, 27, 29, 31, 33, 43, 61, 68, 80, 92, 101, 119
      –, armies 93
   –, monument 29
Shandong, China 45

ship
   –, hull 72
   –, Scandinavian Gokstad 72
shooting mechanism 145
Sicily 46
side pole 95, 97, 101, 103–105, 107, 115, 134, 135, 143
   –, brace 143
   –, distance 143
   –, thickness 143
siege
   –, artillery 57
   –, design 32
   –, engine 19, 35, 51, 52, 107
      –, Roman 48
      –, technology 45
      –, trajectory 116
   –, machinery 43
   –, of Jerusalem 52
   –, technology, turning point 53
   –, warfare, European 53
   –, weapon 5, 35, 43, 66
      –, giant 37
Signorelli, Luca 29, 56, 120
sinew 46, 51, 53, 65, 73
Sistine Chapel 29, 120
   –, dimensions 121, 122
skein bow 55
sling 22, 23, 145, 147
   –, engine 36, 37
   –, giant 37
   –, multiple 37
small cubit (length unit) 144
soldier
   –, Persian 51
   –, with shield 32
soldo (length unit) 44, 144
Solmi, Edmondo 37
span
   –, definition 86
   –, proportions 83, 84
spanning 48, 52
   –, bench 116
   –, length 53
   –, mechanism 53
   –, screw 73
   –, tensioning bench 53
spear thrower, handheld 43
spoke 107
   –, thickness 143
spring 61
   –, capabilities 57
   –, design 63
   –, non-torsion 51
springald 6, 8, 9, 53, 56, 68
   –, giant bent-frame 16
   –, machine, illustration 54
   –, torsion spring 11
   –, vertical bent-beam 9
stabilisation 105
standard, metrological, Milanese 44
statics 1, 61
stone 72, 76
   –, ball, 100-pound 92, 93
stonebow 76

strategy
- -, defensive 35
- -, navel 35
- -, siege and defence 35
- -, underwater attack 38
string
- -, force 86
- -, propelling 86, 87
  - -, angle 88
- -, withdrawal, angle 87
study
- -, bat wing 37
- -, cannon 39
- -, flying machine 37
- -, personal 43
Study of a Dragon (drawing by Leonardo da Vinci) 70
stylus lines (see *metal stylus incisions*)
Suleiman the Magnificent 49

T

Taccola 53, 97
tail, fin 146
tail-piece, hinged 106
Tassili, Egypt 45
technique, proportional 59, 139
technology, Persian 51
Temple of Solomon 122
tendon 73
tensioning device 115
The Beheading of St. John the Baptist (painting by Caravaggio) 69
The Flagellation (painting by Piero della Francesca) 118
theory
- -, impetus 88
- -, of proportion 79
- -, proportion 1
third portions 124
third-part divisions 13
tilting, chassi 116
Timoleone, Brutus Icilio 33
tombstone 48
- -, of Vedennius 47
torsion screw
- -, adjustable 47
- -, earliest visual evidence 47
- -, engine 46
town wall 146
trabucco (length unit) 144
trajectory 46, 76, 116, 137
- -, low 92, 116
tranubolo 147
Trattato di Architettura (treatise by Francesco di Giorgio) 33, 39, 97, 133
treadmill, giant 145
treatise
- -, arrangement 93
- -, military design 33
Treatise on Flight (treatise by Leonardo da Vinci) 68
trebuchet 17, 51–53, 66, 76, 147
- -, catapult 50
Treviso (Italy) 85
trigger 73, 145
- -, arm, inner 113

- -, bar, iron 114
- -, braces, thickness 143
- -, circular 145
- -, design 57
- -, mechanism 13, 26, 27, 45, 55, 88, 95, 109, 112–116
  - -, adjustment 114
  - -, bronze 45
  - -, height 142
  - -, length 143
  - -, measures 44
  - -, nut 55, 114
  - -, proportions 113
  - -, width 142
- -, release 66
- -, spring 114
trigonometry 102
Trinity (painting by Masaccio) 102
trough 145
truss, building 42
tunnelling 32
Tura, Cosmè 34
Turin Royal Library, military drawings 32

U

U-bolt 102, 103, 106
- -, thickness 143
Uffizi
- -, 446 116
- -, 446ev 53, 146
- -, 446r 116
- -, 9r [447Er] 37
  - -, dating 37
- -, 9v [447Ev] 37
unguent, herbal 65
unit 144
- -, proportional 109

V

Valturio, Roberto 5, 9, 33
Vasari, Georgio 68
Vegetius 50, 97
vehicle
- -, armoured 32
- -, covered 17
Venetian galley, armature 54
Venetian Republic (see *Venice*)
Veneziano, Domenico 102
Venice 19, 33
- -, war against Milan 19
Venturi, Giambattisa 32
Verona 19
Vigevano 97
Vikings 52
Villard de Honnecourt 51
Virgil 105
Virgin of the Rocks (altarpiece by Leonardo da Vinci) 21, 25, 29, 34, 39
Vitruvian
- -, human proportions 118
- -, principle 117
Vitruvian Man (drawing by Leonardo da Vinci) 113, 117
Vitruvius (Marcus Vitruvius Pollio) 47, 48, 97

W

walking on water 37
walnut 74
war
  –, galley 35
  –, Milan vs. Venice, 1483–1484 19
  –, of Ferrara 38
  –, with Venice 33
weapon 43, 146
  –, antique 146
  –, naval 34
weight
  –, distribution 52
  –, units 144
whalebone 73
wheel 61, 95, 103, 104, 109, 110
  –, axle 103
  –, canted 105
  –, diameter 104, 143
  –, height 143
  –, spoke 107
  –, thickness 143
  –, tilt 105, 143
      –, angle 104, 105
wheelbarrow 106
whip-driver (*cacciafrusto*) 68, 147
willow 72, 74
  –, soaking 145
winding
  –, passage 17
  –, screw 88, 109
windlass 51, 145, 146
Windsor 92, 93
  –, cannon study 39
  –, page formats 128
Windsor Castle Royal Library 3, 37, 70
  –, 12275r 47
  –, 12337v 47
  –, 12370r 70
  –, 12418r 29
  –, 12439r 40
  –, 12469r 40
  –, 12469v 146
  –, 12632r 37
  –, 12647r 6, 7, 37, 72, 106
  –, 12647r 146
  –, 12649r 146
  –, 12650r 39, 72, 146
      –, dating 39
  –, 12650r-v 37
      –, dating 39
  –, 12650v 146
  –, 12651r 39, 77
  –, 12651r 39, 146
      –, explosive dart 77
  –, 12652r 39
      –, dating 39
  –, 12652r 146
  –, 12652v 146
  –, 12653r 72
  –, 12653r 146
      –, military drawings 32
  –, 12725r 39
  –, 19134r 94
wing 31, 35
  –, design 66
  –, study 31, 37
wood
  –, acacia 73
  –, olive 73
  –, quality 57
  –, strengths 63
worm screw 145
  –, mechanism 115

Y

yew 53, 73
  –, French 53